Springer Geology

More information about this series at http://www.springer.com/series/10172

Dexin Jiang · Eleanora I. Robbins
Yongdong Wang · Huiqiu Yang

Petrolipalynology

Dexin Jiang
Lanzhou Center for Oil and Gas Resources,
  Institute of Geology and Geophysics
Chinese Academy of Sciences
  (Lanzhou Institute of Geology, CAS)
Lanzhou, Gansu
China

Eleanora I. Robbins
Department of Geological Sciences
  Adjunct Faculty
San Diego State University
San Diego, CA
USA

Yongdong Wang
Nanjing Institute of Geology and
  Palaeontology
Chinese Academy of Sciences
Nanjing, Jiangsu
China

Huiqiu Yang
Lanzhou Center for Oil and Gas
  Resources, Institute of Geology
  and Geophysics
Chinese Academy of Sciences
  (Lanzhou Institute of Geology, CAS)
Lanzhou, Gansu
China

ISSN 2197-9545　　　　　　ISSN 2197-9553　(electronic)
Springer Geology
ISBN 978-3-662-47945-2　　ISBN 978-3-662-47946-9　(eBook)
DOI 10.1007/978-3-662-47946-9

Jointly published with Science Press, Beijing
ISBN 978-7-03-045557-4 Science Press, Beijing

Library of Congress Control Number: 2015945126

Springer Heidelberg New York Dordrecht London
© Science Press, Beijing and Springer-Verlag Berlin Heidelberg 2016
This work is subject to copyright. All rights are reserved by the Publishers, whether the whole or part of the material is concerned, specifically the rights of translation, reprinting, reuse of illustrations, recitation, broadcasting, reproduction on microfilms or in any other physical way, and transmission or information storage and retrieval, electronic adaptation, computer software, or by similar or dissimilar methodology now known or hereafter developed.
The use of general descriptive names, registered names, trademarks, service marks, etc. in this publication does not imply, even in the absence of a specific statement, that such names are exempt from the relevant protective laws and regulations and therefore free for general use.
The publishers, the authors and the editors are safe to assume that the advice and information in this book are believed to be true and accurate at the date of publication. Neither the publishers nor the authors or the editors give a warranty, express or implied, with respect to the material contained herein or for any errors or omissions that may have been made.

Printed on acid-free paper

Springer-Verlag GmbH Berlin Heidelberg is part of Springer Science+Business Media
(www.springer.com)

Not for sale outside the Mainland of China (*Not for sale in Hong Kong SAR, Macau SAR, and Taiwan, and all countries except the Mainland of China*).

Foreword 1

The authors and publishers of this English language revised translation of the Chinese publication *Palynology of Petroleum Source* are to be congratulated for re-introducing and reinvigorating a practical and fascinating aspect of palynology; namely the recovery of acid-resistant microfossils, or palynomorphs, from hydrocarbon fluids.

The technique has been applied from the 1930s in, *inter alia*, the UK, the Former Soviet Union, and Australia, but with the exception of work in China, seems now to be rarely used. With the essential requirement to avoid contamination, and the need for caution when handling hydrocarbons, results from this relatively simple technique can add to understanding the "plumbing" of sedimentary basins.

Acid-resistant plant microfossils (palynomorphs) are sedimentary particles, which also have the inherent ability to record geologic age, paleoenvironment and paleoclimate, and thermal history. Consequently, their recovery from transported crustal fluids has the capacity to contribute to understanding structural history, and fluid migration pathways and mechanisms. The authors note, for example, that plant microfossils are, for the most part, larger than pore throats of reservoir rocks—so what is the mechanism that allows their transport—through microfissures resulting from initial expulsion, or during dynamic dilation of faults (often evident from seismic sections)? Analysis of recovered plant microfossil assemblages may also contribute to understanding origins of associated geochemical biomarkers, which sometimes record contributions from organic matter from different geologic periods. Equally a comparison of the thermal maturity indices derived from biomarkers and from recovered palynomorphs may be useful for analysis of migration pathways.

This book is a *tour de force*. The authors provide detailed geological information from five "inland petroliferous basins": Tarim, Junggar, Turpin-Hami, Qaidam, and West Jiuquan; and three "coastal shelf basins": Liaohe, Beibu Gulf, and Zhujiang Mouth. It includes principal stratigraphic units, thickness, lithology, paleoenvironmental/tectonic setting, and age. This provides the context for understanding the formation of hydrocarbon source rocks, carrier rocks, and reservoirs rocks. As noted by the authors, recovered plant microfossils are derived from one or more of these three sedimentary rock types. Contemporaneous recycling of palynomorphs may complicate the picture—emphasizing that all information must considered in context. There are 48 plates of photomicrographs (36 black and white and 12 color) of the almost 500 named species of spores and pollen recovered from oils from 28 oilfields studied. Using globally determined age ranges for these taxa, the authors show that the recovered assemblage is, respectively, of Carboniferous, Permian, Triassic, Jurassic, Cretaceous, and Paleogene age. And from this information identify potential hydrocarbon source formations. This novel palynological information complements other basin assessment techniques, including geochemistry and structural analysis.

Although focussed on recovery of acid-resistant microfossils from hydrocarbons, this technique could be readily adapted to hydrological studies.

The book should be of great interest to explorationists, geochemists, and palynologists.

April 2015

Clinton Foster, Ph.D.
Chief Scientist
Geoscience Australia, Canberra
Adjunct Professor
School of Earth and Environment
The University of Western Australia

Foreword 2

Palynology has been an important discipline in paleontology and biostratigraphy, studying mainly fossil spores and pollen grains from reproductive organs of plants preserved in sedimentary rocks. The field has been an effective approach for better understanding of plant taxonomy, stratigraphic correlation, paleovegetation, and interpretation of paleoclimate and paleoecology. Previously, palynology was regarded as a branch of paleobotany, but now it has become such a useful tool and is often studied independent of plant megafossil remains, especially in geology for stratigraphic correlations and for assessing maturation of oil-rich strata. This monograph shows that fossil spores and pollen in crude oils are also useful indicators for determining petroleum source rocks, studying petroleum origin, migration and accumulation, and forecasting target strata for exploration.

Although palynology of petroleum sources has become accepted as a sub-discipline of palynology, there has been no special academic international monograph on this important subject. Petrolipalynology by Profs. D.X. Jiang, E.I. Robbins, Y.D. Wang, and H.Q. Yang, is the first academic monograph in a new field of interdisciplinary science between palynology and petroleum geology. It represents an achievement in this field and offers a new window for investigating fossil spores and pollen in crude oils from oil fields within ten petroliferous basins of China. In this unique monograph, the authors have defined petroleum sporo-pollen assemblages, introduced the method for extracting fossil spores and pollen from crude oil samples, and explained the approach based on oil source rock correlations for judging hydrocarbon source rocks. In addition, the character, geochronological and geographical distribution of petroleum source rocks in the inland, and coastal shelf petroliferous basins of China are expounded. Based on the ecological characteristics of the original plants that produced the spores and pollen extracted from crude oils, the authors assessed that deep lacustrine deposition under a warm or hot and wet climate was the most favorable condition for formation of petroleum source rocks. They also explained that spores and pollen in crude oils can provide information about passages, phases, directions, routes, and distances of petroleum migration.

This is a well-illustrated book with fossil spores and pollen having wide geographical ranges. It reveals important theoretical significance and practical significance in exploration and development of oil and gas fields. There is no doubt that the publication of this book by Springer will be greatly conducive to international academic exchanges. I firmly believe that this book will have important reference value and serve as a useful scientific work in the fields of petroleum geology, paleontology, and stratigraphy for scientific research, teaching, and talent training.

February 2015

Ge Sun, Ph.D.
Professor, Vice President of Paleontological
Society of China
Dean of College of Palaeontology
Shenyang Normal University
Director, Palaeontological Museum of Liaoning

Foreword (Chinese Edition)

Palynology of Petroleum Source is an academic monograph based on the study on fossil spores and pollen in crude oils from several dozen oil fields within ten petroliferous basins of China. This unique work is the first monograph in this field. Based on more than 20 years of palynological datum accumulation, the authors have produced a book that will be used as both significant guidance for commercial petroleum exploration and undergraduate and graduate student textbook.

Within this monograph, a petroleum sporo-pollen assemblage is defined; an approach based on the oil source rock correlations of spores and pollen for judging potential source rocks is explained; and the character, geochronological, and geographical distribution of petroleum source rocks in the inland and coastal shelf petroliferous basins of China are expounded. The ecological characteristics of the original plants that produced the spores and pollen in crude oils indicate that deep lacustrine deposition under a warm or hot and wet climate is the most favorable for formation of petroleum source rocks. The authors explain that spores and pollen in crude oils can provide information about passages, phases, directions, routes, and distances of petroleum migration. A significant finding is that microfissures in source rocks are probably the features that allowed the primary migration of petroleum. According to fossil spores and pollen in crude oil, the authors demonstrate that petroleum in the igneous rock petroleum reservoir of the Beisantai Oil Field in the Junggar Basin originates from organic materials of the surrounding sedimentary rocks, thus adding supporting evidence for the organic petroleum origin theory.

Academicians in both geology and botany fields have spoken highly of the accomplishments of the study on spores and pollen in crude oils that lead to this monograph. The famous petroleum geologist Prof. Xia Zhu wrote "Research on palynology of petroleum sources has established a bridge of discipline infiltration between palynology and petroleum geology, opening up a new research field." Another renowned palynologist, Prof. Jen Hsü wrote "The work on spores and pollen in crude oils represents the advanced study of Chinese palynologists in this field." Scientists at home in China and abroad have paid attention to the study of palynomorphs in petroleum since the early 1980s. Fleet and others (1988) edited "Spores and pollen in oils as indicators of lacustrine source rocks" in **Lacustrine Petroleum Source Rocks**. Robbins (1990) edited "Palynological evidence for identification of nonmarine petroleum source rocks, China" in **Palynology of Ore Deposits**. Jansonius and McGregor (1996) edited "Fossil pollen and spores in crude oil from an igneous reservoir" in **Palynology: Principles and Applications**. It is considered that these studies merit wider attention. Finally, "Mesozoic nonmarine petroleum source rocks determined by palynomorphs in the Tarim Basin, Xinjiang, northwestern China" was published by the British journal **Geological Magazine** (Jiang et al. 2008).

I sincerely hope that this monograph will be published as early as possible. I believe that this work will be helpful to petroleum exploration and will surely promote development of the discipline.

February 2012

Yongchang Xu
Professor
Lanzhou Center for Oil and Gas Resources
Institute of Geology and Geophysics
Chinese Academy of Sciences
(formerly the Lanzhou Institute of Geology
Chinese Academy of Sciences)
Director
State Key Laboratory of Gas Geochemistry

Preface

Scientists have paid attention to studies about palynomorphs in petroleum since the early 1980s. Traverse (1988) in his unique work **Paleopalynology** wrote "A number of studies have turned up the interesting fact that spores/pollen are capable of being swept along with migrating petroleum as it moves through porous sedimentary rocks." McGregor (1996) reviewed studies of palynomorphs in petroleum and considered that these studies merit wider attention, because the results and interpretations of researchers working on this subject have achieved credibility.

A Chinese monograph **Palynology of Petroleum Source** (in Chinese with English Summary) was published by Science Press in Beijing in 2013. Some palynologists and petroleum geologists suggest that English edition of the monograph should be published. This suggestion is supported by the leadership of Lanzhou Center for Oil and Gas Resources, Institute of Geology and Geophysics, Chinese Academy of Sciences (formerly the Lanzhou Institute of Geology, Chinese Academy of Sciences).

Petrolipalynology is the English version of the Chinese monograph, with additions that advance **Palynology of Petroleum Source**. This book is the first English monograph in this field. The principles and methods for determining petroleum source rocks based on the fossil spores and pollen are explained, and the character and distribution of the petroleum source rocks in the inland petroliferous basins and the coastal shelf petroliferous basins of China are expounded within the monograph. In accordance with the study on palynomorphs in petroleum, the authors discuss how microfissures in source rocks should be the passages for primary migration of petroleum and then expound on the mechanisms of petroleum migration in detail. Based on more than 20 years of palynological data accumulation, the authors have produced a book of use to commercial petroleum exploration and undergraduate and graduate students alike. Just as Prof. Xia Zhu wrote "Research on palynology of petroleum sources has established a bridge of discipline infiltration between palynology and petroleum geology, opening up a new research field." In a word, Petrolipalynology is a seminal discipline with bright prospects.

This English monograph is written by Prof. Dexin Jiang (Lanzhou Center for Oil and Gas Resources, Institute of Geology and Geophysics, Chinese Academy of Sciences), Dr. Eleanora I. Robbins (San Diego State University), and Dr. Yongdong Wang (Nanjing Institute of Geology and Palaeontology, Chinese Academy of Sciences), with plates edited by Prof. Huiqiu Yang.

This project is supported by Lanzhou Center for Oil and Gas Resources and the Open Fund Program of the Key Laboratory of Petroleum Resources Research, Institute of Geology and Geophysics, Chinese Academy of Sciences, and the Key Laboratory Project of Gansu Province (Grant No. 1309RTSA041). This work was also jointly supported by State Key Programme of Basic Research of Ministry of Science and Technology, China (Grant No. 2012CB822003), the Innovation Project of CAS (Grant No. KZCX-2-YW-154), and the Team Program of Scientific Innovation and Interdisciplinary Cooperation of CAS.

The authors acknowledge academicians from the Chinese Academy of Sciences, Prof. Shu Sun, Prof. Ziyuan Ouyang, and Prof. Xu Chen for their encouragement and advice. The authors express their deep gratitude to Prof. Clinton Foster and Prof. Ge Sun for heartily writing Foreword for this book. Special thanks are due to the renowned palynological professor at the Pennsylvania State University, USA, Dr. Alfred Traverse for his encouragement and inestimable help. We are very grateful to Prof. Lianjie Guo, Prof. Yanqing Xia, Prof. Yongchang Xu, Prof. Junchao Wei, Prof. Xianbin Wang (Lanzhou Center for Oil and Gas Resources, Institute of Geology and Geophysics, Chinese Academy of Sciences), Prof. Shu Ouyang (Nanjing Institute of Geology and Palaeontology, Chinese Academy of Sciences), and Prof. Mingcheng Li (China University of Geosciences) for their valuable advice and assistance. The authors also express hearty thanks to Mr. Feng Sun, Ms. Jiang Wei, Ms. Jine Du, Ms. Changyu Lai, Dr. Ning Tian, Dr. Chong Dong, Mr. Ping Wu, Mr. Zheng Shi, Ms. Yuxiao Dong, Ms. Liqin Li, and Ms. Xiaoqing Zhang for making their important contribution to this project.

February 2015

Dexin Jiang
Eleanora I. Robbins
Yongdong Wang

Preface (Chinese Edition)

As energy sources, oil and natural gas are very important for the development of national economies, especially China's. Coal of course is the foremost energy source used in China. The world's petroleum industry developed rapidly in the span between 1950 and 2000; it reached the height of development in the 1980s. Owing to the rapid development of petroleum exploration in the inland petroliferous basins of China, many new petroleum fields such as the Kekeya and the Yakela oil fields in the Tarim Basin, and the Huonan and the Beisantai oil fields in the Junggar Basin were discovered since the 1970s. The development of this petroleum industry offered an opportunity for science and technology to blaze new trails.

An apparatus for extraction of spores and pollen from crude oil samples was set up in the palynological laboratory in Lanzhou Institute of Geology, Academia Sinica, in 1965. Since then, abundant fossil spores, pollen, and algae have been extracted from more than 200 crude oil and natural gas samples associated with oil samples from the Jiuquan, Ordos, Junggar, Turpan, Tarim, Qaidam, Liaohe, Beibu Gulf, Sanshui, and Zhujiang Mouth basins. The initial findings of spores and pollen in crude oils were confirmed by renowned scientists, academicians of Chinese Academy of Sciences. The first appointed director of the original Institute of Geology, Academia Sinica, Prof. Defeng Hou said "Fossil spores and pollen found in crude oils can provide reliable information for judging and dating petroleum source beds, which can also offer evidence for the generation and migration of petroleum." The renowned sedimentologist Prof. Lianjun Ye said "Study on spores and pollen in crude oils serve the production of petroleum by means of palynology as a tool. It is a creative work and is proved to be effective." The renowned palynologist Prof. Jen Hsü said "Spores and pollen in crude oils can provide valuable evidence for petroleum origin and petroleum sources." The famous petroleum geologist Prof. Xia Zhu said "Research on palynology of petroleum sources has established a bridge of discipline infiltration between palynology and petroleum geology, opening up a new research field. The academic accomplishments will be possessed of important function of guidance to petroleum exploration." Thus, the study of spores and pollen in crude oils was listed in the National 1986 to 1990 and 1991 to 1995 Science and Technique Major Research Programs.

The present monograph is a summary of the palynological achievements in the National 1986 to 1990 Program "Geological theory and exploratory technique of petroleum fields" and the National 1991 to 1995 Program "Petroleum resources of the Tarim Basin" as well as the Program of National Natural Science Foundation of China "Principles and methods of petroleum source rock identification by means of spores and pollen" (Grant No. R 850879). The character and distribution of petroleum source rocks in the inland petroliferous basins including the Tarim, Junggar, Turpan, Qaidam, and Jiuquan basins and the coastal shelf petroliferous basins including the Liaohe Basin of East China Sea and the Beibu Gulf and Zhujiang Mouth basins of South China Sea are expounded in this monograph. The work is based on the palynological data that have been accumulated for more than 20 years. The main achievements of this monograph are as follows:

1. The definition and classification of petroleum sporo-pollen assemblage are provided. An approach based on the oil source rock correlations of fossil spores and pollen species and color for judgment of petroleum source rocks is explained. By way of the application in eight petroliferous basins, the approach is proved to be effective.
2. The authors expound that microfissures in source rocks are the important passages for the primary migration of petroleum, allowing the passage of pollen and spores, and showing that fossil spores and pollen in crude oils are capable of dating petroleum source rocks.
3. The ecological characteristics of the original plants that shed the spores and pollen act as indicators of petroleum source rocks, thereby indicating that the lacustrine and swamp/marsh sedimentary environments under warm/hot and humid/wet climatic conditions are favorable for the formation of nonmarine petroleum source rocks.
4. Fossil spores and pollen in crude oils show reliable information about passages, phases, directions, routes, and distances of petroleum migration. Microfissures in source rocks formed by abnormal high pressure and undercompaction during the process of diagenesis are supported as the passageways for primary migration of petroleum. The passageways for secondary migration include connective pore spaces, bedding voids, joints, fissures, faults, and unconformities in the carrier bed and the reservoir bed. The phase state of primary migration includes the oil phase, gas phase, water-soluble phase, oil-soluble phase, gas-soluble phase, and diffusion phase. The phase state of secondary migration generally inherits the phase state of primary migration. The directions of petroleum migration are from low porosity and permeability rocks to high porosity and permeability rocks, following either vertical migration or lateral migration pathways. The routes of migration are from petroleum source beds to traps. The distances of migration are dependent on the distances between petroleum source beds and traps.
5. Ninety-six species of fossil spores and pollen referred to 52 genera were found in crude oil from an igneous petroleum reservoir in the East Junggar Depression. Igneous rocks cannot yield biological fossils; thus, the spores and pollen in crude oil must have been carried by oil, gas, and water from the surrounding sedimentary petroleum source rocks to the

igneous reservoir during petroleum migration. This discovery is convincing evidence for the organic petroleum origin theory.
6. One hundred and eighty-three species of fossil spores and pollen referred to 89 genera were found in crude oils from the Tarim Basin. The original plants of the spores and pollen are continental plants, which have their sources from the terrestrial environment. These spores and pollen bear witness to petroleum generation from the continental facies.
7. The results of the study on fossil spores and pollen in crude oils indicate that in China, the Carboniferous, Permian, Triassic, Jurassic, Cretaceous, and Tertiary Systems of the inland petroliferous basins contain petroleum source rocks, and the Paleogene System of the coastal shelf petroliferous basins contains excellent petroleum source rocks. The spore/pollen exine colors indicate that these petroleum source rocks are mature.

In the course of the performance of the present project, the authors obtained encouragement, advice, and assistance from the leadership of Division of Earth Science of Academia Sinica, the leadership of Lanzhou Center for Oil and Gas Resources, Institute of Geology and Geophysics, Chinese Academy of Sciences (the original Lanzhou Institute of Geology, Academia Sinica), Academician Shu Sun, Prof. Jindong Zhang, Prof. Yanqing Xia, Prof. Yongchang Xu, Prof. Difan Huang, Prof. Digang Liang, Prof. Zhuosheng He, and Prof. Kailin Dong. To the above leaders and specialists, the authors express their deep gratitude.

The Chinese monograph was written by Dexin Jiang, while the plates were edited by Huiqiu Yang. Mr. Feng Sun, Mr. Yongdong Wang, Ms. Jine Du, Ms. Changyu Lai, Mr. Ping Wu, and Mr. Zheng Shi, all of whom participated in this project.

The authors acknowledge the careful review of Prof. Junchao Wei. Special thanks are due to Ms. Jiang Wei for computer technical assistance. Appreciation is extended to Ms. Yuxiao Dong for her obtaining some necessary references in libraries.

January 2012   Dexin Jiang

Contents

1 **Introduction**.. 1
   1.1 Early History... 2
   1.2 On Petrolipalynology..................................... 3
      1.2.1 Objectives and Purposes......................... 4
      1.2.2 Principles and Methods.......................... 4
      1.2.3 Review and Outlook.............................. 5
   References... 7

2 **Geological Background**..................................... 11
   2.1 Introduction.. 11
   2.2 Geological Conditions for Oil–Gas Field Formation....... 11
      2.2.1 Complex of Source Rock, Reservoir Rock, and Seal Rock.................................. 11
      2.2.2 Generation, Migration, and Accumulation of Petroleum.................................... 12
      2.2.3 Petroleum Deposit and Its Classification........ 13
   2.3 Inland Petroliferous Basins.............................. 13
      2.3.1 The Tarim Basin................................. 13
      2.3.2 The Junggar Basin............................... 23
      2.3.3 The Turpan-Hami Basin........................... 28
      2.3.4 The Qaidam Basin................................ 31
      2.3.5 The West Jiuquan Basin.......................... 33
   2.4 Coastal Shelf Petroliferous Basins...................... 35
      2.4.1 The Liaohe Basin................................ 35
      2.4.2 The Beibu Gulf Basin............................ 37
      2.4.3 The Zhujiang Mouth Basin........................ 39
   References.. 41

3 **Fossil Spores and Pollen in Crude Oils**................... 43
   3.1 Materials and Methods................................... 43
      3.1.1 Materials....................................... 43
      3.1.2 Methods... 44
   3.2 Tarim Basin... 44
      3.2.1 North Tarim Upheaval............................ 44
      3.2.2 Kuqa Depression................................. 51
      3.2.3 Southwest Tarim Depression...................... 53

3.3	Junggar Basin		57
	3.3.1	East Junggar Depression	57
	3.3.2	South Junggar Depression	66
3.4	Turpan Basin		70
	3.4.1	Qiktim Oil Field	70
	3.4.2	Shengjinkou Oil Field	73
3.5	Qaidam Basin		75
	3.5.1	North Border Block-fault Zone	75
	3.5.2	Mangnai Depression	78
3.6	West Jiuquan Basin		82
	3.6.1	Laojunmiao Anticlinal Zone	82
	3.6.2	Baiyanghe Monoclinal Zone	82
3.7	Liaohe Basin of Bohai Gulf		83
3.8	Shelf Basins of South China Sea		85
	3.8.1	Beibu Gulf Basin	85
	3.8.2	Zhujiang Mouth Basin	87
References			89

4 Petroleum Sporo-pollen Assemblages and Petroleum Source Rocks 95

4.1	Definition and Classification of Petroleum Sporo-pollen Assemblages		96
	4.1.1	Definition	96
	4.1.2	Classification and Character	96
4.2	Tarim Basin		96
	4.2.1	North Tarim Upheaval	96
	4.2.2	Kuqa Depression	100
	4.2.3	Southwest Tarim Depression	103
4.3	Junggar Basin		105
	4.3.1	East Junggar Depression	105
	4.3.2	South Junggar Depression	109
4.4	Turpan Basin		110
	4.4.1	Qiktim Petroliferous Region	110
	4.4.2	Shengjinkou Petroliferous Region	112
4.5	Qaidam Basin		112
	4.5.1	North Border Block-fault Zone	112
	4.5.2	Mangnai Depression	114
4.6	West Jiuquan Basin		114
	4.6.1	Laojunmiao Anticlinal Zone	114
	4.6.2	Baiyanghe Monoclinal Zone	116
4.7	Liaohe Basin		117
4.8	Beibu Gulf Basin		117
4.9	Zhujiang Mouth Basin		117
References			120

5	Spore/Pollen Fossil Coloration and Petroleum Source Rock Quality		123
	5.1	Spores/Pollen Fossil Coloration and Maturity of Organic Material	123
	5.2	Organic Material Type and Maturity with Hydrocarbon Generation Potential	126
	References		127
6	Palynological Evidence for Organic Petroleum Origin Theory		129
	6.1	Sporopollenin and Petroleum Origin	130
	6.2	Fossil Spores and Pollen in Crude Oils from Sedimentary Rock Petroleum Reservoirs	131
	6.3	Fossil Spores and Pollen in Crude Oils from Igneous Rock Petroleum Pools	132
	References		133
7	Environment for the Formation of Petroleum Source Rocks		135
	7.1	Botanical Relationship of Dispersed Spores and Pollen	135
	7.2	Paleoecology and Paleoclimate	140
	7.3	Paleoecology and Paleogeography	144
	7.4	Sedimentary Environment and Petroleum Source	147
	References		150
8	Mechanisms of Petroleum Migration		153
	8.1	Introduction	154
	8.2	Primary Migration	154
		8.2.1 Passageways for Primary Migration	154
		8.2.2 Phase States of Primary Migration	155
	8.3	Secondary Migration	155
		8.3.1 Passageways for Secondary Migration	156
		8.3.2 Phase States and Flow Types of Secondary Migration	156
		8.3.3 Directions, Routes, and Distances of Secondary Migration	157
	8.4	Period of Petroleum Deposit Formation	158
	References		158
9	Geochronic and Geographic Distribution of Nonmarine Petroleum Source Rocks		159
	9.1	Inland Petroliferous Basins	160
	9.2	Coastal Shelf Petroliferous Basins	161
	References		162

Concluding Remarks 163

Explanation of Plates and Plates 167

Introduction 1

Abstract

The primary purpose for studying palynology of petroleum sources, *Petrolipalynology*, is to determine geological ages and stratohorizons of petroleum source rocks. In accordance with theories on origin, migration, and accumulation of petroleum, fossil spores and pollen in crude oils should be from the source bed, the carrier bed, and the reservoir bed, thereby forming a three-part assemblage. The reservoir rock of an oil field is always known, so fossil spores and pollen deriving from the reservoir bed itself can easily be separated from the three-part assemblage. The remainder of the assemblage indicates source and carrier beds. Studies of the incorporated pollen and spores show that source rocks can be younger or older than the reservoir rocks. Although the geological circumstances are often complicated, petroleum source rocks can be judged by correlations between fossil spores and pollen in oils and those in potential petroleum source rocks. The fossil spores and pollen of petroleum source rocks can be carried to petroleum traps during the course of petroleum migration, because they are minute, light, flexible, and have a strong exine which can withstand such transport. Laboratory experiments on petroleum migration have demonstrated that microfossils entrapped by oil and gas from source rocks and carrier rocks can migrate together with oil and gas and that they can also be separated from oil and gas without any appreciable change in structural features.

Keywords

Petroleum source rocks · Spores and pollen · Correlation of oil and source

1.1 Early History

Palynology is a field that embraces all pollen and spore studies. Both spores of cryptogams and pollen of spermatophytic plants are genital cells. Microspores of heterosporous plants are male, having diameters from 20 to 100 μm, and megaspores of heterosporous plants are female, having diameters over 200 μm. Pollen grains are male genital cells, having diameters from 20 to 50 μm, while coniferous pollen grains can be over 100 μm. Pollen and spore fossils are important microfossils because they are abundant in sediments; furthermore, special structures and ornamentation of the outer, exine layer are often dependable for identification to genus and sometimes species.

In the middle to late nineteenth century, C.A. Weber and others discovered pollen and spore fossils from late-glacial peat in Sweden. This discovery made Scandinavia the center of pollen and spore studies. The publication of the dissertation of Swedish scholar Von Post (1916) on quantitative analysis of tree pollen from a marsh is considered as the beginning of pollen analysis. Studies of morphological descriptions and systematic classification of spores and pollen proceeded widely in Europe and America. In 1944, H.A. Hyde and D.W. Willems proposed a new word "palynology" to express the newly emerging discipline (Traverse 1988). The first international meeting of palynologists (The International Association of Palynology) was held at the 7th International Botanical Conference in Stockholm, Sweden, in 1950. The 1st International Palynological Conference was held at the University of Arizona in Tucson in 1962. Since then, every four or five years an International Conference has been held for exchange of palynological studies. These conferences discuss morphology, taxonomy, evolution, ecology, physiology, biochemistry, biostratigraphy, paleoclimatology, paleogeography, aerobiology, melissopalynology, and data processing models.

Due to the tremendous development of the petroleum and coal industries and funded geological exploration programs, palynology in China has rapidly progressed since 1949. About 100 palynological laboratories were established in China in the 1950s–1970s. Three to four hundred people are engaged in palynology, and courses are offered either in the geological or botanical departments of many universities. The 1st Chinese Palynological Conference was held in Tianjin in 1979. Also in 1979, the Palynological Society of China was founded and the academician (Chinese Academy of Sciences) and palynological pioneer Professor Jen Hsü was elected as its president at that conference. About 200 palynologists attended the conference where discussions included morphology, biostratigraphy, oceanic palynology, and techniques of pollen analysis. Since then, the Chinese Palynological Conference has been held every four years.

In 1953, academician and Professor Fuhsiung Wang (Director of Division of Morphology, Institute of Botany, Chinese Academy of Sciences) arranged an important research project for palynology. Under the leadership of Professor Fuhsiung Wang, the Institute of Botany (Chinese Academy of Sciences) became the center of modern spore/pollen study in China. Based on the observations of 450 species of modern pollen, morphological descriptions of spores and pollen have been refined. A well-illustrated book *Pollen Flora of China* (in Chinese) by Fuhsiung Wang et al. was published in 1960; its 2nd edition was in 1995. More than 1400 species of modern pollen in China belonging to 900 genera and 118 families were described in detail, and the geographic distribution and ecological environment of the plants were added in the 2nd edition (Wang et al. 1995). Another well-illustrated book *Sporae Pteridophytorum Sinicorum* (in Chinese) was published in 1976 by the Institute of Botany (Chinese Academy of Sciences). About 1000 species of pteridophytic spores belonging to 174 genera and 52 families were described systematically; the book included 119 figures of spores, 89 plates of modern spores, and 12 plates of fossil spores (Beijing Institute of Botany, Academia Sinica 1976). These two books have provided an important basis for the identification of fossil spores and pollen in China. Furthermore, these compendiums have offered palynological foundations for research into botanical taxonomy, evolution, paleogeography, paleoclimatology, and paleovegetation. In addition, the journals *Acta Botanica Sinica* (in Chinese) and the *Chinese Journal of Botany* publish palynological papers.

Paleopalynology has flourished under the leadership of Professor Jen Hsü (Institute of Palaeontology, Chinese Academy of Sciences, and Institute of Geology, Chinese Academy of Geological Sciences). In the 1950s, palynologists began to research Paleozoic, Mesozoic, and Cenozoic palynology at these institutes. Results from these studies are primarily published in the journals *Acta Palaeontologica Sinica* (in Chinese) and *Acta Micropalaeontologica Sinica* (in Chinese). Some systematic palynological studies have been published in *Palaeontologia Sinica* (in Chinese with English summary). Important works in Chinese include the following: "Sporo-pollen complex and geological age of the red beds of Wenmingsze, Jucheng of southern Hunan" (Hsü 1958), "Microflora and geological age of the basal part of the Lower Huihuipou Formation of the Chiuchüan Basin of western Kansu" (Hsü and Chow 1956a), "Microflora and geological age of the uppermost part of the Lower Huihuipou Formation of the Chiuchüan Basin of western Kansu" (Hsü and Chow 1956b), "Tertiary spore and pollen complexes from the red beds of Chiuchuan, Kansu and their geological and botanic significance" (Sung 1958), "Palynology of Upper Permian and Lower Triassic strata of Fuyuan district, eastern Yunnan" (Ouyang 1986), "Late Triassic spores and pollen from central Sichuan" (Zhang 1984), "Triassic and Jurassic sporopollen assemblages from the Kuqa Depression, Tarim Basin of Xinjiang, NW China" (Liu 2003), and "Cretaceous and Early Tertiary sporo-pollen assemblages from the Sanshui Basin, Guangdong province" (Song et al. 1986). Nationally known works in Chinese include the following: "Mesozoic Stratigraphic Paleontology of the Shaanxi-Gansu-Ningxia Basin" (Institute of Geology, Chinese Academy of Geological Sciences 1980), "Permian and Triassic strata and fossil assemblages in the Dalongkou area of Jimsar, Xinjiang" (Institute of Geology, Chinese Academy of Geological Sciences, and Institute of Geology, Xinjiang Bureau of Geology and Mineral Resources 1986), "Palynology of the Carboniferous and Permian strata of northern Xinjiang, northwestern China" (Ouyang et al. 2003), "Cretaceous-Tertiary palynological assemblages from Jiangsu" (Song et al. 1981), and "A research on Tertiary palynology from the Qaidam Basin, Qinghai province" (Research Institute of Exploration and Development, Qinghai Petroleum Administration, and Nanjing Institute of Geology and Palaeontology, Academia Sinica 1985). Tens of hundreds of species of fossil spores and pollen including new genera and species were described, and the sporo-pollen assemblage sequences of each geological age were established in the above-mentioned works.

These publications and others provide an important basis for stratigraphic division and correlation; they also have offered a strong foundation for research into paleovegetation, paleoclimatology, paleogeography, development and evolution of flora, and phytogeographical zonation. Such palynological accomplishments have laid the foundation for research into fundamental theories and applied sciences, as well as national economic endeavors.

1.2 On Petrolipalynology

The development of a long-lasting discipline must include ever-branching developments. Thus, palynology has led to paleopalynology, aeropalynology, melissopalynology, and so on. Paleopalynology is the study of fossil spores and pollen from sedimentary rocks of all geological ages; its focus is on geochronology, biostratigraphy, and paleoecology (Traverse 1988). Aeropalynology is the study of spores and pollen in the air; its focus is on analysis and improvement of the ecological environment and prevention and cure of pollen allergy (Hyde 1969). Melissopalynology is the study of pollen contained in honey; its purposes are improvement of honey quality and increase in production of honey (Lieux 1972).

Palynology has played a major role in age dating and ecological information of use to the petroleum industry (De Jersey 1965; Jiang and Yang 1992; Jiang et al. 2008). Petrolipalynology is a subdiscipline that has evolved gradually as the practice of evaluation and prediction of petroleum resources (Jiang and Yang 1983b).

1.2.1 Objectives and Purposes

Palynomorphs found in liquid petroleum is the focus of the study of petrolipalynology. This includes spores, pollen, fungi, and algae such as dinoflagellates and acritarchs being carried by crude oils, natural gas, and oil field waters. Furthermore, palynomorphs in the solid rocks that are potential petroleum source rocks are also subject to intense study because they are useful for many aspects in that industry.

Purposes for studying petrolipalynology include helping to determine petroleum source rocks, forecasting target strata for further exploration, and providing a scientific basis for petroleum prospect evaluation and exploration. Research includes age determination of petroleum source rocks, theory of petroleum origin, and mechanisms of petroleum migration. While reservoir rocks are directly known, petroleum-generating source rocks are inferred. Exploration into petroleum sources and source rocks is a wide-open field. Correctly identifying petroleum source rocks is important for prospect evaluation and further exploration in petroliferous region. Therefore, knowledge of all aspects of petroleum sources and source rocks is an important precondition and key for oil–gas field exploration.

1.2.2 Principles and Methods

Petroleum usually contains fossil spores, pollen, and algae (Timofeev and Karimov 1953; De Jersey 1965; Jiang and Yang 1992, 1994). These microfossils are providing a great deal of reliable information for petroleum source research. Spores, pollen, and algae themselves can be the original materials from which petroleum is generated; furthermore, their exines produce well-preserved fossils. These two characteristics are important in judging petroleum source rocks. In specific, spores, pollen, and algae that were originally buried in the source rocks have contributed their waxy, fatty, and oily secretions to the eventual petroleum complex, leaving only their acid-resistant, alkali-resistant, and microbial decay-resistant exine remains (Sanders 1937). The exines of spores and pollen are strong enough to undergo the thermal metamorphism during petroleum generation and the high pressure and temperature of petroleum migration, and yet retain primary textures and ornamentation (Jiang and Yang 1992; Jiang et al. 2008).

Laboratory experiments on petroleum migration by Chepikov and Medvedeva (1971) proved that plant microfossils can migrate not only together with oil, but also with natural gas and that spores and pollen can undergo migration without any appreciable change in structural features. Additional support for this interesting observation about palynomorph migration in petroleum is the fact that fossil spores and pollen having different geological ages from the reservoir beds themselves occur in crude oils (Jiang and Yang 1980, 1986, 1994). Therefore, by means of studies on genus and species identification and geological distribution of palynomorphs in crude oils, it is possible to determine petroleum source rock ages and stratigraphic position, as well as to deduce the evolution of changes in petroleum provinces.

The above studies stand in direct contradiction to Hunt (1979) who stated "Spores and pollen have periodically been suggested for age dating but their diameters, which are mostly over 15 μm, prevent their migration with liquid or gaseous hydrocarbons. Particles of that size are unable to migrate through shales or siltstones." The key to understanding why this was not accurate is due to unimagined complex processes and the unknown presence of ever-changing passageways used by liquids and associated solid particles during primary migration of petroleum. Although spores and pollen cannot be expelled from source rocks via rock pores, they can be expelled from source rocks via microfissures.

Microfissures are formed by abnormal high pressure during the process of diagenesis, and they are common (Tissot and Welte 1978). Furthermore, hydrocarbon generation itself is an important cause of abnormal high pressures (Hua and Lin 1989; Li 2004). Such microfissures are considered to be the important passages for primary migration of petroleum (Li 2004). Moreover, fossil spores and pollen have the ability to pass through various migration passageways because they are very thin, light, and flexible. When hydrocarbon fluid pressure is high, fossil spores and pollen can wrinkle and pass through narrow passageways; surprisingly, they can subsequently recover their original state when the space around them increases. This flexibility of fossil spores and pollen has been observed under microscope (Jiang and Yang 1991, 1992).

In accordance with theories on origin, migration, and accumulation of petroleum, palynomorphs in oil pools must be either from petroleum source beds, carrier beds, or reservoir beds. These three different environments can be considered as petroleum source rock series, carrier rock series, and reservoir rock series. The reservoir bed of an oil pool is always known, so fossil spores and pollen deriving from the reservoir bed itself can be easily identified. Those not deriving from the reservoir bed are indicators for either source bed or carrier bed. In general, the oldest palynomorphs in crude oils indicate the source bed, those of intermediate age indicate the carrier bed, and the youngest palynomorphs indicate the reservoir bed. Sometimes, the three parts are coeval, indicating that the source bed, carrier bed, and reservoir bed belong to the same stratum. It is worth bearing in mind that the carrier rock series usually contains source rocks. Therefore, in order to correctly judge all of the petroleum source rocks, it is necessary to compare the palynomorphs in crude oils with those in the potential petroleum source rocks.

Similar to fingerprint compounds being used by geochemists as indicators for correlation between oil and its supposed source, fossil spores and pollen may be used as indicators by palynologists for correlation between oil and its potential source to determine petroleum source rocks. The approach to determine petroleum source rocks by means of studies of palynomorphs in crude oils has proved to be an effective determination model. In addition, fossil spores and pollen are useful for identification of geological ages and ecological environments of petroleum source rocks.

1.2.3 Review and Outlook

Biot (1835) proved that some petroleum fractions exhibit rotary polarization, which is known to be formed only by organisms. Therefore, most scientists consider that petroleum originates from the remains of organisms. For example, White and Stadnichenko (1923) discovered the sporangia of *Protosalvinia*, megaspores of Lycopodiales, and thalluses of algae from the black and gray shales of the Upper Devonian Ohio Shale in Kentucky, USA, and chemical experiments elucidated that these plants should be the mother materials of petroleum generation there. Sanders (1937) extracted fossil spores, algae, and fungi from crude oils in Cretaceous and Tertiary oil fields of Mexico and the Tertiary oil field of Romania; he also stated that the spores and algae had provided waxy, fatty, and oily secretions for petroleum generation there, leaving only their decay-resistant exines. Sanders (1937) also extracted carapaces of fossil insects that resemble Acarina from crude oils in a Cretaceous oil field in Mexico; such well-preserved fossil insects had been found in Jurassic rock cores in the region before (Sanders 1937). Therefore, he supported the idea that petroleum from Mexico, Cuba, Colombia, Venezuela, and southwestern Texas, USA, originated from the Jurassic limestones and shales (Wright and Sweet 1924). Sanders is the first scientist who associated microfossils in crude oils with petroleum source rocks, thereby creating this field of study. His contribution is rarely acknowledged.

In order to determine the petroleum source rocks of the Colorado USA oil field, Waldschmidt (1941) extracted diatoms and plant fossil fragments from Permian crude oils using Sanders' method. Subsequently, Timofeev and

Karimov (1953) extracted spores, pollen, algae, and plant fossil fragments from crude oil samples taken from the Cambrian oil field of East Siberia, from the Devonian, Carboniferous, and Permian oil fields of Volga-Ural, from the Cretaceous oil field of Emba, and from the Tertiary oil field of North Gorgaso and Kamchatka. Banerjee (1965) researched microflora in crude oils from the Assam oil field of India. De Jersey (1965) published on the palynoflora in crude oils from the Moonie oil field in Queensland, Australia; because petroleum in the producing sandstones of the Moonie oil field contained a characteristically Jurassic palynoflora, a Jurassic source for the petroleum was assumed.

In China, crude oil sample processing uses a Soxhlet apparatus to extract fossil spores and pollen from crude oils (Jiang et al. 1974). Fossil spores and pollen produced in the black shales of the Lower Cretaceous lower Xinminbu Formation were found in crude oil samples taken from the Tertiary reservoir bed and the Silurian reservoir bed of the Laojunmiao oil field and the Yaerxia oil field in the West Jiuquan Basin of Gansu; therefore, the Lower Cretaceous black shales were considered to be the petroleum source rocks of the Yumen petroliferous region (Jiang and Yang 1980, 1981a, b). Since then, abundant fossil spores and pollen have been found in crude oils from the petroliferous regions in the Tarim, Junggar, and Turpan basins of Xinjiang, the Qaidam Basin of Qinghai, and the Shaanxi-Gansu-Ningxia Basin (Jiang 1988, 1990, 1991). In addition to these continental petroliferous basins, various fossil spores, pollen, and algae have been found in crude oils from the petroliferous regions in the offshore shelf basins of the East China and the South China Seas. Studies on fossil spores and pollen in crude oils show that there are multiple-period petroleum source rocks in the Chinese continental petroliferous basin; for example, the Carboniferous, Permian, Triassic, Jurassic, and Paleogene all contain producing petroleum source rocks in the Junggar Basin (Yang and Jiang 1989; Jiang 1991; Jiang and Yang 1994). These studies also show that the offshore shelf basins of the East China and the South China Seas are rich in Paleogene petroleum sources (Yang and Jiang 1981; Jiang 1991; Jiang and Yang 1999a). Jurassic spores and pollen are abundant in crude oils from the Kuqa Depression and the Yecheng and Kashi Sags in the Tarim Basin, reflecting the fact that Jurassic petroleum sources are important for the Tarim Basin (Jiang and Yang 1983a, 1986, 1991, 1996). Most of the spores and pollen in crude oils from the North Tarim Upheaval petroliferous region are Mesozoic palynomorphs; mainly Triassic and Jurassic rocks are the primary sources (Jiang and Yang 1991, 1992, 1999b). According to the results of correlation between spores and pollen in crude oils from the Jurassic reservoir bed of the Qiktim oil field of the Turpan Basin and those in the Lower Jurassic Badaowan and Sangonghe formations and the Middle Jurassic Xishanyao and Toutunhe formations, Lower Jurassic and Middle Jurassic dark-colored mudstones should be the petroleum source rocks in the Turpan Basin (Jiang and Yang 1989). The results of correlation between spores and pollen in crude oils from the Lenghu oil field in the Qaidam Basin and those in potential source rocks indicate that the dark-colored mudstones of the Lower Jurassic Xiaomeigou Formation and the Middle Jurassic Dameigou Formation should be the petroleum source rocks of this famous petroliferous region (Jiang and Yang 1997).

Traverse (1988) in his unique work *Paleopalynology* gave examples from Australia and China on palynomorphs in petroleum and their significance. Fleet and others in *Lacustrine Petroleum Source Rocks* introduced "Spores and pollen in oils as indicators of lacustrine source rocks" (Jiang 1988) and considered this study as a novel technique for studying the origin and migration history of oils (Fleet et al. 1988). Robbins (1990) incorporated "Palynological evidence for identification of nonmarine petroleum source rocks, China" (Jiang 1990) in her special edition of *Palynology of Ore Deposits* and suggested that this evidence should be useful in both new and partially explored regions. McGregor (1996) reviewed studies about palynomorphs in petroleum and considered that such work merited wider attention, because the results and interpretation by researchers working on this subject have achieved credibility. Furthermore, a case study of palynomorphs in petroleum from the Junggar Basin of China "Fossil pollen and spores in crude oil from an igneous reservoir" (Jiang 1996) was published in Jansonius and McGregor (1996) *Palynology: Principles and Applications*. Also, the British journal, *Geological Magazine* published "Mesozoic non-marine petroleum source rocks determined by palynomorphs in the Tarim Basin, Xinjiang, northwestern China" (Jiang et al. 2008).

In spite of additional new energy sources including nuclear, solar, and wind energy, petroleum is still the primary energy resource used around the world. So exploration and development of petroleum resources shoulder heavy responsibilities. The determination of petroleum source rocks is the foundation for prediction and reserve estimations of petroleum resources. When a petroleum source rock is determined, it becomes possible to calculate oil-generating quantity on the bases of the area, thickness, and oil-generating potential of the source rock. Therefore, quality and distribution of petroleum source rocks are the groundwork for evaluating prospective petroliferous regions. In contrast, when a petroleum source rock is not determined, wildcat drilling usually results in large economic losses. Therefore, to locate specific target strata for exploration, determination of geological age, and stratigraphic position of petroleum source rocks is needed before exploration is deployed.

Palynomorphs in petroleum itself can be used as indicators for correlation between oil and source in order to identify petroleum source rocks. As fossils, they can also indicate geological ages and ecological environments of petroleum source rocks. Therefore, petrolipalynology should be a valuable tool in guidance for exploration and development. In a newly explored region, if fossil spores and pollen or algae can be found from crude oil samples in a timely manner, it will be possible to guide further exploration. In partially explored regions, studies of this subject have a bright future. There are more than 100 sedimentary basins on the continent of China; some of them are proved petroliferous basins, but exploration is still comparatively light. Future research projects can be predicted. For example, we can predict (1) distribution and oil-generating potential of Jurassic petroleum source rocks in the petroliferous basins of northwestern China, (2) the evolved conditions of petroleum pools in multiple-source, superposed petroliferous basins, and (3) the environment for formation of petroleum source rocks within continental facies in northwestern China will produce positive contributions to the exploration and development of petroleum resources in China and elsewhere.

In oceanic areas, the value of such insights is great. According to the results of correlations between fossil spores and pollen in crude oils and those in potential source rocks, the Bohai Gulf of the East China Sea and the Beibu Gulf of the South China Sea are rich in Paleogene petroleum sources (Yang and Jiang 1981; Jiang and Yang 1999a). With reserves of 600 million tons, the Penglai oil field found in the Bohai Gulf shows a bright prospect for ocean petroleum exploration (Huang 2001). Research projects, such as (1) correlations between fossil dinoflagellates in crude oils and those in potential source rocks, (2) the mother materials of petroleum source rocks and the organic material maturity, and (3) the environment for formation of petroleum source rock series in alternating facies of marine and nonmarine strata, can be predicted to be very significant for petroleum exploration under the ocean. In other petroliferous regions of the world, if palynomorphs are sought in crude oil samples, *Petrolipalynology* will be able to provide useful scientific bases for determination and deployment of exploration.

References

Banerjee, D. (1965). Microflora in crude petroleum from Assam India. *Transactions of the Geological, Mining and Metallurgical Society of India, 62*, 51–66.

Beijing Institute of Botany, Academia Sinica. (1976). *Sporae Pteridophytorum Sinicorum* (pp. 1–451). Beijing: Science Press (in Chinese).

Biot, M. (1835). Memoir sur la polarisation circulaire et sur ses applications à la chimie organique. *Mem. Acad. Roy. Sci. Inst. France, 13*, 39.

Chepikov, K. R., & Medvedeva, A. M. (1971). Spores and pollen in oil and gas as migration indices. *Journal of Palynology, 7*, 56–59.

De Jersey, N. J. (1965). Plant microfossils in some Queensland crude oil samples. *Geological Survey of Queensland Publication, 329*, 1–9.

Fleet, A. J., Kelts, K., & Talbot, M. R. (1988). *Lacustrine petroleum source rocks* (pp. 1–391). London: Blackwell Scientific Publications.

Hsü, J. (1958). Sporo-pollen complex and geological age of the red beds of Wenmingsze, Jucheng of southern Hunan. *Acta Palaeontologica Sinica, 6*(2), 141–158. (in Chinese with English abstract).

Hsü, J., & Chow, H. I. (1956a). Microflora and geological age of the basal part of the Lower Huihuipou Formation of the Chiuchüan Basin of western Kansu. *Acta Palaeontologica Sinica, 4*(4), 491–507. (in Chinese with English summary).

Hsü, J., & Chow, H. I. (1956b). Microflora and geological age of the uppermost part of the Lower Huihuipou Formation of the Chiuchüan Basin of western Kansu. *Acta Palaeontologica Sinica, 4*(4), 509–524. (in Chinese with English summary).

Hua, B. Q., & Lin, X. X. (1989). Discussion on some problems of oil migration in Jiuxi Basin. *Acta Sedimentologica Sinica, 7*(1), 39–47. (in Chinese with English abstract).

Huang, D. F. (2001). The task and prospect of petroleum geochemistry in China in Early 21st Century. *Acta Sedimentologica Sinica, 19*(1), 1–6. (in Chinese with English abstract).

Hunt, J. M. (1979). *Petroleum geochemistry and geology* (pp. 1–617). San Francisco: W. H. Freeman and Company.

Hyde, H. A. (1969). Aeropalynology in Britain—An outline. *New Phytology, 68*, 579–590.

Institute of Geology, Chinese Academy of Geological Sciences. (1980). *Mesozoic stratigraphic paleontology of the Shaanxi-Gansu-Ningxia Basin* (pp. 1–230). Beijing: Geological Publishing House (in Chinese).

Institute of Geology, Chinese Academy of Geological Sciences, Institute of Geology, Xinjiang Bureau of Geology and Mineral Resources. (1986). *Permian and Triassic strata and fossil assemblages in the Dalongkou area of Jimsar, Xinjiang* (pp. 1–262). Beijing: Geological Publishing House (in Chinese).

Jiang, D. X. (1988). Spores and pollen in oils as indicators of lacustrine source rocks. In A. J. Fleet, K. Kelts & M. R. Talbot (Eds.), *Lacustrine petroleum source rocks* (pp. 159–169). London: Blackwell Scientific Publications.

Jiang, D. X. (1990). Palynological evidence for identification of nonmarine petroleum source rocks, China. *Ore Geology Reviews, 5*, 553–575.

Jiang, D. X. (1991). Fossil spores and pollen in petroleum and their significance. *Chinese Journal of Botany, 3*(1), 62–76.

Jiang, D. X. (1996). Fossil pollen and spores in crude oil from an igneous reservoir. In J. Jansonius & D. C. McGregor (Eds.), *Palynology: Principles and applications* (Vol. 3, pp. 1123–1128). Dallas, Texas: American Association of Stratigraphic Palynologists Foundation.

Jiang, D. X., & Yang, H. Q. (1980). Petroleum sporo-pollen assemblages and oil source rock of Yumen oil-bearing region in Gansu. *Acta Botanica Sinica, 22*(3), 280–285. (in Chinese with English abstract).

Jiang, D. X., & Yang, H. Q. (1981a). Spores and pollen in crude oils and Judgment of petroleum source rocks. *Mem. of Lanzhou Inst. of Geol., Acad. Sinica, 1*, 99–109 (in Chinese).

Jiang, D. X., & Yang, H. Q. (1981b). Spores and pollen in crude oils from West Jiuquan Basin. *Petroleum Geology, 5*(1), 1–9. (in Chinese).

Jiang, D. X., & Yang, H. Q. (1983a). Petroleum sporo-pollen assemblages and oil source rock of Kuche depression in Xinjiang. *Acta Botanica Sinica, 25*(2), 179–186. (in Chinese with English abstract).

Jiang, D. X., & Yang, H. Q. (1983b). On palynology of petroleum sources. *Chinese Bulletin of Botany, 1*(1), 32 (in Chinese).

Jiang, D. X., & Yang, H. Q. (1986). Petroleum sporo-pollen assemblage and oil source rocks of Yecheng Seg in Xinjiang. *Acta Botanica Sinica, 28*(1), 111–116. (in Chinese with English abstract).

Jiang, D. X., & Yang, H. Q. (1989). Spores and pollen from crude oils of Turpan Basin, Xinjiang. *Acta Botanica Sinica, 31*(6), 477–483. (in Chinese with English abstract).

Jiang, D. X., & Yang, H. Q. (1991). Spores and pollen in crude oils and petroleum sources of Tarim Basin. *Science in China (Series B), 21*(12), 1313–1318. (in Chinese).

Jiang, D. X., & Yang, H. Q. (1992). Spores and pollen in crude oils and petroleum sources of Tarim Basin. *Science in China (Series B), 35*(8), 1005–1012.

Jiang, D. X., & Yang, H. Q. (1994). Spores and pollen in oil from igneous rock petroleum pool and petroleum origin of Junggar Basin. *Science in China (Series B), 24*(7), 774–778, *37*(12), 1499–1505 (in Chinese).

Jiang, D. X., & Yang, H. Q. (1996). Spores and pollen from crude oil of Kashi Seg, Xinjiang. *Acta Botanica Sinica, 38*(10), 809–813. (in Chinese with English abstract).

Jiang, D. X., & Yang, H. Q. (1997). Palynological evidence for Jurassic petroleum source of Qaidam Basin. *Acta Botanica Sinica, 39*(12), 1160–1164. (in Chinese with English abstract).

Jiang, D. X., & Yang, H. Q. (1999a). Spores and pollen from crude oils of Weizhou oil-field in Beibu Gulf of South China Sea. *Acta Botanica Sinica, 41*(1), 102–106. (in Chinese with English abstract).

Jiang, D. X., & Yang, H. Q. (1999b). Petroleum sporo-pollen assemblages and petroleum source rocks of North Tarim Upheaval in Xinjiang. *Acta Botanica Sinica, 41*(2), 213–218. (in Chinese with English abstract).

Jiang, D. X., Yang, H. Q., & Du, J. E. (1974). Method of extraction of spores/pollen from crude oils. *Journal of Botany, 1*(1), 31–32. (in Chinese).

References

Jiang, D. X., Wang, Y. D., Robbins, E. I., Wei, J., & Tian, N. (2008). Mesozoic non-marine petroleum source rocks determined by palynomorphs in the Tarim Basin, Xinjiang, northwestern China. *Geological Magazine, 145*(6), 868–885.

Li, M. C. (2004). *Petroleum and natural gas migration* (3rd ed., pp. 1–350). Beijing: Petroleum Industry Press (in Chinese).

Lieux, M. H. (1972). A melissopalynological study of 54 Louisiana (USA) honeys. *Review of Palaeobotany and Palynology, 13*, 95–124.

Liu, Z. S. (2003). *Triassic and Jurassic Sporo-pollen assemblages from the Kuqa Depression, Tarim Basin of Xinjiang, NW China*. Palaeontologia Sinica, Whole Number 190, New Series A, No. 14 (pp. 1–244). Beijing: Science Press (in Chinese with English summary).

McGregor, D. C. (1996). Palynomorphs in petroleum and formation water: A review. In J. Jansonius, D. C. McGregor (Eds.), *Palynology: Principles and applications* (Vol. 3, pp. 1115–1121). Dallas, Texas: American Association of Stratigraphic Palynologists Foundation.

Ouyang, S. (1986). *Palynology of Upper Permian and Lower Triassic strata of Fuyuan district, eastern Yunnan*. Palaeontologia Sinica, Whole Number 169, New Series A, No. 9 (pp. 1–122). Beijing: Science Press (in Chinese with English summary).

Ouyang, S., Wang, Z., Zhan, J. Z., & Zhou, Y. X. (2003). *Palynology of the Carboniferous and Permian strata of northern Xinjiang, northwestern China* (pp. 1–700). Hefei: University of Science and Technology of China Press (in Chinese with English summary).

Research Institute of Exploration and Development, Qinghai Petroleum Administration, Nanjing Institute of Geology and Palaeontology, Academia Sinica. (1985). *A research on tertiary palynology from the Qaidam Basin, Qinghai Province* (pp. 1–297). Beijing: Petroleum Industry Press (in Chinese with English abstract).

Robbins, E. I. (1990). *Palynology of ore deposits* (pp. 387–578). Amsterdam: Elsevier Science Publishers B V.

Sanders, J. M. (1937). The microscopical examination of crude petroleum. *Journal of Institute of Petroleum Technology, 23*(167), 525–573.

Song, Z. C., Zheng, Y. H., Liu, J. L., Ye, P. Y., Wang, C. F., & Zhou, S. F. (1981). *Cretaceous—Tertiary palynological assemblages from Jiangsu* (pp. 1–268). Beijing: Geological Publishing House (in Chinese with English abstract).

Song, Z. C., Li, M. Y., & Zhong, L. (1986). *Cretaceous and Early Tertiary sporo-pollen assemblages from the Sanshui Basin, Guangdong Provice*. Palaeontologia Sinica, Whole Number 171, New Series A, No. 10 (pp. 1–170). Beijing: Science Press (in Chinese with English summary).

Sung, T. C. (1958). Tertiary spore and pollen complexes from the red beds of Chiuchuan, Kansu and their geological and botanical significance. *Acta Palaeontologica Sinica, 6*(2), 159–167. (in Chinese with English summary).

Timofeev, B. V., & Karimov, A. K. (1953). Spores and pollen in mineral oil. *Dokl. Akad. Nauk. SSSR, 92*(1), 151–152 (in Russian).

Tissot, B. P., & Welte, D. H. (1978). *Petroleum formation and occurrence* (pp. 1–314). New York, Berlin: Springer.

Traverse, A. (1988). *Paleopalynology* (pp. 1–600). Boston: Unwin Hyman.

Von Post, L. (1916). Om Skogsträdpollen i sydsvenska torfmosslagerfölyder. *Geol. Fören. Stockholm Förh, 38*, 384–390.

Waldschmidt, W. A. (1941). Progress report on microscopic examination of Permian crude oils. *American Association of Petroleum Geologists Bulletin, 25*, 934.

Wang, F. H., Qian, N. F., Zhang, Y. L., & Yang, H. Q. (1995). *Pollen flora of China* (2nd ed., pp. 1–461). Beijing: Science Press (in Chinese with English abstract).

White, D., & Stadnichenko, T. (1923). Some mother plants of petroleum in the Devonian black shales. *Economic Geology, 18*, 238–252.

Wright, J. A., & Sweet, P. W. K. (1924). The Jurassic as a source of oil in western Cuba. *American Association of Petroleum Geologists Bulletin, 8*(4), 516–519.

Yang, H. Q., & Jiang, D. X. (1981). Pollen and spores extracted from petroleum of Liaohe oil-field and their significance. *Acta Botanica Sinica, 23*(1), 52–57. (in Chinese with English abstract).

Yang, H. Q., & Jiang, D. X. (1989). Spores and pollen from crude oil of Dushanzi oil-field in Xinjiang. *Acta Botanica Sinica, 31*(12), 948–953. (in Chinese with English abstract).

Zhang, L. J. (1984). *Late Triassic spores and pollen from central Sichuan*. Palaeontologia Sinica, Whole Number 167, New Series A, No. 8 (pp. 1–100). Beijing: Science Press (in Chinese with English abstract).

2. Geological Background

Abstract

Crude oil and natural gas deposits are considered to be stratabound ore deposits, because of similar mechanisms of formation. The necessary geological conditions for the formation of oil and gas deposits are petroleum source rock, reservoir rock, caprock, and trap, all of which require a favorable time and space association. Details about these geological conditions, tectonic factors, and locations of oil fields are delineated for the inland petroliferous basins of west China, such as the Tarim, Junggar, Turpan, Qaidam, and West Jiuquan basins, and the coastal shelf petroliferous basins in the East and South China Seas, such as the Liaohe, Beibu Gulf, and Zhujiang Mouth Basins.

Keywords

Stratigraphy and lithofacies · Tectonic units · Potential petroleum source rock series · Oil–gas fields

2.1 Introduction

The geological background to understand why the inland and coastal shelf basins of China are so rich in hydrocarbons is quite complex. It requires analysis of three essential parts. The first is a clear description of the source rocks, reservoirs, and caprocks. The second is an explanation of the generation, migration, and accumulation of oil and gas. Finally, the geological processes that created and still act in the inland and coastal shelf basins must be explained.

2.2 Geological Conditions for Oil–Gas Field Formation

2.2.1 Complex of Source Rock, Reservoir Rock, and Seal Rock

The formation of oil and gas fields in general has essential conditions including oil source rocks (generating rocks), reservoir rocks, seal rocks, and traps, while at the same time having favorable configurations for these numerous factors in time and space. Among sedimentary rocks, carbonates and dark argillaceous rocks form the most productive hydrocarbon source rocks. Carbonate rocks that have developed holes and sandstones having well-developed pore structures are ideal reservoir rocks. Mudstone, shale, and halite layers as well as carbonate rock layers can form caprocks. In addition, igneous and metamorphic rocks with developed fissure structures can also be regarded as oil and gas reservoir rocks. Crustal movements and climate change participate in the process with regression and transgression of sea levels; such kinds of tectonic movements and regression–transgression may result in different types of hydrocarbon associations, as well as traps for source rocks, reservoit rocks, and caprocks.

In general, reservoir rocks are typically underlain by source rocks and overlain by caprocks. When source rocks and reservoir rocks are formed in the same period without depositional breaks, then in situ generation and in situ deposit formation are created. When source rocks and reservoir rocks are formed in different periods separated by depositional breaks, then the new complexes of source rock, reservoir rock, and seal rock are formed. Complexes including igneous or metamorphic reservoir rocks associated with sedimentary Mesozoic and Cenozoic petroleum source rocks have been found.

2.2.2 Generation, Migration, and Accumulation of Petroleum

Oil and gas are both fluids. They change their motion from a dispersed state in the source rocks to aggregation in reservoir rock traps, during which a migration process is also necessary. Oil and gas generation, migration, and accumulation represent a continuous evolution process. When buried organic materials are introduced into the threshold of oil generation and begin forming petroleum in the source rocks, the oil–gas generated from this source bed may begin to migrate through the carrier bed or the reservoir bed.

Hydrocarbon generation is an important cause of the abnormal high pressure and formation of microfissures. Microfissures are important channels for primary migration of oil and gas. Thus, hydrocarbon generation and hydrocarbon expulsion participate in a continuous process. The power that drives oil and gas migration includes abnormal stratigraphic pressure, tectonic stress, compaction water power, gravity, buoyancy, molecular diffusion forces, and molecular penetration (Li 2004).

Tectonic forces produce two essential structural features important in producing large oil and gas fields. One is the formation of depressions into which sediment can accumulate and the second is uplift. Therefore, depressions of hereditary nature are favorable for oil and gas generation, and uplift is the target of oil and gas migration. However, oil and gas loss in the course of oil and gas migration is absolute, while the accumulation of oil and gas that occurs in the process of migration is relative and conditional.

Trap and caprocks are basic conditions for oil and gas accumulation and preservation. The trap represents the boundary space that prevents oil and gas from escaping, resulting in the accumulation of oil and gas in reservoir beds. Trap formation mainly depends on structural factors; anticline trap and fault trap are common structural trap types. In addition, stratigraphic overlap, stratigraphic unconformity between strata, as well as lithologic pinch out and lenticular beds are also important factors for trap formation. The caprock bed is a relatively impermeable bed that can prevent upward or up direction motion of hydrocarbon migration and leakage. Whether the oil–gas gathered in such traps can be preserved largely depends on the sealing capacity of the trap and the sealing ability of the caprocks.

Common caprocks include dense shale, shale in general, salt deposits, and carbonate rocks; both the lithology and thickness of the caprock determine caprock capability. The ideal caprocks are anhydrite, gypsum, rock salt, and compacted mudstone. Such caprocks do not require great thickness in order to prevent leakage of oil and gas.

2.2 Geological Conditions for Oil–Gas Field Formation

2.2.3 Petroleum Deposit and Its Classification

A petroleum deposit (petroleum pool) is the volume of oil and gas accumulated in a trap that encloses a single reservoir bed. The petroleum deposit is the smallest unit of oil–gas accumulation in a petroliferous basin. Oil and gas fields occur where a trap encloses two or several reservoir beds. According to the causes of trap formation, petroleum deposits can be divided into the following types: structural trap petroleum deposits, including anticline trap and fault trap; stratigraphic–lithologic trap petroleum deposits, including stratigraphic trap and lithologic trap; and complex trap petroleum deposits, including lithology–anticline trap and strata–structural trap. According to the phases of the hydrocarbon, petroleum deposits can be divided into four categories, including gas deposit, boundary state deposit, oil deposit, and asphaltic oil deposit (Li 2004).

2.3 Inland Petroliferous Basins

There are many inland oil- and gas-bearing basins in China. In this chapter, we highlight and discuss the most important basins in great detail because of their significance to the thesis that palynology plays a major role both in understanding the hydrocarbon potential of such basins and in providing information about other exploration targets. They include the Tarim, Junggar, Turpan-Hami, Qaidam, and West Jiuquan basins.

2.3.1 The Tarim Basin

The Tarim Basin is located in between the Tianshan, Kunlun, and Aljin mountains, with a cover area of 560,000 km². The basin depth is as much as 13,000–15,000 m. It is a large-sized cratonic, superimposed, composite basin.

The geological development history of the basin is divided into three stages, namely the Proterozoic geosynclinal stage, the Paleozoic platform developmental stage, and the Cenozoic platform fault block developmental stage. The Jinning Orogeny is equivalent in time to the Tarim Orogeny participated in the Proterozoic geosynclinal folding, thus resulting in the creation of the ancient Tarim platform and its crystalline basement. During the Paleozoic, the basin sank and was subject to five transgressions and a huge thickness of marine carbonate sediments. These sediments constitute the bottom cover of the platform. Late Hercynian movements included folding which also affected the peripheral Tianshan Mountains and Kunlun geosyncline. Within the platform, the sediments were subject to relatively uneven subsidence, and in the mountain front depression and sag regions, continental clastic sediments were deposited during the Mesozoic. The Himalayan Orogeny created sharply rising peripheral mountains, and finally, today's Tarim Basin was formed. A great thickness of Cenozoic continental sediments was next deposited. The Mesozoic and Cenozoic deposits form the upper caprock of the platform. The Tarim Basin is thus a Mesozoic and Cenozoic fault depression basin formed above Pre-Sinian crystalline basement and the Paleozoic stable platform. Therefore, it represents a typical cratonic, superimposed, composite basin.

2.3.1.1 Stratigraphy and Lithofacies (Table 2.1)

The Sinian System: In the northwest margin of the Tarim Basin, there is a set of continuous transgressive deposits in the Keping area. These are about 1940 m in total thickness and composed of conglomeratic sandstones, sandstones, and sandy shales that gradually transition to carbonates. The lower part of the Sinian System in the Keping area is represented by the Yourmeilake Formation, which is composed of conglomerates, conglomeratic mudstones, with intercalations of mudstones. The middle part is the Sugeitblake Formation which is mainly composed of shallow marine and coastal clastic rocks of terrestrial origin; it also contains a basalt in the lower stratigraphic member, as well as muddy limestone, dolomitic limestone, and dolomite in the upper stratigraphic member. The upper part of the Sinian System in the Keping

Table 2.1 Potential petroleum source rock series of the Tarim Basin

Geological age	Stratigraphic formation	Code	Lithology	Facies
Late Cretaceous	Yingjisha Group	K_{2y}	Gray limestone, dolomite, and sandy mudstone	Shallow sea facies
Middle Jurassic	Qiakemake (Taergan) Formation	J_{2q}	Gray sandstone, siltstone, and mudstone	Lacustrine
	Kezilenuer (Yangye) Formation	J_{2k}	Gray sandstone, mudstone, carbonaceous shale, coal beds	Lacustrine to swamp facies
Early Jurassic	Yangxia (Kansu) Formation	J_{1y}	Gray mudstone, sandstone, and carbonaceous shale with coal beds	Lacustrine to swamp facies
	Ahe formation	J_{1a}	Gray sandstone, conglomerates with intercalations of mudstone	Fluvial–lacustrine facies
Late Triassic	Taqilike Formation	T_{3t}	Gray mudstone, sandstone and carbonaceous shale, coal beds	Lacustrine to swamp facies
	Huangshanjie Formation	T_{3h}	Gray mudstone, sandstone and carbonaceous shale, coal beds	Lacustrine to swamp facies
Middle Triassic	Karamay Formation	T_{2k}	Gray mudstone, sandstone, carbonaceous shale	Lacustrine to swamp facies
Early Triassic	Ehuobulake Formation	T_{1e}	Conglomeratic sandstones alternated with mudstone	Fluvial–lacustrine facies
Carboniferous-Permian	Muziduke Group	C-P	Biologic reef limestone, black shale	Shallow sea facies
Late Carboniferous	Bijingtawu Formation	C_{2b}	Limestone, sandstone, black shale	Shallow sea facies
Middle Ordovician	Yinggan Formation	O_{2y}	Black shale, mudstone, containing graptolites	Littoral facies
	Saergan Formation	O_{2s}	Black shale, rich in graptolites	Littoral facies
Early Ordovician	Qiulitage Group	O_{1q}	Limestone, dolomite, geode seepage	Shallow sea facies
Early Cambrian	Wusongge'er Formation	ε_{1w}	Limestone, dolomite	Marine facies
	Xiaoerbulake Formation	ε_{1s}	Limestone, dolomite; contains trilobites	Marine facies

area of the Tarim Basin is represented by the Qigebulake Formation, which is mainly composed of gray algal dolomite carbonate deposits, with well-developed algal laminations and a few columnar laminated beds.

In the Kuluketage region, which is at the margin at the northeast of the Tarim Basin, there is a set of massive, thick deposits of coastal to shallow marine clastic, volcanic, and pyroclastic rocks, as well as tillite and sandy limestone deposits. These deposits are as much as 5750 m in total thickness. In the Kuluketage region, the Qigebulake Formation is unconformably underlain by Pre-Sinian System.

The Cambrian System: The Cambrian System in the Keping region of the Tarim Basin is represented by a series of shallow marine platform carbonate deposits, about 600 m in thickness. The lower part of the Lower Cambrian is dominated by dark gray, phosphatic lenticular siliceous rocks and phosphates with intercalations of dolomitic limestones. The middle part is characterized by black carbonaceous shales, gray green, and purple shales with intercalations of dolomites. And the upper part is a gray white, thin-layered dolomicrite, and knotty dolomitic shales, about 8–35 m in thickness. The middle part of the Lower Cambrian is represented by the Xiaoerbulake Formation which consists mainly of limestone and dolomite, yielding fossil trilobites and brachiopods, about 150 m in thickness. The upper part is defined as the Wusongger Formation, which is composed of dolomitic limestone, nodular limestone, as well as limy dolomite with cherts, about 150 m in thickness. Of Middle Cambrian age, the lower part of the Shayilike Formation is dominated by biolimestone, leopard skin limestone, bamboo-like limestone, and conglomeratic limestone; the upper part of this formation is characterized by thin chert strip limestone with muddy dolomite, about 90 m in thickness. The Awatage Group of the Middle to Upper Cambrian is dominated by a series of gray and brick red gypsum, muddy dolomite with chert strips, and algal dolomite deposits representing a shallow marine lagoon facies, about 200 m in thickness.

In the Kuluketage region, the lower part of the Lower Cambrian is characterized by the Xishanbulake Formation, which at its base is composed of black siliceous and phosphatic rocks; the middle section is volcanic rocks; and the upper part is black siliceous rocks, and the entire formation is 37–130 m in thickness. The upper part of the Lower Cambrian is represented by the Xidashan Formation which is composed of gray black mudstone, sandy limestone with intercalations of cherty limestone, nodular limestone, and purple shale; it contains fossil trilobites and brachiopods and is 13–88 m in thickness. The Middle Cambrian is defined as the Mohe'er Group, which consists of dark gray and gray black sandy and muddy limestones with intercalations of cherty limestone, leopard skin limestone, and marls and shales; the formation is 120–300 m in thickness. The Upper Cambrian in the Kuluketage region is represented by the Tuershaketage Group, which manly consists of dark gray, thick limestones with intercalations of thin limestones and has a thickness of over 250 m.

The Ordovician System: The Ordovician of the Tarim Basin is mainly dominated by shallow marine carbonate deposits; in the Keping area, it is about 1400 m in thickness, and in the Kuluketage area, the thickness varies from 2600 to 2800 m. In the Keping area, the Lower Ordovician Qiuliketage Group is a gray to dark gray dolomite and limestone with cherty masses or thin beds; it contains fossil trilobites and brachiopods and has a thickness of 178–1300 m. The Middle Ordovician Saergan Formation consists of black, gray black, and gray green shales with intercalations of thin layers or nodular muddy limestones; it contains fossil graptolites, trilobites, and conodonts and has a thickness of 16 m in the Dalanggou Section. The Kanling Formation of the lower part of the Middle Ordovician is composed of purple calcilutite limestone, with intercalations of gray green silty sandstone; it contains fossil conodonts, cephalopods, and trilobites and has a thickness of 17–36 m. The Qilang Formation of the middle part of the Middle Ordovician consists of gray, thin-layered calcilutite limestone interbedded with gray green siltstone, mudstone, and muddy limestones; it contains fossil graptolites, trilobites, and cephalopods and ranges in thickness from 158 to 177 m. The upper part of the Middle Ordovician is represented by the Yingan Formation which consists of black and dark gray carbonaceous shale and mudstones with intercalations of calcilutite limestone; it contains fossil graptolites, trilobites, and brachiopods and has a thickness of 35 m. The Late Ordovician is missing because the platform was uplifted during the latest Middle Ordovician.

The Silurian System: In the Keping area of the Tarim Basin, the Silurian System is characterized by clastic rocks deposited in a littoral facies; a disconformity separates it from the underlying Ordovician. The Lower Silurian Kepingtage Formation is composed of gray green sandstone, siltstone, and mudstone; it contains fossil graptolites, trilobites, and plants and has a thickness of 390–445 m. The Middle and Upper Silurian Tataertage Formation consists of purple and gray green sandstones, siltstones, and silty shales; it varies from 160 to 270 m in thickness. In the Kuluketage area, only the Lower Silurian Series is exposed, and is represented by the Tushibulake Formation, which consists of gray green sandstones, siltstones, dark gray to black shales, marls, and limestones, with a thickness of 720–2271 m. It lies disconformably above rocks of the Ordovician System.

The Devonian System: In the Keping area of the Tarim Basin, the Devonian is of littoral to continental facies origin; the rocks are red clastics which grade conformably into underlying Silurian rocks. The Lower Devonian Yimugantawu Formation consists of purple sandstones and siltstones, including gray white greywacke; the formation varies from 450 to 500 m in thickness. The Middle to Upper Devonian Keziertage Formation consists of brick red sandstones, fine sandstones, and siltstones, with a thickness of 920–1200 m. In the Kuluketage area, the Lower Devonian Shugouzi Formation is composed of gray green and interbedded red sandstones; the sandstones of the upper part are red. The unit disconformably overlies the Silurian and varies from 300 to 400 m in thickness. The Middle to Upper Devonian Series is represented by the Aertemeishibulake Formation which consists of purple, course sandstones, and gray white sandstones in the lower part; and red conglomerates and course sandstones in the upper part, and the formation has a thickness of 1200–1300 m.

The Carboniferous System: During the Carboniferous period, a shallow sea and lagoon facies developed in the Tarim Basin. Carboniferous strata are widely exposed and disconformably cover Devonian rocks. In the Tiekelike area at the southwestern margin of the basin, the basal part of the Lower Carboniferous, the Kelitage Formation, consists mainly of dark gray limestone with intercalations of gray dolomite and a few gray green sandstone and siltstone beds, with a thickness of 190–380 m. The upper part of the Lower Carboniferous Heshilapu Formation consists of the conglomerate, sandstone, siltstone, biolimestone, and carbonaceous shale with coal seams, about 818–1900 m in thickness. The lower part of the Upper Carboniferous, the Kalawuyi Formation, consists mainly of gray green and gray black mudstone, and light gray sandstone alternating with dark gray limestone; the formation has locally developed coal seams and contains fusulinids, brachiopods, and plant fossils, about 140–190 m in thickness. The middle part is represented by the Azigan Formation, consisting of gray limestone with intercalations of a minor sandstone and gray green mudstone; it contains fossil fusulinids, brachiopods, and corals and is 170–445 m in thickness. The upper part is defined by the Tahaqi Formation, which is composed of gray limestone with intercalations of minor purple silty mudstone and dark gray mudstone; it contains fusulinid and brachiopod fossils and is about 130–175 m in thickness.

In the northwestern margin of the Tarim Basin, in the Keping area, the Lower Carboniferous Series Kongtaiaikenggou Formation consists of gray conglomerate, sandstone, and siltstone with intercalations of gray limestone, marl, and purple mudstone and siltstone; total thickness is 346 m. The lower part of the Upper Carboniferous Bijingtawu Formation consists mainly of dark gray limestones alternating with sandstones and siltstones; the middle part of the series is represented by biolimestone, and sandstone alternating with black shales and coal seams. The upper part is sandstone having intercalations of black shales and limestone; fossils include brachiopods and corals, and the total thickness of the Bijingtawu Formation is 1660 m at the southern end of the Wushi area. The upper part of the Upper Carboniferous Kangkelin Formation consists of black shales, grayish brown sandstones, and gray siltstones intercalated with dark gray limestones; the upper part contains limestones yielding fossil fusulinids, brachiopods, and corals. The total thickness of the Kangkelin Formation varies from 86 to 663 m (Ma and Wen 1991).

In the western margin of the Tarim Basin where the Yanguan Series is absent, the Lower

Carboniferous is mainly dominated by carbonate deposits. At the southern margin, the Lower Carboniferous Datang Series and the Upper Carboniferous deposits consist mainly of clastic rocks deposited in a shallow sea along with carbonates; volcaniclastic rocks are also present. At the eastern margin of the basin, in addition to the shallow sea clastics and carbonates, there are a large number of volcanic intrusive and eruptive rocks in the Carboniferous. The entire Carboniferous has a maximum thickness of more than 5000 m (Zhou and Chen 1990).

The Permian System: The Permian System is mainly distributed in the northwest and southwest margin of the Tarim Basin, conformably covering the Carboniferous System. In the Keping area at the northwestern basin margin, the Lower Permian is bounded by the Yinggan Shan Mountain Range. The western part is of marine origin, whereas the eastern side is a terrestrial deposit. In western part of the Yinggan Mountain Range, the basal part of the Lower Permian Balikelike Formation is a marine deposit, consisting of dark gray and black limestone, and shelly limestone with intercalations of carbonaceous shale; the formation contains brachiopods, bivalves, and fusulinid fossils and is 340 m in thickness. The upper part of the Lower Permian, the Kalundaer Formation, is of terrestrial facies origin and is composed of variegated sandstone and mudstone as well as carbonaceous mudstone and siltstones, yielding fossil spores and pollen grains, and is 22–340 m thick. In the eastern Yinggan Mountain Ranges, the basal Lower Permian is the Kupukuziman Formation which consists of a series of variegated sandstone, siltstone, and mudstone with intercalations of volcanic tuff. The upper part of this formation is composed of basalt with intercalations of grayish green and grayish purple silty mudstone, tuffaceous sandstone, volcanic breccia, with a thin coal seam at the top of the formation yielding plant fossils; the entire formation is 451–470 m thick. The upper part of the Lower Permian Series is represented by the Kaipaizileike Formation. In this formation, the lower part is characterized by variegated sandstone and dark mudstone with coal seams, and the upper part consists of basalt with variegated sandstone, siltstone, mudstone, and coal seams. The formation contains plant and sporo-pollen fossils, and the entire formation is 580–660 m thick. For the Upper Permian Series, the Shajingzi Formation is of terrestrial origin, and consists of variegated sandstone, siltstone, carbonaceous mudstone with thin coal seams, about 700 m in thickness.

In the southwestern margin of the Tarim Basin, which is the western part of the Pishan River region, the Lower Permian Qipan Formation is a marine deposit, and consists of gray and grayish black calcareous mudstone, carbonaceous mudstone, grayish green as well as mousy gray sandstone and siltstone with intercalations of biolimestone and shelly limestone, yielding fossil brachiopods and lamellibranchs, as well as bryozoans and spores and pollen; the formation is 476 m thick. The Upper Permian Daliyue'er Formation is a continental facies, consisting of purple, maroon, and grayish green sandstone, siltstone, and mudstone; it contains plant (*Calamites*) and ostracode fossils and is 191 m thick. In the eastern part of the Pishan River region, the Lower Permian Pusige Formation is a continental facies, consisting of brownish red, maroon, and grayish green sandstone, sandy mudstone, siltstone with thin-layered limestone, and containing plant fossils; it is about 1791 m in thickness. The Upper Permian Duwa Formation is a continental facies, consisting of maroon and taupe conglomerates, coarse sandstone, and siltstone, with grayish green mudstone and a thin-layered limestone at the top of the formation, bearing lamellibranch and conchostracan fossils; the thickness varies from 418 to 1268 m.

In addition, the Upper Permian Biyoulebaoguzi Group in the Kuqa area consists of purple and brownish red conglomerate, sandstone, and dark sandy mudstone with intercalations of grayish green sandstone and carbonaceous shale. The unit disconformably covers Carboniferous rocks and has a thickness of 286 m.

The Triassic System: The Triassic System is mainly exposed in the northern portion of the Tarim Basin and contains terrestrial sediments. In the Kuqa–Baicheng region, the Lower Triassic

Ehuobulake Formation disconformably covers the Lower Permian Biyoulebaoguzi Group. The unit is dominated by purple, sandy conglomerates that alternate with grayish green sandy mudstone and is underlain by a basal conglomerate. The fossils in the formation are plants, conchostracans, charophytes, and ostrocodes; the thickness varies from 191 to 592 m. The Middle Triassic "Karamay Formation" consists of grayish green and beige conglomerates, sandstone intercalated with purple sandy mudstone, and siltstone; it also contains black carbonaceous shale at the top of the formation with fossil plants, spores, pollen, and conchostracans. Total thickness of the formation is 572–885 m.

The lower Upper Triassic Huangshanjie Formation consists of grayish green sandstones and siltstones intercalated with dark gray mudstone, carbonaceous mudstone, and coal beds; it contains various fossils including plants, spores and pollen, charophytes, and conchostracans, with a thickness varying between 168 and 279 m. The uppermost Triassic Taliqike Formation includes gray conglomerates, sandstones, grayish green siltstone, sandy shale, as well as black carbonaceous shale and coal beds, yielding plants, spores and pollen, and conchostracans, with a thickness of 544–837 m.

In the Avati region, Triassic strata are encountered in the drilled boreholes including Yuecan #1, Shacan #1, and Acan #1. The Triassic Manjiaer Group in the drill samples is composed of basal grayish green, beige mudstones, siltstone, and fine sandstone alternating with fine conglomerate and thin-layered limestone, with a thickness of 239 m. The middle part consists of grayish green and beige mudstones, and siltstone with intercalations of dark gray carbonaceous mudstone and coal beds, with a thickness of 241 m. The upper part is composed of gray, beige, and reddish brown mudstones, siltstone with intercalations of black carbonaceous shale and thin coal beds, and the lower part with variegated conglomerate with a thickness of 190 m (Ma and Wen 1991).

The Jurassic System: Jurassic strata are mainly distributed in the Kuqa–Baicheng and Kashi areas in the northern Tarim Basin and Tiekelike area in the southwestern Tarim Basin. The sediments are mainly of a lacustrine and swamp facies with coal-bearing clastic deposits. The Jurassic lies disconformably or unconformably in contact with the underlying strata. In the Kuqa–Baicheng area, the lower part of the Lower Jurassic, the Ahe Formation, disconformably lies above the underlying Triassic Series; consists of grayish white, yellowish gray conglomerates and sandy conglomerates, and sandstones with some thin dark gray mudstone layers; and is 300–500 m thick. The upper part of the Lower Jurassic is represented by the Yangxia Formation, which consists of gray and yellowish gray course sandstones, siltstones, grayish black mudstones, and black carbonaceous shale with coal beds, including fossil plants, pollen, and bivalves; total formation thickness is 200–400 m. The Middle Jurassic Kezilenuer (Yangye) and Qiakemake (Taerga) formations include gray sandstones, dark gray and greenish gray mudstones, black carbonaceous mudstones, coal beds, and oil shales, containing miospores, megaspores, estherids, ostracodes, and bivalves; formation thickness ranges from 872 to 1245 m. The Upper Jurassic Qigu Formation consists of brown and brownish red mudstones intercalated with sandstones, and the thickness is 256 m; fossils include charophytes, ostracodes, and bivalves.

In the Kashi area, the lower part of the Lower Jurassic Series is represented by the Shaliqike Formation, which unconformably overlies Permian strata, and is composed of grayish green conglomerates and sandstones, siltstones, and black carbonaceous shales, containing plants and sporo-pollen fossils, with a thickness of 1314 m. The upper part of the series is the Kansu Formation, consisting of grayish green sandstones and sandy conglomerates with alternating dark gray mudstones, black carbonaceous mudstones, and coal beds, yielding plants, pollen, and bivalves; the thickness is 1160 m. For the Middle Jurassic Series, the lower part is the Yangye Formation which consists of gray, grayish black mudstones, with intercalations of sandstones, shale, and marl, yielding fossil spores and pollen, ostracodes, and bivalves, with a thickness of

1030 m. The upper part of the Middle Jurassic, the Taergan Formation, consists of variegated clastic rocks dominated by grayish green mudstone, yielding spores and pollen, ostracodes, bivalves, and conchostracans; and is 215 m thick. The Upper Jurassic Kuzigongsu Formation is composed of red conglomerates with intercalations of sandstone and mudstone, and 423 m in thickness.

In the Tiekelike area, the basal part of the Lower Jurassic, the Shalitashi Formation, consists of conglomerates, whereas the upper part of the Lower Jurassic Formation, the Kansu Formation consists of grayish green and grayish yellow sandstones with black carbonaceous shales. The basal part of the Middle Jurassic Yangye Formation consists of grayish green and yellowish green mudstones with intercalations of sandstones, conglomerates, and coal beds. The upper part of the Middle Jurassic, the Taerga Formation, consists of purple mudstones and siltstones. The Lower to Middle Jurassic Ye'erqiang Group has a total thickness of 850–2759 m and lies disconformably over the underlying Permian strata. The Upper Jurassic strata are missing in this region (Zhou and Chen 1990; Ma and Wen 1991; Zhou 2001).

The Cretaceous System: The Cretaceous System is of terrestrial origin, a clastic facies in most of the Tarim Basin, except in the southwestern part, where the Upper Cretaceous is a marine facies. In the Kuqa–Baicheng area, the Lower Cretaceous Kapushaliang Group lies unconformably above the Upper Jurassic Qigu Formation. The basal part of the Lower Cretaceous, the Yageliemu Formation, consists largely of grayish purple conglomerates, conglomeratic sandstones, and sandstones, about 120–290 m thick. The middle part Shushanhe Formation consists of brownish red, reddish brown, and blue gray mudstones with intercalations of yellowish green fine sandstones and siltstones, yielding fossils of sporo-pollen and ostracodes, with a thickness of 141–1099 m. The upper part of the Lower Cretaceous, the Baxigai Formation, consists of beige and red brown conglomerates and sandstones, with mudstones and siltstones, yielding fossil sporo-pollen, ostracodes, and conchostracans; the formation is 240–400 m thick. The Upper Cretaceous Bashijiqike Formation consists of purple red conglomerates and sandstones and lies disconformably over the Lower Cretaceous; it is about 115–200 m in thickness.

In the Shache–Kashgar area, the Lower Cretaceous Kezlesu Formation lies unconformably or disconformably over Lower Jurassic strata and consists of brownish red and grayish green sandstones and conglomerates intercalated with mudstones and siltstones; it yields fossil spores and pollen, and ostracodes and has a thickness of 500–1300 m. The Upper Cretaceous Yingjisha Group is marine facies that is divided into Kukebai, Wuyitage, Yigeziya, and Tuyiluoke formations in ascending order; the group lies conformably over the Kezilesu Group. The lower part of the Kukebai Formation consists of reddish brown and grayish green mudstones having intercalations of dolomite and thin-layered gypsum; the base of this formation consists of white sandy dolomicrites, and the top consists of micrites and aphanitic limestones, yielding foraminifera and bivalve fossils, about 62 m thick. The middle part of this formation consists of dark gray and grayish green mudstones with oyster-shell limestones and aphanitic limestones, yielding fossil foraminifera and bivalves, about 69 m thick. The upper part of this formation consists of gray mudstones with limestones, yielding bivalves and ostracode fossils, about 26 m thick. The Wuyitage Formation consists of grayish green and dark purple mudstones and siltstones with intercalations of white gypsum layers, about 87 m in thickness. The Yigeziya Formation is composed of grayish green, gray, and purple dolomite and limestone with mudstones in the upper part, yielding fossil bivalves, echinoderms, foraminifera, bryozoans, and other fossils, about 162 m thick. The Tuyiluoke Formation is composed of brownish red mudstone and gypsum-bearing mudstone and is about 67 m in thickness.

In addition, in the Jianggeshayi Section of Qiemo County, southwestern Tarim Basin, the Lower Cretaceous Kezlesu Group consists of brownish red sandstones, siltstones, and mudstones with intercalations of green sandstones and mudstones, and the base of the formation is composed of light gray, fine conglomerates that lay disconformably over the Upper Jurassic Kuzigongsu Formation; the Kezlesu Group has a thickness of 498 m.

The Tertiary System: In the southwest Tarim Basin, the Lower Tertiary (Paleogene) is a marine facies; in other places, the Lower Tertiary deposits are terrestrial facies, lying disconformably or unconformably above the underlying strata. In the Shache–Hetian area, the Paleocene Alta Formation is marine in the upper part, consisting of white, massive gypsum, about 214–450 m thick. The Paleocene to Eocene Series Qimugen Formation consists of marine limestones and grayish green mudstones in the lower part, and of a lagoon facies in the upper part with red mudstones, gypsum-bearing mudstone, and gypsum, all about 86–194 m in thickness. The middle part of the Eocene Kalataer Formation consists of red gypsum-bearing mudstone, gypsum, and limestones in the lower member, and marine limestones in the upper member, all about 78–98 m in thickness. The Upper Eocene Series Wulagen Formation is a shallow marine and littoral facies; the middle to lower parts are composed of grayish green mudstone with limestone, whereas the upper part is red mudstones with intercalations of grayish green sandstones, mudstones, and a thin shell layer, all about 98–130 m thick. The Oligocene Bashibulake Formation is a lagoon facies, consisting mainly of brownish red mudstones, grayish red siltstones, and fine sandstones with intercalations of grayish green mudstones, siltstones, sandstones, shell layers, and white gypsum, all with a thickness of 315–800 m. The lower part of the Miocene Keziluoyi Formation consists of brownish red mudstones with grayish beige siltstones and sandstones, with a basal layer of gypsum and gypsum-bearing mudstone, lying conformably over the Paleogene; the formation is about 422–450 m in thickness. The Middle Miocene Anju'an Formation consists of gray and grayish green mudstones, siltstones, sandstones, and a gypsum layer, all about 486–668 m in thickness. The Upper Miocene Pakabulake Formation consists largely of brown, sandy mudstones, with alternations of siltstones and sandstones, about 1100–2168 m thick. The lower part of the Pliocene Atushi Formation consists of gray and brown sandstones and conglomerates with intercalations of yellow, sandy shale, and siltstone, all of which is 1750–3497 m thick. The Pliocene Xiyu Formation consists of dark gray conglomerates; the thickness varies from 100 to 3000 m.

In the Kuqa–Baicheng area, the Lower Tertiary Kumugeliemu Group is a tidal flat and lagoon facies composed of gypsum-bearing mudstones. The Paleocene to Eocene Talake Formation consists of variegated clastic rocks and dolomites with a basal conglomerate, whereas the upper part is gypsum-bearing mudstone and siltstone containing ostracode and bivalve fossils, all with a thickness of 150–210 m. The middle part of the Eocene Xiaokuzibai Formation consists of brick red and purple sandy mudstone, siltstone with a thin layer of gypsum and halite, and yields ostracode and foraminifera fossils; this sequence is 230–500 m thick. The Upper Eocene Awate Formation is composed of purple and brown mudstone, and sandy mudstone with intercalations of grayish green siltstone and halite; the formation varies from 259 to 500 m in thickness.

The Neogene System is mainly a terrestrial deposit which lies disconformably over the underlying strata. The basal Miocene Suweiyi Formation consists of maroon conglomerate, sandstone, siltstone, and mudstone with gypsum and halite, all about 200–400 m in thickness. The Middle Miocene Jidike Formation consists of red and maroon conglomerate, sandstone and conglomeratic sandstone, and mudstone with gypsum-bearing mudstone, varying from 600 to 800 m in thickness. The Upper Miocene Kangcun Formation is composed of gray sandstone alternating with brown mudstone that includes a banded, grayish green siltstone; total thickness is 300–800 m. The lower part of the Oligocene Kuqa Formation is composed of beige conglomerate that changes to the south into gray to brownish gray sandstone and siltstone alternating with conglomerate, all of which is about 300–

700 m in thickness. The Upper Oligocene Xiyu Formation consists of dark gray conglomerate with intercalations of beige sandy mudstone and conglomeratic sandstone; the thickness is 50–600 m in general, with the maximum thickness of 1366 m in the Tarim Basin.

2.3.1.2 Tectonic Characteristics and Tectonic Units

Tectonically, the Tarim Basin (Fig. 2.1) is located in the Tarim platform area. The basin is bounded to the north by the Tianshan Mountain Range, to the southwest by the Kunlun Mountain Range, and to the southeast by the Altun Mountain Range. The Tarim Basin has experienced seven tectonic movements and is characterized by six unconformity surfaces. The Tarim Orogeny is equivalent to the Jinning Orogeny that inverted folds in the Paleozoic geosyncline and formed crystalline basement. The Sinian System forms caprocks that lie unconformably over the Pre-Sinian System. Middle Ordovician Caledonian Orogeny resulted in uplifting the platform and creating exposed land. Erosion of the Upper Ordovician sediments thus formed an unconformity surface between the Silurian and Ordovician systems. Early Hercynian Orogeny during the latest Devonian sank the platform, thereby forming an unconformity surface between the Carboniferous and Devonian systems. Late Hercynian Orogeny formed an unconformity surface between the Lower and Upper Permian Series. The Indosinian Orogeny at the end of the Triassic produced an unconformity surface between the Jurassic and Triassic. At the end of the Cretaceous period, the Late Yanshanian Orogeny produced an unconformity surface between the Tertiary and the Cretaceous. The Himalayan Orogeny formed the basic pattern seen in the modern basin. In addition to stratigraphic unconformities produced in different periods, all previous tectonic movements determine the shape, structure, and tectonic characteristics of the basin. The rigid crystalline basement of the Pre-Sinian System is cut by the two deep faults bearing west-northwest and east-northeast, thus forming rhombohedral blocks. These two set of deep faults control the basic tectonic framework of the basement and cover, thereby forming three upheavals, four

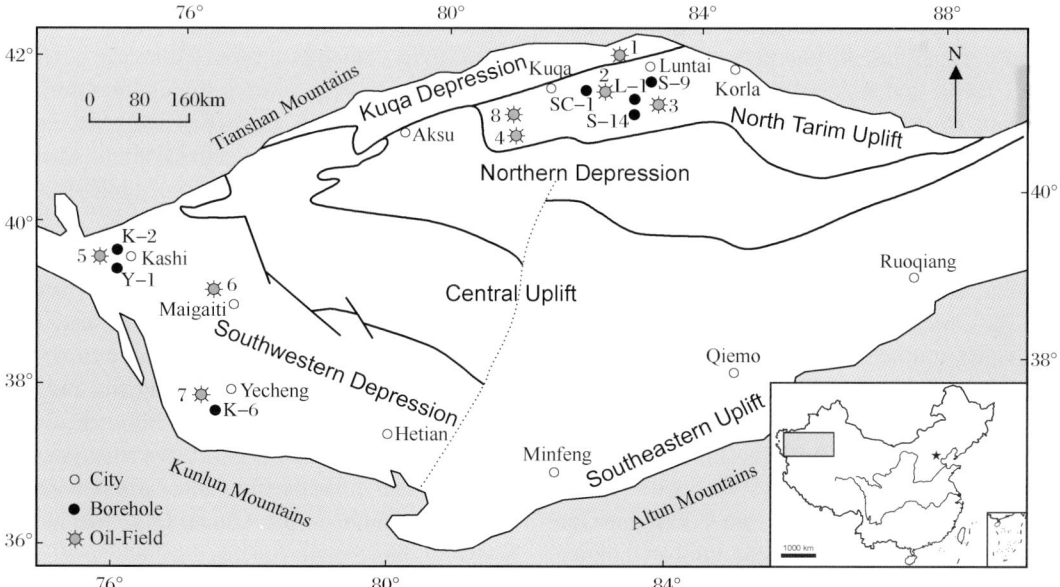

Fig. 2.1 Sketch map showing tectonic subdivisions and locations of oil fields and boreholes in the Tarim Basin. *1* Yiqikelike oil field. *2* Yakela oil field. *3* Lunnan oil field. *4* Yingmaili oil field. *5* Kelatu oil field. *6* Maigaiti oil field. *7* Kekeya oil field. *8* Donghetang oil field

depressions, and various types of folds and faults. Inside the Tarim Basin, seven tectonic units are recognized, including Central Upheaval, North Depression, North (Tarim) Upheaval, Southeast Upheaval, Kuqa Depression, Southwest (Tarim) Depression, and the Southeast Depression. The North Depression can be further subdivided into four secondary tectonic units, such as the Avati Sag, Manjiaer Sag, Yingjisu Sag, and the Kongque River Slope. The Southwest (Tarim) Depression can be subdivided into five secondary tectonic units, including the Kashi Sag, Yecheng Sag, Tanggu Ancient Sag, and Meigaiti Slope (Zhou and Zheng 1990) (Fig. 2.1). Fault activity plays an important role in the formation of the Tarim Basin tectonic framework and petroleum reservoir formation; all of the above tectonic units demonstrate block fault features. The formation and evolution of each block are controlled by deep fault activity that formed basin basement fault blocks in the west-northwest and east-northeast directions. Thus, the tectonic axis and fracture direction are mostly in the west-northwest and east-northeast directions.

2.3.1.3 Potential Petroleum Source Rock Series and Oil–Gas-Bearing Prospects

The massive, thick sedimentary sequence in the Tarim Basin contains many sets of carbonates and dark mudstones; these are potential hydrocarbon source rocks in a variety of geological periods (Table 2.1). The grayish black carbonate rocks are well developed in the Lower Cambrian Xiaoerbulake and Wusongger formations, which are fossiliferous and exhibit good potential for oil and gas generation. The Lower Ordovician Qiulitage Group contains neritic facies limestone and dolomite, and liquid crude oil is found in the geodes of the miarolitic limestones, showing that these carbonate rocks should be a good oil source rocks. The black shales and mudstones in the Middle Ordovician Saergan and Yinggan formations are rich in fossil graptolites and pyrites nodules, representing a deoxygenated environment of stagnant conditions and anoxic sediment. These are favorable for enrichment and preservation of organic matter and thus are also a potential oil source rock series. In the Carboniferous and the Lower Permian, black and carbonaceous shales with intercalations of minor sandstone, siltstone, and limestone should also be ideal hydrocarbon rock series. In the Keping–Aheqi area, the organic reef facies limestone of the Upper Carboniferous to Lower Permian Muziduke Group contains abundant reef-forming organisms, such as fusulinids, corals, brachiopods, bryozoans, algae, and sponges; these thrived in the warm water environment of the platform edge. Such reef facies are often hydrocarbon source rocks. The Middle Triassic "Karamay Formation" and the Upper Triassic Huangshanjie and Taliqike formations are deposits of inland lake and swamp deposits, consisting of dark mudstones with intercalations of thin layers of siderite and coal beds; these units are rich in organic matter and thus are ideal continental hydrocarbon source rocks. The Lower Jurassic Yangxia (Kangsu) Formation and Middle Jurassic Kezillenuer (Yangye) Formation are a coal-bearing series, containing rich organic matter deposited in lacustrine and swamp facies. The Middle Jurassic Qiakemake (Taergan) Formation is another lacustrine deposit (Zhong et al. 2003) and is also an ideal continental hydrocarbon source rock. The Upper Cretaceous Yingjisha Group contains marine sediments, yielding ostracodes, sea urchins, foraminifera, dinoflagellates, and algal fossils, reflecting a shallow bay or lagoon environment that is favorable for the formation of hydrocarbon source rocks.

Apart from the above potential hydrocarbon source rock series, there is no lack of reservoir rocks and caprocks in the Tarim Basin. Cambrian–Ordovician carbonate rocks were exposed to the sea surface in the Late Ordovician and then eroded; the holes and cracks formed during long-term dissolution provided space to create oil and gas reservoirs, and to form carbonate reservoir rocks such as the Qiulitake Group in the Yakela oil field. The sandstones with well-developed pores from a variety of horizons in the Upper Paleozoic and Mesozoic–Cenozoic strata are also ideal oil and gas reservoir rocks. Additionally, the Ordovician and Permian all have mudstones and shales that can serve as caprocks; finally, the compact mudstone, gypsum-bearing mudstone, and gypsum in the Mesozoic to Cenozoic strata also provide good caprock potential for oil and gas preservation.

Depressions in basins are favorable zones for organic matter accumulation and preservation; multiple superimposed tectonic events resulted in superimposed composite depressions that are particularly favorable for organic matter accumulation and preservation. The four depressions of such a hereditary nature, the Kuqa, North, Southwest, and Southeast depressions, occur in the Tarim Basin. Oil and gas fields are already known in some of these basins. The North Tarim Upheaval is bounded to the north by the Kuqa Depression and to the south by the North Depression. Being surrounded by depressions, there is a potentially rich oil source supply. Indeed, it has the most favorable conditions to form large oil and gas fields. The discovery of the Yakela and Lunnan oil fields, as well as other rich oil and gas fields, demonstrates that the Tarim Basin is a petroliferous basin having great prospective potential.

2.3.2 The Junggar Basin

The Junggar Basin, which is located in the intersection of the Kazakhstan, Siberian, and Tarim plates, is a continental intraplate bounded on three sides by Paleozoic sutures. Its western part is the Hercynian fold belt that is the leading edge of the NE–NEE Kazakhstan Plate, composed of the West Junggar anticlinorium belt. At the northeastern and eastern margins of the Junggar Basin, the Altai and East Junggar fold belts were created by motion at the southern edge of the Siberian Plate. At the southern end of the Junggar Basin, the Tianshan fold belt formed at the northern edge of the Tarim Plate.

According to aeromagnetic and seismic data, the Junggar Basin has a rigid Precambrian crystalline basement, where the general trend of its basement relief is such that the central western region is higher, surrounded by a deep depression. The formation of the Junggar Basin was greatly influenced by the Hercynian Orogeny, and the basin is a fault depression basin within the plate developed in the Late Hercynian. The basin was initially formed in the Permian period. The Mesozoic was a period that involved integrated development of the basin, and thus, the continental sedimentary sequence was formed by the combination and transformation of both uplift and depression. By Cenozoic time, most orogeny had ceased except along the southern edge of the basin where thick continental sediments were deposited.

The Junggar Basin is a large-sized superimposed composite basin with an overlapping area about 145,000 km^2 and a superimposed sedimentary thickness of 12,000–15,000 m. The basin contains six petroliferous strata from Carboniferous to Tertiary in age.

2.3.2.1 Stratigraphy and Lithofacies (Table 2.2)

The Carboniferous System: The lower part of the Lower Carboniferous in the East Junggar subarea is characterized by the Tamugang Formation, which is mainly composed of gray and black argillaceous siltstone, gray conglomerate, and carbonaceous shale, with a thickness of 1550 m. The upper part of the Lower Carboniferous is the Dishuiquan Formation, which unconformably overlies the underlying strata. The Dishuiquan Formation is grayish green and taupe conglomerate, sandstone, and siltstone tuff, with a thickness of about 1600 m. The lower part of the Upper Carboniferous is the Batamayineishan Formation (=Liushugou Formation), which is composed of variegated volcanic and pyroclastic rocks, tuff and tuff breccia, with a thickness of 176–1011 m, in angular unconformity with the underlying Dishuiquan Formation. The middle part of the Upper Carboniferous, the Shiqiantan Formation, is grayish yellow siltstone, mudstone, and limestone with conglomerate, and about 480 m in thickness. The upper part of the Upper Carboniferous is the Liukeshu Formation (=Ao'ertu Formation), consisting of grayish green and gray siltstone, shale with intercalations of limestone, and siltstone glutenite, about 268–380 m in thickness. In the Carboniferous, tectonic

Table 2.2 Potential petroleum source rock series of the Junggar Basin

Geologic age	Lithostratigraphic units	Code	Lithology	Lithofacies
Oligocene	Anjihaihe Formation	E_{3a}	Gray, gray green sandstone, and mudstone	Lacustrine facies
Middle Jurassic	Toutunhe Formation	J_{2t}	Variegated mudstone alternating with sandstone, interbedded with carbonaceous shale and coal beds	Lacustrine–swamp facies
	Xishanyao Formation	J_{2x}	Gray conglomerate, sandstone, interbedded with clay and coal seams	Fluvial–swamp facies
Early Jurassic	Sangonghe Formation	J_{1s}	Gray green sandstone, mudstone, and siltstone	Lacustrine facies
	Badaowan Formation	J_{1b}	Gray mudstone, sandstone, interbedded with conglomerate and coal seams	Fluvial–swamp facies
Late Triassic	Haojiahe Formation	T_{3ha}	Gray sandstone and mudstone, interbedded with coal seams and siderites	Lacustrine–swamp facies
	Huangshanjie Formation	T_{3h}	Gray mudstone, shale, and black carbonaceous shale	Lacustrine–swamp facies
Middle Triassic	Kelamayi Formation	T_{2k}	Gray green mudstone, sandstone, and carbonaceous shale seams	Lacustrine–swamp facies
Early Triassic	Shaofanggou Formation	T_{1s}	Purple sandstone and mudstone, interbedded with gray mudstone and siltstone	Lacustrine facies
	Jiucaiyuan Formation	T_{1j}	Sandstone alternating with mudstone, interbedded with dark mudstone, siltstone	Lacustrine facies
Late Permian	Wutonggou Formation	P_{2w}	Gray glutenite alternating with mudstone, with intercalation of carbonaceous shale and coal seams	Lacustrine–swamp facies
Middle Permian	Quanzijie Formation	P_{2q}	Lower part with conglomerate, the upper part of mudstone, carbonaceous shale, and coal seams	Fluvial–swamp facies
	Hongyanchi Formation	P_{2h}	Gray, gray green mudstone, sandstone, interbedded with marl	Lacustrine facies
	Lucaogou Formation	P_{2l}	Gray black mudstone, oil shale, sandstone, siltstone, interbedded with dolomite	Lacustrine facies
Late Carboniferous	Ao'ertu Formation	C_{2a}	Gray green, dark gray siltstone, interbedded with glutenite	Continental facies
	Batamayineishan Formation	C_{2b}	Andesitic porphyry, basalt porphyry, tuffaceous sandstone	Continental facies
Early Carboniferous	Dishuiquan Formation	C_{1d}	Gray clastic interbedded with tuff	Marine facies
	Tamugang Formation	C_{1t}	Gray siltstone and mudstone, carbonaceous shale, glutenite	Marine and terrestrial alternating facies

and volcanic activities were intense with frequent marine transgressions and regressions, so the sedimentary combination of marine and continental facies as marine–terrestrial alternating deposits was coeval. In the Kelameily region, volcanic and pyroclastic rocks are well developed.

The Permian System: The Permian System is well developed in the Bogda foreland area of the

Junggar Basin. The Lower Permian Xiajijicao Group, which lies uncomformably or disconformably over the underlying strata, is composed of littoral or paralic dark clastic sediments in the Manas–Jimusaer region. The lower part of the Xiajijicao Group is the Shirenzigou Formation, and the upper part is the Tashikula Formation. The former is grayish green tuffaceous conglomerate, sandstone, siltstone, and sandy mudstone with tuff layers, containing brachiopods, bivalves, plants, spores and pollen, and other fossils, about 205–770 m in thickness. The latter formation is gray to grayish green fine-grained sandstone, siltstone, silty mudstone, limestone, and black shale, including bivalves, plants, pollen, and other fossils, about 1102–2593 m in thickness. The Upper Permian Shangjijicao and Xiacangfanggou groups consist of continental sediments, conformably overlying the underlying strata. The Shangjijicao Group has been subdivided into Wulabo, Jingjingzigou, Lucaogou, and Hongyanchi formations, in ascending order. The Xiacangfanggou Group is subdivided into Quanzijie, Wutonggou, and Guodikeng formations, in ascending order. The Wulabo Formation is as much as 1065–2543 m thick, consisting of gray and grayish green sandstone, siltstone, and silty mudstone, yielding bivalves, plant, spores and pollen, and other fossils. The Jingjingzigou Formation is composed of blue gray tuff, gray sandstone, siltstone, and mudstone, yielding ostracodes, plants, spores and pollen, and other fossils, about 319–1654 m thick. The Lucaogou Formation is composed of grayish black mudstone, shale, oil shale, sandstone, and siltstone with intercalations of dolomite and dolomitic limestone, yielding fossils of bivalves, ostracodes, fish, plants, spores, and pollen, with a thickness of 1100–1300 m. The Hongyanchi Formation is as much as 735 m thick and is composed of gray to grayish green mudstone, shale, and sandstone with intercalations of marl, yielding fossils of bivalves, ostracodes, plants, as well as spores and pollen. The Quanzijie Formation is 196–273 m in thickness; its lower part consists of brownish red conglomerate sandstone and mudstone, and the upper part consists of gray and dark gray mudstone, and carbonaceous shale interbedded with coal beds, yielding bivalves, ostracodes, plants, pollen, and other fossils. The Wutonggou Formation is 87–286 m thick, consisting of grayish green glutenite and mudstone interbedded with grayish black carbonaceous shale and thin coal beds, yielding fossil bivalves, ostracodes, and spores and pollen. The Guodikeng Formation consists of dark red and grayish green mudstone, and sandy mudstone intercalated with fine sandstone and carbonaceous mudstone, yielding bivalves, ostracodes, conchostracans, pollen, and other fossils; formation thickness is 68–128 m (Hu et al. 1991; Ouyang et al. 2003).

The Triassic System: The Triassic deposits crop out around the Junggar Basin. These are well developed in the Bogda piedmont sag, which is represented by inland fluvial and lacustrine facies. The Triassic mostly lies conformably over the underlying Permian strata; however, there are also unconformity or disconformity contacts. The Lower Triassic Shangcangfanggou Group is red clastic sediments, the lower part of which is the Jiucaiyuanzi Formation, composed of grayish green and dark red sandstone, mudstone interbedded with gray and grayish green mudstone, silty mudstone, and muddy siltstone, yielding fossils of reptiles, and spores and pollen, about 237–376 m thick. The upper part of the Shangcangfanggou Group is the Shaofanggou Formation, which consists of light purple and grayish green sandstones and red mudstones intercalated with gray mudstones and pelitic siltstones, yielding fossils of ostracodes, and spores and pollen, and having a total thickness of about 258–352 m. The Middle Triassic Karamay Formation is composed of grayish green to reddish brown and grayish black mudstones and grayish green sandstones intercalated with carbonaceous shales and thin coal beds, yielding fossils of plants and spores and pollen; the total thickness is about 465 m. The Lower–Upper Triassic Huangshanjie Formation consists of dark gray and yellowish gray mudstones and shales with black carbonaceous shales at the base, yielding plants, and spores and pollen fossils, about 225–450 m thick. The upper part of the Upper Triassic is the Haojiagou Formation, which consists of gray and grayish yellow sandstones and mudstones intercalated with thin coal and siderite beds, yielding fossils of plants, and spores and pollen, and a total thickness of 184–380 m.

The Jurassic System: Jurassic outcrops are exposed along the Junggar Basin margins and in mountain depressions and are well developed at the southern edge of the basin and the Bogda piedmont depression. Jurassic deposits are 976–1237 m thick, consisting of fluvial, lacustrine, and swamp facies. The Lower and Middle Jurassic are represented by coal-bearing strata. The Lower Jurassic is divided into the Badaowan Formation at the base and the Sangonghe Formation above it. The Baodaowan Formation contains coal and clastic sediments of fluvial, lacustrine, and swamp facies, consisting of gray, grayish green, and yellowish green mudstone, and sandstone with intercalations of coal and conglomerate; and includes fossils of bivalves, plants, and spores and pollen; the thickness is 300–670 m. The Sangonghe Formation is composed of yellowish green and dark red sandy mudstone interbedded with sandstone, conglomerate, carbonaceous, and siderite lenticular bodies, yielding plant fossils; total thickness is about 234 m. The Lower–Middle Jurassic Xishanyao Formation consists of gray conglomerate, and sandstone intercalated with clay and coal beds; the Upper and Middle Jurassic Toutunhe Formation is grayish green and purple mudstone and sandstone intercalated with carbonaceous shale and coal beds. The total thickness of the Middle Jurassic in the basin is as much as 377 m. The Upper Jurassic Qigu Formation is brownish red mudstone intercalated with grayish green and grayish brown sandstone and conglomerate, about 168–352 m thick.

The Cretaceous System: The sedimentary deposits of Cretaceous age in the Junggar Basin are mainly fluvial and lacustrine facies; these lay unconformably or disconformably on the underlying strata. The Lower Cretaceous Tugulu Formation consists of red sandstone and mudstone, with a basal conglomerate, and includes fossils of fish, ostracodes, conchostracans, and charophytes, about 100–260 m thick. The Upper Cretaceous Donggou Formation (Honglishan Formation–Wulungu Formation) is composed of red coarse clastic sediments, mainly consisting of conglomerate including ostracodes, charophytes, and vertebrate fossils; the unit lies unconformably on the underlying Tugulu Group.

The Tertiary System: Tertiary strata are well developed at the southern edge of the Junggar Basin and unconformably cover the underlying Upper Cretaceous Donggou Formation. The Paleocene–Eocene Ziniquanzi Formation crops out in foothills where it is a fluvial facies, consisting of red sandstone and conglomerate intercalated with sandy mudstone; it contains charophytes and ostracodes and is about 13–854 m thick. The Oligocene Anjihai Formation is a lacustrine sediment and consists of gray and grayish green sandstone and mudstone, yielding bivalves, ostracodes, spores and pollen, and other fossils, with a thickness of about 167–763 m. The Miocene Taxihe Formation consists of gray, gray brown, and grayish green sandstone, mudstone, and conglomerate, yielding fossils of bivalves, ostracodes, and charophytes; the total thickness is 500–2300 m. The Pliocene Dushanzi Formation contains fossil teeth of the three-toed horse *Hipparion* and is equivalent to the "three-toed horse laterite" horizons in North China.

The potential oil source rock series in the Junggar Basin are shown in Table 2.2.

2.3.2.2 Tectonic Units

The Junggar Basin (Fig. 2.2) can be subdivided into five tectonic units, including the Manashu Depression, the Wulungu Depression, the Luliang Uplift, and the East and South Junggar depressions. In addition, two units can be distinguished in the basin subsurface using magnetic data that show two structures named the Mosuowan Uplift and the Moqu Depression (Division of Earth Science of Academia Sinica and Xinjiang Petroleum Administration 1989) (Fig. 2.2).

The Manashu Depression is located in the northwestern edge of the basin, about 250 km in length and 40 km in width, extending in a northeast direction. It is the locality of the Karamay–Urho oil field region. The northwestern margin of the basin was subject to a break in subsidence during

2.3 Inland Petroliferous Basins

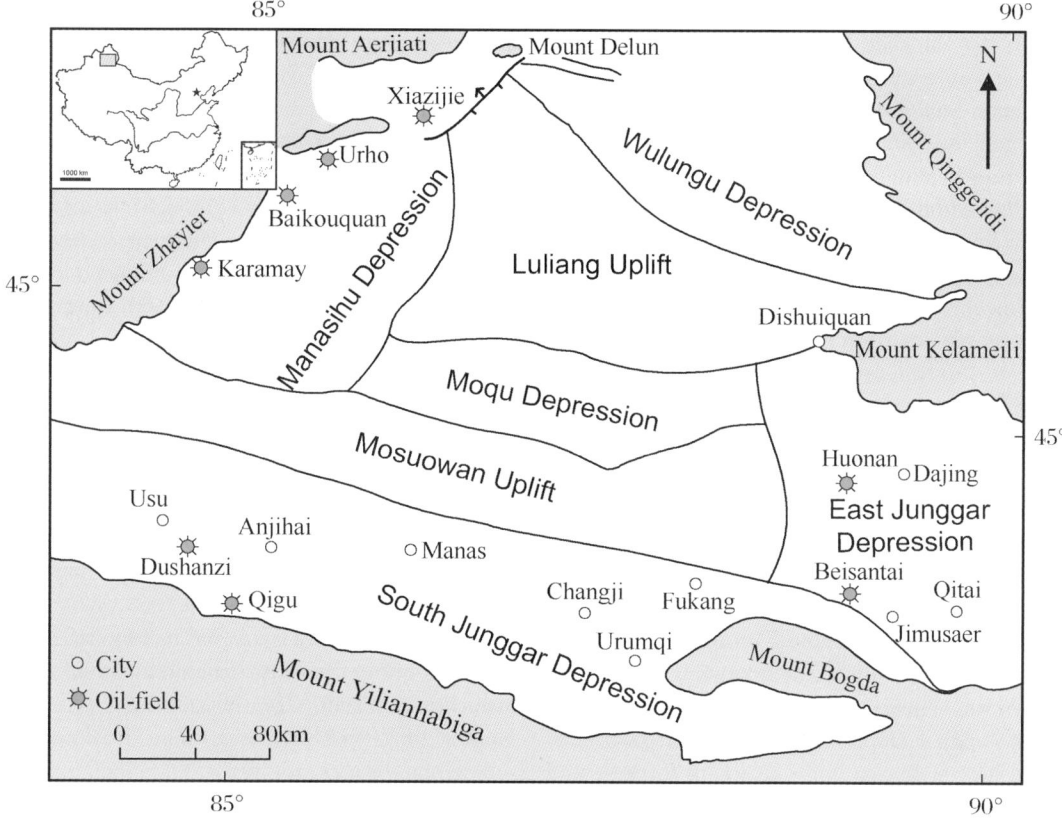

Fig. 2.2 Sketch map showing tectonic subdivisions and locations of oil fields in the Junggar Basin

the Permian, thereby forming two deposition centers: the larger one of which is in the Manashu area, which has sediments about 3000 m thick; and the smaller depo-center is in the Urho region, having sediments about 2900 m thick. After the Triassic period, the depression continued to sink rapidly, and the total thickness of sedimentary rocks in the depression is as much as 8000 m. So far, about 13 anticline structures and nose-shaped folds have been discovered in the basin.

The Wulungu Depression is located in the northeast edge of the basin, about 235 km in length and 40–60 km in width; it strikes NWW, the eastern end of which converges and narrows to the north of Kelameili Mountain. The depression is inherited from the Permian and continued sinking into the Triassic. So while the thickness of Permian sediments is about 1500 m, the thickness of Triassic sediments is as much as 4500 m. The total thickness of sedimentary rock in the depression is therefore about 6000 m.

The Luliang Uplift is located in the north-central part of the Junggar Basin, and is bounded at the NW by the Luliang fracture, at the north by the Wulungu Depression, at the west by the Manashu Depression, and at the south by the Moqu Depression. The uplift extends NW, and it about 120 km long and 80 km wide, with an area of approximately 9000 km^2. The uplift is composed of Permian, Triassic, Jurassic, and Cretaceous sequences. Drilling and seismic data show two unconformities between the Permian and Triassic, and the Jurassic and Cretaceous in this area.

The East Junggar Depression of the Junggar Basin is located in the Kelameili piedmont and the east–west "Corridor" of the north Jimusaer–Qitai areas, with an area of about 3500 km^2. The depression was formed by intense sinking in the Late Permian, followed by further sinking in the Triassic, and the sedimentary thickness is about 6000 m. The East Junnggar Depression is complex and is divided into a series of sags and uplifts named the Wucaiwan Fault Sag, the Shaqiu Fault Uplift, the Huoshaoshan Fault Sag, the Zhangpenggou Fault Uplift, the Dajing Fault Sag, the Beisantai–Shaqiu Fault Uplift, the Jimusaer Fault Sag, and the Qitai Fault Sag. The Huoshaogou Fault Sag is mainly Permian in age and contains the Huonan oil field. At the north end of the Beisantai–Shaqiu Fault Sag, the Wucaiwan, Huoshaoshan, and Dajing fault sags lay at the north, whereas the Jimusaer Fault Sag lays to the south, where it represents an inherited fault uplift of the east sag. The Beisantai–Shaqiu Fault Sag contains direct evidence for oil and gas migration and accumulation into its Beisantai oil field.

The South Junggar Depression (Changji Depression) of the Junggar Basin is located in the Mesozoic fold belt of the piedmont of Changji–Fukang and Mount Bogda and in the piedmont of Mount Yilianhabiga. This is a large depression that began sinking in the Permian where total thickness of the enclosed sediments is as much as 16,000 m. There are two depo-centers in the South Junggar Depression: one in the piedmont of the Mount Bogda, with sediment thickness of about 4000 m, and the other near the Changji area, where sediment thickness is about 3000 m. During the Mesozoic era, the depositional center moved westward; the Jurassic and Cretaceous depositional center is in the Changji to Manas areas. The Tertiary depositional center shifted westward to the Anjihai area. Seismic data show a significant unconformity between Jurassic and Cretaceous strata. The Jurassic Qigu oil field and the Tertiary Dushanzi oil field are both located in the west part of the South Junggar Depression.

2.3.3 The Turpan-Hami Basin

The Turpan-Hami Basin (Fig. 2.3) is an intermontane basin in the Northern Tianshan Eugeosynclinal fold belt. It is bounded to the south by the Aqikuduke break and adjacent to the uplift belt of the central Tianshan Mountains. The basin is bounded to the north by buried deep faults and is adjacent to the Bogda–Kaerlike fold belt. Surrounded by mountains, the basin is bounded on the north by Mount Bogda, Mount Balikun, and Mount Karlike where it is adjacent to the Junggar Basin. It is bounded on the south by Mount Jueluotage and adjacent to the Tarim Basin.

The Turpan-Hami Basin has its long axis in an east–west direction; its structure is that of an extensional basin. Its area is approximately 50,000 km^2 with the maximum thickness of sedimentary rocks as much as to 8600 m. The basement of the basin belongs to the easternmost part of the Kazakhstan Plate, so it should be composed of Precambrian crystalline rocks. Collision of the Kazakhstan and Tarim plates caused Jueluotage and Bogda–Kaerlike Orogeny activities, which is the main factor affecting the development and evolution of the basin's structure. The Turpan-Hami Basin is one of the three major oil and gas basins in Xinjiang, where the

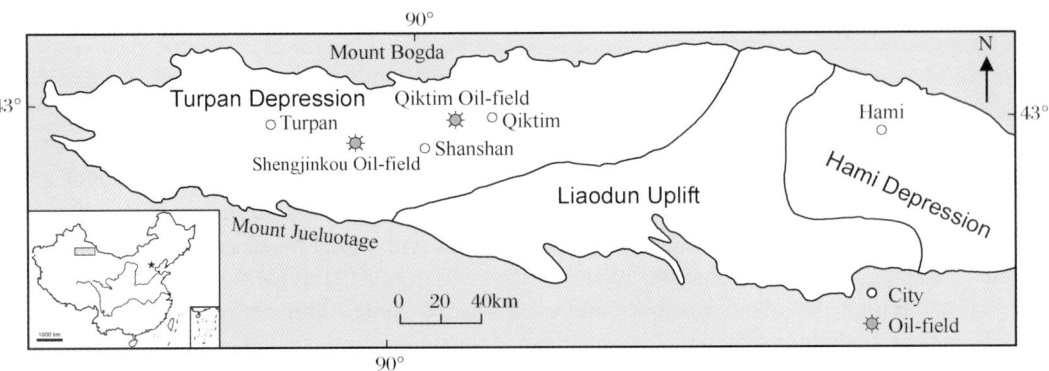

Fig. 2.3 Sketch map showing tectonic subdivisions and locations of oil fields in the Turpan-Hami Basin

Shengjinkou, Qiketai, and Shanshan oil and gas fields were discovered. Moreover, the basin has large-scale oil and gas deposits, enhanced by Jurassic coal seams, and thus, additional gas prospects can be predicted (Hu et al. 1991; Wang et al. 1993).

2.3.3.1 Stratigraphy and Lithofacies (Table 2.3)

The Permian System: The Lower Permian is lacking in the western part of the Turpan-Hami Basin. The lower part of the Upper Permian is the Taodonggou Group, which is quite similar to the Shangjijicao Group in Junggar Basin. The upper part is the Xiacangfanggou Group and has lithofacies that can be correlated with that of the Xiacangfanggou Group in the Junggar Basin. The Daheyan Formation in the Lower Taodonggou Group consists of purple conglomerate, pebbly coarse sandstone, sandstone interbedded with minor siltstone, tuff, and limestone lenses, with a thickness of 163 m; these sediments are in angular unconformity with the underlying strata. The Tarlang Formation in the Upper Taodonggou Group consists of variegated mudstone, marl, siltstone, sandstone, and fine-grained conglomerates, containing fish, ostracodes, and plant fossils and having a thickness of 352 m. The Upper Permian in the eastern basin is composed of alternating coastal marine and terrestrial facies and is composed of variegated clastic sediments. The Upper Permian can be divided into Daheyan, the Tarlang, and the Wutongou Formations, in ascending order. The Daheyan Formation consists of purple and gray fine conglomerate, sandstone, siltstone, and mudstone, with a thickness of 165 m; the formation lays unconformably above the underlying Lower Permian strata. The Tarlang Formation is composed of variegated mudstone, silty mudstone, siltstone, and sandstone, with a thickness of 520 m. The Wutonggou Formation consists of grayish green and pale yellow coarse sandstone and conglomerate, interbedded with purple fine silty mudstone and mudstone, with a thickness of 205 m; it lays unconformably above the underlying Tarlang Formation. The Late Permian deposits of the basin in mountain foothills are fluvial–lacustrine sediments.

The Triassic System: Triassic rocks are mainly exposed in the western part of the basin adjacent to the Junggar Basin where they are inland fluvial–lacustrine deposits. The Shangcangfanggou Group in the Lower Triassic is mainly red clastic sediment. The lower part of this group is the Jiucaiyuan Formation and consists of red sandstone, conglomerate, and mudstone marl lenses. The total thickness of this

Table 2.3 Potential petroleum source rock series of the Turpan-Hami Basin

Geological age	Lithostratigraphic units	Code	Lithology	Lithofacies
Middle Jurassic	Toutunhe Formation	J_{2t}	Sandstone, siltstone, mudstone, intercalated with coal seams	Lacustrine–swamp facies
	Xishanyao Formation	J_{2x}	Siltstone, mudstone, carbonaceous mudstone, intercalated with coal beds	Lacustrine–swamp facies
Early Jurassic	Sangonghe Formation	J_{1s}	Sandstone, mudstone intercalated with coal beds	Lacustrine–swamp facies
	Badaowan Formation	J_{1b}	Mudstone, carbonaceous mudstone, sandstone, intercalated with coal beds	Lacustrine–swamp facies
Late Triassic	Haojiagou Formation	T_{3ha}	Sandstone alternated with mudstone, with intercalations of coal beds and siderite	Lacustrine–swamp facies
	Huangshanjie Formation	T_{3h}	Mudstone, shale, carbonaceous shale, containing siderite nodules	Lacustrine facies
Late Permian	Tarlang Formation	P_{2t}	Variegated mudstone, siltstone, fine sandstone, and conglomerate	Lacustrine facies

formation is 128–235 m. The upper part of the Shangcangfanggou Group is the Shaofanggou Formation, which is composed of red and green sandstone with alternations of mudstone; the thickness is 83–169 m. The Middle Triassic Karamay Formation consists of gray green sandstone, sandy mudstone, and red mudstone; the basal part is a yellow green thick conglomerate. The fossils include reptiles, amphibians, ostracodes, and plants, and the thickness is about 300 m. The lower part of the Upper Triassic is represented by the Huangshanjie Formation, consisting of dark gray, gray, yellow green mudstone, and sandstone interbedded with coal beds that also include siderite nodules, as well as fossils of plants and bivalves, with a thickness of 193–289 m. The upper part of the Upper Triassic is Haojiagou Formation, which consists of grayish yellow glutenite and dark gray mudstone, and carbonaceous mudstone with coal beds that bear siderite nodules and plant fossils; formation thickness is 92–157 m.

The Jurassic System: The Jurassic System is mainly exposed in the west part of the basin and is represented by inland fluvial, lacustrine, and swamp facies deposits having a thickness of 1340–2178 m. The lower and middle parts of the Jurassic are coal-bearing strata, while the middle part also contains hydrocarbon-bearing strata. The lower part of the Lower Jurassic is the Badaowan Formation, consisting of grayish black mudstone, carbonaceous mudstone, and yellowish green sandstone interbedded with thin coal seams; the basal part of the formation is conglomerate. The formation also has siderite nodules and fossil plants and pollen and spores, with a thickness of 43–540 m. The upper part of the Lower Jurassic, the Sangonghe Formation, consists of gray sandstone with conglomerate and gray mudstone, and silty mudstone with coal beds, yielding plant and pollen fossils; the thickness is 18–246 m. The lower part of the Middle Jurassic is represented by the Xishanyao Formation, consisting of gray and grayish yellow siltstone and black mudstone with coal beds and siderite nodules, yielding plant and pollen fossils; the thickness is 412 m. The upper part of the Middle Jurassic, the Toutunhe Formation, consists of grayish green and brownish red sandstone, siltstone, and silty mudstone with coal beds; the thickness is 455 m. The Upper Jurassic is represented by the Qigu Formation, which consists of purple and grayish green sandstone and mudstone with intercalations of conglomerate, yielding fossil ostracodes, with a thickness of 24–295 m.

The Cretaceous System: The Cretaceous is represented mainly by alternative clastic and lacustrine sediments deposited in inland fluvial, lacustrine, and fluvial–lacustrine environments. The Lower Cretaceous Tugulu Group is divided into Sanshilidadun, the Shengjinkou, and the Lianmuqin formations, in ascending order. The Upper Cretaceous is divided into the Kumtag Formation and the Subashi Formation. The Sanshilidadun Formation consists of red sandstone and sandy mudstone, with a basal fine-grained gravel; these lie unconformably or disconformably over the underlying strata, and the thickness is 768 m. The Shengjinkou Formation consists of grayish green mudstone, sandy mudstone, and siltstone with the fish fossils *Turfanichths* sp. and *Hunyshania* sp., and the thickness is 49–62 m. The Lianmuqin Formation consists of brownish red sandy mudstone and grayish green and taupe fine sandstone, yielding fossil ostracodes, with a thickness of 213–307 m. The Kumtag and Subashi formations consist of hyacinth and orange fine-grained sandstone containing ostracodes and vertebrate fossils, with a thickness of 57–120 m.

The Tertiary System: Tertiary rocks were deposited in an oxygenated environment that left numerous inland lacustrine ostracode fossils. The Paleogene consists of orange, yellow, and green mudstone and sandstone; it has a basal conglomerate, yields ostracodes and vertebrate fossils, and has a thickness of 170–701 m. The Neogene consists of purple and pale yellow mudstone, siltstone, and conglomerate bearing a gypsum layer; fossils are ostracodes and vertebrates, and the thickness is 57–120 m.

The potential petroleum source rock system in the Turpan-Hami Basin is shown in Table 2.3.

2.3.3.2 Tectonic Units

The Turpan-Hami Basin (Fig. 2.3) can be subdivided into three tectonic units, including the Turpan Depression, the Liaodun Uplift, and the Hami Depression. Bordered by the Central Flaming Hill Fault, the Turpan Depression is bounded to the north by the Taibei Sag and to the south by the Aidinghu Slope and the Tuokexun Sag. There are seven anticline belts in the Taibei Sag, including the Taoshuyuan, the Kekeya, the Tudunzi, the Kushuijing, the Gedatai, the Flaming Hill, and the Yanmuxi anticline belts. The Shengjinkou and Qiktim oil fields are located in the west and east part of the Flaming Hill Anticline Belt, respectively (Hu et al. 1991) (Fig. 2.3).

2.3.4 The Qaidam Basin

The Qaidam Basin (Fig. 2.4) is located in the northeastern Qinghai–Tibetan plateaus. It is bounded at the northwest by the Altun Mountains, at the northeast by the Qilian Mountains, and at the south by the Kunlun Mountains. It is a Mesozoic–Cenozoic intermontane basin having an area of 120,000 km^2 and was formed after the Indosinian Orogeny. The basement of the basin consists of Pre-Sinian crystalline rocks and the Caledonian fold belt. The Qaidam Basin has a Mesozoic–Cenozoic continental sedimentary thickness of more than 10,000 m.

The Qaidam Basin is a large inland petroliferous basin located in western China, where more than 100 oil-bearing structures, excellent trap conditions, 17 oil fields, and 5 gas fields have been found so far. Boreholes reveal a thickness greater than 4000 m in the Lower–Middle Jurassic petroliferous rock series in the Lenghu area; total thickness of the Tertiary potential petroleum source rock series is probably 3600 m (Research Institute of Exploration and Development, Qinghai Petroleum Administration, Nanjing Institute of Geology and Palaeontology, Academia Sinica, 1985; Huang et al. 1993). Zhu (1986) pointed out that, in a comparatively broader tilted fault depression, especially in a semi-graben, because of the constraint of geological factors, coarse clastic rocks should be deposited on one side of the fault, while variegated sandstone and shale intercalated with dark pelites should be developed on the slope side. The transition zone between these two kinds of deposits can be regarded as of great significance for both generation and the accumulation of oil and gas. The Qaidam Basin is mainly controlled by fault depression even as it

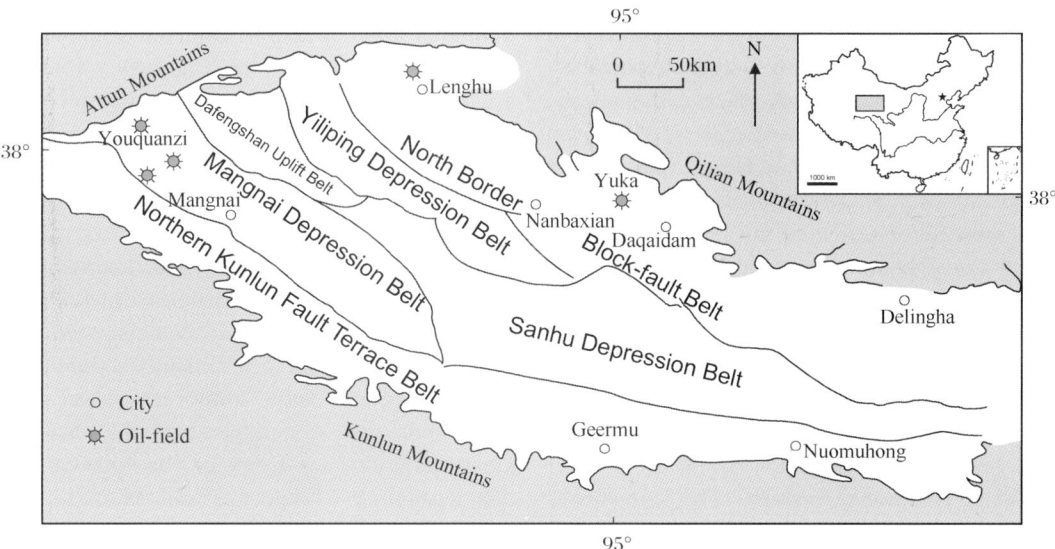

Fig. 2.4 Sketch map showing tectonic subdivisions and locations of oil fields in the Qaidam Basin

also possesses a combination of fault depression and depression. Near the Qilian Mountains and Altun Mountains, Tertiary coarse clastic rocks developed, while fine-grained clastic rocks represented by dark pelites alternate with sandstones that are well developed in the center of basin. Dark pelites favor oil generation, and alternations of sandstones and shales provide oil–gas accumulation and preservation conditions.

2.3.4.1 Stratigraphy and Lithofacies (Table 2.4)

The Jurassic System: The Jurassic System in the Qaidam Basin is composed of fluvial, lacustrine, and swamp facies which are mainly distributed in the piedmont fault depression of the Altun and Qilian mountains at the northern margin of the basin. Most of the strata are well exposed in the Dameigou and Xiaomeigou areas in the southeastern Daqaidam area and have a thickness of 1260 m. The Lower Jurassic Xiaomeigou Formation consists of grayish yellow and grayish green conglomerates and sandstones interbedded with dark gray mudstones and black carbonaceous shales; these are also intercalated with coal beds and yield siderite concretions as well as fossil plants, and spores and pollen. The thickness is 141 m in the Xiaomeigou Section. Grain size of borehole sediments becomes finer in the Lenghu area, where thickness is more than 1000 m. The Middle Jurassic Dameigou Formation consists of grayish green and grayish yellow sandstones, conglomerates, and grayish black carbonaceous shales intercalated with coal and siderite beds; contains fossils of conchostracans, plants, and pollen; and has a thickness of 1029 m at the Dameigou Section, and in the Lenghu area, the grain size of borehole sediments also becomes finer, where thickness is more than 3000 m. The Upper Jurassic Caishiling Formation is composed of purple mudstones intercalated with sandstones and gypsum and has a thickness of 90 m. Lower–Middle Jurassic dark pelites are the main source rocks in the northern Qaidam Basin.

The Cretaceous System: The Cretaceous Quanyagou Group in the Qaidam Basin is a red clastic deposit that consists of purple, brownish red, and grayish brown sandstones and mudstones, yielding fossils of ostracodes, conchostracans, and charophytes; the thickness is 300–2100 m.

The Tertiary System: The Tertiary is widely distributed in the Qaidam Basin, where the depositional environments left inland fluvial and lacustrine facies. Sedimentary thickness of the Tertiary is 7000 m in the Mangnai Depression of the western basin, and as much as 13,000 m in the Yiliping Depression of the central basin. The Paleo-Eocene Lulehe Formation is mainly composed of fluvial deposits and consists primarily of dark brown and brownish red conglomerates and psephitic sandstones intercalated with brownish red and yellowish green sandstones, mudstones, and siltstones; total thickness is 200–1000 m. The Oligocene Xiaganchaigou Formation in the interior basin is a lacustrine facies that consists mainly of gray and dark gray pelites intercalated with siltstones; the thickness is 300–2800 m. The Lower Miocene Shangganchaigou Formation in the central basin is composed mainly of a lacustrine facies, consisting of gray mudstones intercalated with grayish green sandstones and siltstones; the thickness is 300–1200 m. The Upper Miocene Xiayoushashan Formation in the central basin is mainly a shallow lacustrine facies that is primarily composed of grayish green, greenish gray, and brownish red sandstones interbedded with mudstones that intercalate with light gray marlites and variegated mudstones; the thickness is 300–2000 m. The Lower Pliocene Shangyoushashan Formation at the margin of the basin is a fluvial facies; in the central basin, it forms shallow lacustrine deposits with gypsum, consisting mainly of dark brown and yellowish green sandy mudstones intercalated with grayish green sandstones, siltstones, gray marlites, and thin-bedded gypsum. Formation thickness is 150–1800 m. The Upper Pliocene Shizigou Formation is a fluvial facies, mainly composed of yellowish gray sandy mudstones intercalated with sandstones, conglomerates, and siltstones; the thickness varies from 300 to 2000 m. The thickness of the Oligocene and Miocene dark mudstones in the Mangai Depression is over 2000 m, and the dark mudstones are the main source rocks in the basin.

2.3 Inland Petroliferous Basins

Table 2.4 Potential petroleum source rock series of the Qaidam Basin

Geologic Age	Lithostratigraphic units	Code	Lithology	Lithofacies
Pliocene	Shangyoushashan Formation	N_2^1	Sandy mudstones intercalated with sandstones, siltstones, and gypsum	Fluvial and lacustrine facies
Miocene	Xiayoushashan Formation	N_1^2	Sandstones interbedded with mudstones, intercalated with marlites	Shallow lacustrine facies
	Shanganchaigou Formation	N_1^1	Gray mudstones intercalated with sandstone and siltstones	Lacustrine facies
Oligocene	Xiaganchaigou Formation	E_3	Dark gray mudstones and siltstones	Lacustrine facies
Middle Jurassic	Dameigou formation	J_{2d}	Sandstones, conglomerates, carbonaceous shales intercalated with coal beds and siderite	Lacustrine–swamp facies
Early Jurassic	Xiaomeigou formation	J_{1x}	Sandstones, conglomerates, mudstones, carbonaceous shales intercalated with coal beds	Lacustrine–swamp facies

Note The dark pelites in the Paleo-Eocene Lulehe formation (E_{1-2}) have oil generation potential

The potential petroleum source rock series of the Qaidam Basin are shown in the Table 2.4.

2.3.4.2 Tectonic Units

The Qaidam Basin (Fig. 2.4) can be subdivided into six tectonic units, i.e., the North Border Block-fault Belt, the Mangnai Depression Belt, the Dafengshan Uplift Belt, the Yiliping Depression Belt, the Northern Kunlun Fault Terrace Belt, and the Sanhu Depression Belt. The North Border Block-fault Belt formed in the Jurassic and can be subdivided, from the west to the east, into four secondary tectonic units, i.e., the Lenghu Tectonic Fault Zone, the Mahainanbaxian Anticlinal Zone, the Yuka Sag, and the Delingha Sag. The Lenghu oil field is located in the Lenghu Tectonic Fault Zone and the Yuka oil field is located in the Yuka Sag. The Mangnai Depression subsided during the Tertiary from the Oligocene to the Pliocene and is the main oil-bearing area in the basin. The Youquanzi and the Xianshuiquan oil fields were discovered in this depression. The Dafengshan Uplift Belt consists of the Jianshan, the Jiandingshan, and the Dafengshan Anticlinal Zones, and the Dafengshan oil field is located in the Dafengshan Anticlinal Zone. The Yiliping Depression Belt possesses the thickest Mesozoic–Cenozoic deposits, where the thickness is as much as 15,500 m. The Sanhu Depression Belt subsided during the Quaternary and is thus the Quaternary gas-bearing area in the basin. The Sebei#1 and Sebei#2 gas fields were discovered there (Huang et al. 1993) (Fig. 2.4).

2.3.5 The West Jiuquan Basin

The West Jiuquan Basin (Fig. 2.5) is located in the western piedmont depression of the Qilian Mountains. It is a Mesozoic–Cenozoic continental petroliferous basin having an area of 2700 km² and a sedimentary thickness of 7000 m. Tectonically, the Qilian Caledonian fold belt is located between the Alxa Block and the Qaidam Block. The Qilian Mountains are developed on Caledonian fold basement rocks. In the northern piedmont depression of the Qilian Mountains, Carboniferous to Mesozoic–Cenozoic rocks are exposed. The Jiuquan Basin in the western part of the depression contains valuable oil and gas resources. The Carboniferous deposition was in shallow seas where marine and terrestrial rocks alternate and are only exposed in the southern part of the Jiuquan Basin. The grayish green and red clastic sediments of the Permian and the Triassic are not developed in the Jiuquan Basin, and the Upper Triassic is lacking. Due to the influence of the Indosinian and Yanshan orogenies, the depression was tilted so that the rocks are uplifted

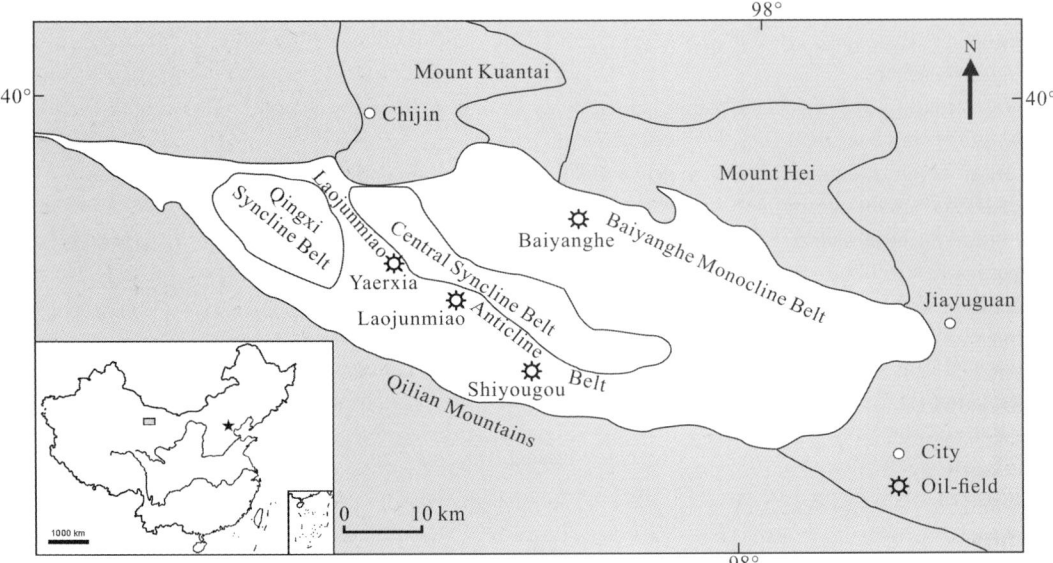

Fig. 2.5 Sketch map showing tectonic subdivisions and locations of oil fields in the West Jiuquan Basin

in the east and down dropped in the west. Jurassic and Cretaceous sediments in the basin have a total thickness of about 3600 m. The Himalayan Orogeny influenced deposition in the basin where Paleocene to Eocene rocks are thin because of regional uplift, but Late Cenozoic deposits are thick due to strong subsidence. Red clastic sediments of Cenozoic age in the basin are about 2500 m in total thickness.

2.3.5.1 Stratigraphy and Lithofacies

The Jurassic System: The Jurassic of the Western Jiuquan Basin is represented by fluvial and swamp facies, which lay disconformably over the underlying strata. The southern part of the basin contains the Longfengshan Group, which is about 1350 m in thickness, and composed of grayish green and purple sandy shale, and sandstone intercalated with conglomerate, with coal beds at the base containing plant fossils. In the northern part of the basin, the Jurassic is represented by the Chijinbao Group, which consists of grayish green and purple brecciated sandstone, and sandstone with alternations of black shale; it yields fossil plants, conchostracans, ostracodes, and bivalves and has a thickness of about 250–1440 m.

The Cretaceous System: The Cretaceous System is widely distributed in the West Jiuquan Basin, where it crops out in the Hongliuxia and Hanxia areas. The Lower Cretaceous is represented by lacustrine sediments having a total thickness as much as 1400 m, and the Upper Cretaceous is represented by a fluvial facies having a thickness of 100–1100 m. The Lower Cretaceous and Lower Huihuibao groups (Lower Xinminpu Formation) are composed of grayish black shales intercalated with grayish green sandstones and gray marl; the base of the formation is conglomerate. The formation has pyrite nodules and fossil plants, spores and pollen, conchostracans, bivalves, ostracodes, and fish and has a thickness of 300–1140 m. The Lower Xinminpu Formation lays unconformably over Precambrian or Paleozoic metamorphic rocks at the edge of the basin, but disconformably over the underlying Jurassic Chijinbao Formation in the central part of the basin. Lower Cretaceous organic-rich strata are the potential petroleum source rock series in this basin. The lithology of the Upper Cretaceous and Upper Huihuibao groups are red, gray, yellow, and green conglomerate, sandstone, and shale, with the thickness of about 100–1100 m; it lies disconformably over the underlying Lower Huihuibao Group.

The Tertiary System: The Paleogene Huoshaogou Formation is distributed in the northern part of the West Jiuquan Basin, where it is a red clastic, fluvial facies, composed of brownish red conglomerate intercalated with gray sandstone and red sandy mudstone, about 930 m thick. The Huoshaogou Formation lies uncomformably over the underlying Cretaceous strata. According to the spore–pollen assemblage, the geological age of the formation is Oligocene (Sung 1958). The Neogene, laying unconformably in contact with the underlying Huoshaogou Formation, is widely distributed in the southern and northern parts of the basin. The Miocene Baiyanghe Formation is a fluvial facies composed of red quartz sandstone and dark red mudstone intercalated with a gypsum layer, and the formation thickness is 410–470 m. The sandstone of the Baiyanghe Formation is well sorted and loosely packed and has high psephicity; it is a reasonable reservoir rock, with the gypsum layer that could form a cover rock. The Pliocene Shulehe Formation is khaki sandstone and brownish yellow sandy mudstone, about 1410 m thick.

2.3.5.2 Tectonic Units

The West Jiuquan Basin (Fig. 2.5) can be divided into four tectonic units: the Laojunmiao Anticline Belt, the Central Syncline Belt, the Baiyanghe Monoclinal Belt, and the Qingxi Syncline Belt. The Laojunmiao Anticline Belt has eight anticlinal structures, including the Qingcaowan, the Yaerxia, the Laojunmiao, the Shiyougou, and the Dahongjuan structures. The Laojunmiao Anticline Belt is a relative uplift in the depression, and therefore a prospective target for petroleum migration. The anticline belt, therefore, is favorable for the accumulation of petroleum and creation of oil and gas fields. In fact, the Laojunmiao, Yaerxia, and Shiyougou oil fields have all been found in this anticline belt. The Central Syncline Belt is located in the northeast side of the Laojunmiao Anticline Belt and is the center of piedmont depression, where depth of basement is 3500–4000 m, and it contains three buried structures. The Baiyanghe Monoclinal Belt is located in the northern part of the piedmont depression. The basement and covering strata of the Baiyanghe Monoclinal Belt are both gently south-dipping with dip angles of 3°–5°; however, the dip steepens greatly in the Central Syncline Belt, where dip angles are as much as 30°–50°. The change in dip reflects the bedrock fault. The Baiyanghe Monoclinal Belt includes the Baiyanghe, Huihuibao, and Nanshan Brachyanticlines; the Baiyanghe oil field is located in the middle part of the Baiyanghe Monoclinal Belt. The Qingxi Syncline Belt is located in the west of the Qingcaowan area, between the Hongliuxia, Hanxia, and Dongdakou areas; the depth of basement is about 4000 m (Lanzhou Institute of Geology, Academia Sinica 1960).

2.4 Coastal Shelf Petroliferous Basins

China has several coastal shelf petroliferous basins, three of which will be discussed here in detail. They are useful because of their significance to explaining the role that palynology plays both in understanding the hydrocarbon potential of such basins and in providing information about other exploration targets. These basins are the Liaohe, the Beibu Gulf, and the Zhujiang Mouth Basins.

2.4.1 The Liaohe Basin

The Liaohe Basin (Fig. 2.6) is a Cenozoic petroliferous basin in the northern Bohai Gulf area of eastern China. The land area of the basin is about 12,400 km^2, and the sedimentary thickness is over 8000 m. The Liaohe Basin was formed by extensional forces caused by complex motion between the Pacific and China plates following onset of the Himalayan Orogeny at the end of the Cretaceous. Complex motion included reverse torsional pressure or pressure torsion stress that twisted the China Plate on which the Liaohe Basin sits. In eastern China, the development of this twist extension formed several basins,

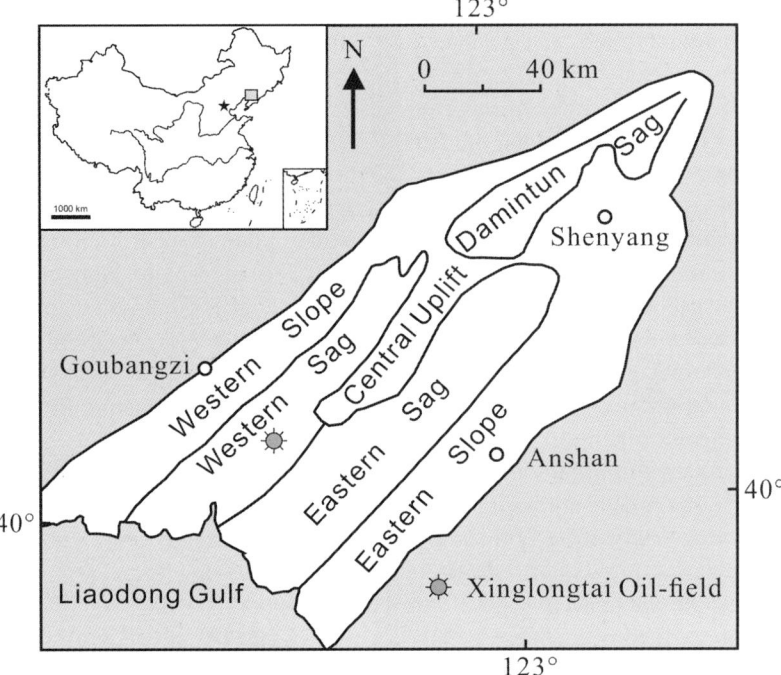

Fig. 2.6 Sketch map showing tectonic subdivisions and location of oil field in the Liaohe Basin

including the Liaohe Basin, Cenozoic sedimentary basins in north China, the Bohai Sea, and the eastern China continental shelf. Indeed, Zhu (1986) argued that interactions between the Pacific and China plates created basins on the China Plate. The location of the Liaohe Basin is close to the eastern "frontline" of the intersection of the Pacific Plate with the China Plate and is thus a Cenozoic rift-depression transformation-type basin.

The basement of the Liaohe Basin is pre-Tertiary in age. While the central parts of the basin were uplifted and the sediment stripped, downwarped edges filled with Tertiary deposits. Within the basin, the maximum thickness of Cenozoic sedimentary rocks is over 8000 m, most of which are Lower Tertiary in age.

2.4.1.1 Stratigraphy and Lithofacies

The Lower Tertiary in the coastal region of the Bohai Sea is dominated by lacustrine deposits. About ten members of three formations are recognized and have been divided in ascending order:

The Eocene Kongdian Formation is composed of three sets of red and gray coarse clastic rocks, divided into three members:

The 3rd Member of the Kongdian Formation is a red sandstone alternating with mudstones, over a basal conglomerate; it lays unconformably over pre-Cenozoic strata, with a thickness of 300–500 m.

The 2nd Member of the Kongdian Formation is a gray and dark gray mudstone with intercalations of carbonaceous shale, oil shale, coal beds, and sandstone, about 500–600 m thick.

The 1st Member of the Kongdian Formation is a red mudstone alternating with sandstone, 300–500 m thick.

The Oligocene Shahejie Formation consists mainly of gray and dark gray mudstones with four members as follows:

The 4th Member of the Shahejie Formation in the lower part is composed of red mudstone with minor thin sandstones, 150–500 m thick. The middle part is composed of bluish gray mudstone with gypsum, 100–300 m thick. The upper part is composed of grey mudstone intercalated with carbonates and oil shale, 100–150 m thick.

The 3rd Member of the Shahejie Formation mainly consists of massive sets of gray mudstones, 500–1000 m thick, and as much as 1600 m in the Xinmin area of Liaoning Province. The lower part is composed of gray mudstone with intercalations of sandstone; the sandstone is mostly concentrated at the base, about 100–150 m thick. The middle part is composed of gray and dark gray mudstone and shale intercalated with lenticular sandy conglomerate, about 400–600 m thick. The upper part consists of gray mudstone alternating with siltstone and fine sandstone, 300–400 m thick.

The 2nd Member of the Shahejie Formation consists mainly of the sandstone with grayish green mudstone and carbonaceous shale in the lower part; the upper part is a grayish green and red mudstone, 100–250 m thick.

The 1st Member of the Shahejie Formation consists mainly of mudstone, with intercalations of shale and carbonates. The lower part is composed of gray mudstone intercalated with shale, shale, limestone, and dolomite, 100–200 m thick. The upper part is composed of gray and dark gray mudstone with a small amount of sandstone, 100–200 m thick.

The Oligocene Dongying Formation is composed of gray and dark gray mudstones intercalated with sandstones; deposition was greatest in the Eastern Hebei Plain where the sediment is 1000–1500 m thick. Away from there, it gradually becomes courser and thinner. Three members are divided as follows:

The 3rd Member of the Dongying Formation consists of gray mudstone with carbonaceous shale and sandstone, and the thickness is as much as 800 m in Eastern Hebei Plain. In the lower distributary of Liaohe River area, the unit is more sandy and thinner.

The 2nd Member of the Dongying Formation is gray mudstone with sandstone, about 300–500 m thick.

The 1st Member of the Dongying Formation is brownish red and grayish green mudstone with intercalations of sandstone, 0–300 m in thickness. It lays unconformably below the overlying Guantao Formation of Neogene age.

Based on investigations on fossil algae, and spore and pollen assemblages, the upper part of the 4th Member of the Shahejie Formation is equivalent to the marine facies in the Kenli, Boxing, and Bingxian areas of Shandong Province; the 3rd Member of the Shahejie Formation is a brackish water facies; the 2nd Member of the Shahejie Formation is a freshwater lacustrine facies; the lower part of the 1st Member of the Shahejie Formation is a semi-brackish water facies, while the upper part of this member is a fresh water facies; the 3rd and 2nd members of the Dongying Formation represent semi-brackish to littoral facies; and the 1st Member of this formation is a continental freshwater facies (Research Institute of Petroleum Exploration and Development, Ministry of Petroleum Chemistry Industry, Nanjing Institute of Geology and Palaeontology, Chinese Academy of Sciences 1978b).

2.4.1.2 Tectonic Units and Oil–Gas Reservoirs

In the Liaohe Basin (Fig. 2.6), six tectonic units can be subdivided, including Damintun Sag, Eastern Sag, Western Sag, Central Uplift, West Slope, and Eastern Slope. In the Western Sag, the 3rd Member of the Shehejie Formation is a deep lake facies, consisting of dark mudstone with turbidite rock, thereby being an excellent example of a reservoir and seal complex favorable for the formation of oil and gas reservoirs. Founded in 1972, the Xinglongtai oil field has a turbidite reservoir that is located in the Western Sag. The Xinglongtai oil field is dominated by oil, with an association to an oil-type gas reservoir. The lacustrine mudstone in the Shahejie and Dongying formations is the ideal gas source rock, and indeed, the gas output is greater than the oil output. The natural gas production in the Liaohe Basin ranks second in China (Xu et al. 1994, 2000).

2.4.2 The Beibu Gulf Basin

The South China Sea is the largest marginal sea of the northwest Pacific Ocean, bearing remarkable oil and gas resources. Along the northern

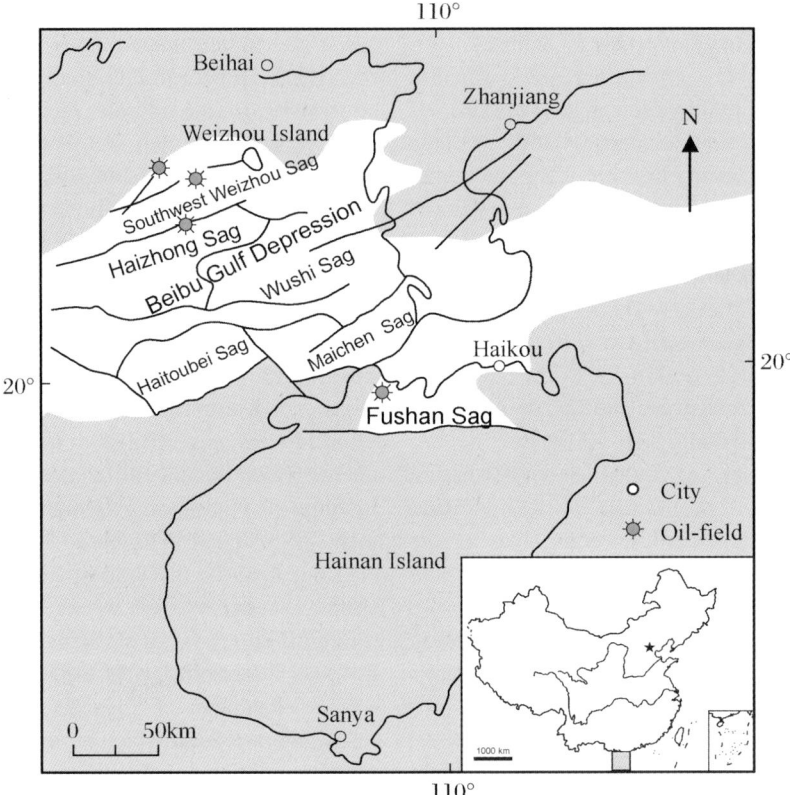

Fig. 2.7 Sketch map showing tectonic subdivisions and locations of oil fields in the Beibu Gulf Basin

continental shelf of South Sea, there are three large Cenozoic petroliferous basins, including the Beibu Gulf Basin, the Yinggehai Basin, and the Zhujiang Mouth Basin.

The Beibu Gulf Basin (Fig. 2.7) lies in the Beibu Gulf Depression of the western North Shelf of South China Sea. It is located east of longitude 108°, offshore of the Leizhou Peninsula, and north of latitude 35°19′ to Qinzhou area of Guangxi Autonomous Region and covers an area of about 40,000 km². The basin is a Cenozoic petroliferous basin, and is developed on Paleozoic basement and Yanshanian granite to form a large fault depression to depression-type basin, with a sedimentary thickness of 7000 m. The embryonic basin was formed in the Late Cretaceous, and the fault tectonic framework formed during Late Yanshanian Orogeny and was inherited and transformed in the Tertiary Himalayan Orogeny, thereby creating the Beibu Gulf Basin in the Tertiary. The Beibu Gulf Basin predominantly filled in the Paleogene when a set of lacustrine facies sediments were deposited, consisting of the Changliu, Liushagang, and Weizhou formations in ascending order, with a total thickness of as much as 6000 m. During the Neogene period, new oceanic crust was formed and further expanded in the South China Sea. As sea level began to rise, coastal and shallow marine facies clastic sediments were deposited in the Beibu Gulf Basin. Deposition in the basin, in ascending order, included the Xiayang, Jiaowei, Dengloujiao, and the Wanglougang formations, with a total thickness of 1200–2000 m.

2.4.2.1 Stratigraphy and Lithofacies

The Paleocene Changliu Formation is characterized by pluvial to alluvial facies deposits, consisting of brownish gray mudstones with alternations of gravel layers, about 40–840 m thick.

The Eocene–Oligocene Liushagang Formation represents a littoral lacustrine deposit, consisting of grayish black and dark gray mudstone, beige shale, and oil shale. The lower part bears an oil-bearing coarse-grained sandstone, and the upper part is a grayish white fine-grained sandstone; formation thickness is as much as 4400 m. The solitary dark argillaceous rock layer in the Liushagang Formation is 500 m thick, where its depositional environment was a deep anoxic basin, and it is the major oil source rock series in the Beibu Gulf Basin. The dark mudstone with intercalations of sandstone may represent turbidites of the deep lake facies and is a potential target layer for oil exploration.

The Middle Oligocene Weizhou Formation represents alternation of marine and terrestrial facies, consisting of variegated mudstone and silty mudstone with alternations of sandstone and gravel; the middle part is composed of dark gray mudstone. The maximum thickness is as much as 2000 m, with 766.5 m in the Wan #2 Well.

The Lower Miocene Xiayang Formation is a coastal shoal facies. The lithology consists of greenish gray gravels, coarse sandstone with intercalations of gray mudstone, and sandy dolomite; the thickness is 235 m in the Wan #2 Well.

The Middle Miocene Jiaowei Formation is a shallow marine facies; the lower part consists of grayish green fine sandstone alternating with mudstone; the upper part is composed of a gray thick mudstone. Formation thickness is 264.5 m in the Wan #2 Well.

The Upper Miocene Dengloujiao Formation is a shallow marine facies. The lower part consists of grayish yellow sandy gravels; the middle part is dominated by gray mudstone with intercalations of muddy gravels. The upper part is marked by grayish yellow coarse-grained sandstone and conglomerates with gray mudstone. Formation thickness is 383 m in the Wan #2 Well.

The Upper Miocene Wanglougang Formation is a littoral facies. The lower part consists of gray mudstone with intercalations of grayish yellow sandstone; the middle part is grayish yellow sandstones with alternations of mudstones. The upper part is grayish yellow, sandy conglomerates. Formation thickness is 282.5 m in Wan #2 Well.

The dark mudstone in the Liushagang Formation is rich in organic matter that creates a thick sedimentary unit that could be the major hydrocarbon source rock in the basin. The 3rd Member of the Liushagang Formation contains massive thick dark mudstone intercalated with sandstone, indicating a good reservoir potential. In addition, in the Weizhou, Xiayang, and Jiaowei formations contain sandstones that could be productive oil reservoirs (Zeng and Guo 1981).

2.4.2.2 Tectonic Units and Oil–Gas Reservoirs

There are six sags in the Beibu Gulf Depression (Fig. 2.7), including the Southwest Weizhou, Haizhong, Wushi, Haitoubei, Maichen, and Fushan Sags. Abundant oil has been extracted by industry from the Southwest Weizhou, Wushi, and Fushan sags since exploration began in the 1970s. For example, the Wan #2 Well in the Southwest Weizhou Sag has produced oil in the sandstones of the 3rd Member of the Liushagang Formation; the Wan #5 Well has produced oil in the sandstone reservoir of the Xiayangjiao and Jiaowei formations; the Wan #11 Well in the Wushi Sag has produced oil in the 1st sandstone reservoir of the Liushagang Formation; and the Fu #23, Fu #28, and Fu #29 wells in the Fushan Sag have produced oil in the sandstone reservoirs of the Weizhu Formation (Zeng and Guo 1981).

2.4.3 The Zhujiang Mouth Basin

The Zhujiang Mouth Basin (Fig. 2.8) is one of three Cenozoic basins on the northern continental shelf of the South China Sea. The basin is bounded at the north by Guangdong Province, at the south by the Xisha Islands, at the west by

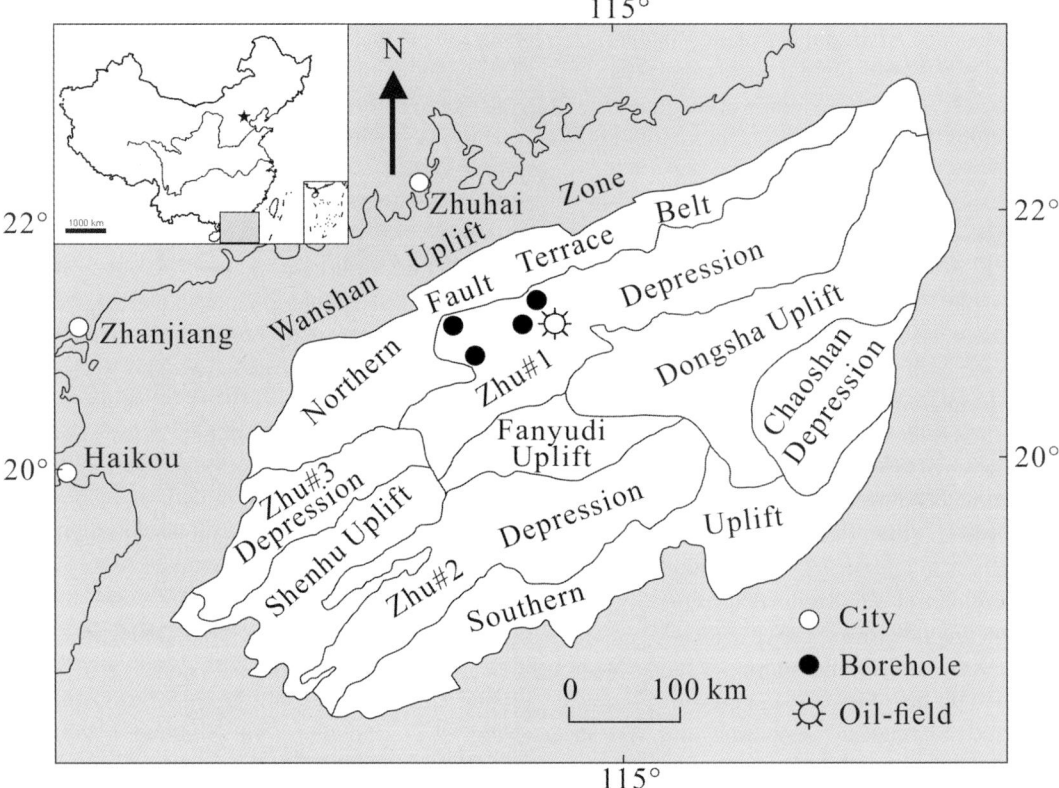

Fig. 2.8 Sketch map showing tectonic subdivisions and locations of oil fields and boreholes in the Zhujiang Mouth Basin

Hainan Island, and adjacent to the Dongsha Islands at the east. It is located in between longitude 110°–117°E and latitude 19°–22°N and covers an area of 150,000 km².

Tectonically, the basin lies in the Zhujiang Mouth Depression of the eastern North Shelf of the South China Sea. It is bounded at the north by the Wanshan Uplift Zone, at the south by the North Xisha Fault Belt, at the west by the Hainan Uplift Zone, and at the east by the Dongsha Uplift. This basin developed on Caledonian, Hercynian, and Yanshanian fold basement; sedimentary thickness is about 8000–10,000 m. The basin was formed in the Middle Eocene due to the influence of Himalayan Orogeny. At that time, a series of graben fault depressions were formed in the south Wanshan Uplift area; they filled with lacustrine facies deposits that are Eocene to Oliogocene in age. In the Early Miocene, north–south differences in uplift and the activity of a NE fault resulted in creating land to the north of the Wanshan Uplift; to the south, downdropping laid a foundation for unification of the individual basins creating the Zhujiang Mouth Basin following the Miocene when there was large-scale marine transgression.

The maximum thickness of the Tertiary in wells in the Zhujiang Mouth Basin is as much as 4620 m. Paleogene deposition was mainly terrestrial, consisting of the Shenhu, Wenchang, Enping, and Zhuhai formations, in ascending order. Neogene deposition was marine, consisting of the Zhujiang, Hanjiang, Yuehai, and Wanshan formations, in ascending order.

2.4.3.1 Stratigraphy and Lithofacies

The Paleocene to Lower Eocene Shenhu Formation is a fluvial facies, consisting mainly of grayish white and brownish red sandstone with alternations of mudstone, with a maximum thickness of 95 m.

The Middle Eocene Wenchang Formation is a lacustrine facies deposit, consisting mainly of dark gray and grayish black mudstone and shale, with grayish white sandstone. The maximum drilled thickness is 1110 m.

The Upper Eocene to Middle Oligocene Enping Formation is a lacustrine and swamp facies, consisting mainly of light gray sandstone with alternations of gray and dark gray mudstone and siltstone with thin coal beds. The maximum drilled thickness is 2055 m.

The Upper Oligocene Zhuhai Formation is a lacustrine and littoral facies, consisting mainly of light gray and gray white fine conglomerates, argillaceous sandstone, dolomitic fine sandstone, argillaceous siltstone, and dark gray silty mudstone with intercalations of multilayers of fine sandstone, oil sands, oil shale, bituminous shale and lignite, and occasional glauconite. The drilled thickness is 730–817 m.

The Lower Miocene Zhujiang Formation is a littoral and shallow marine facies, consisting mainly of light gray sandstone, siltstone with intercalations of dark gray mudstone, bituminous shale and oil shale, and containing minor glauconite. The drilled thickness is 210–1180 m.

The Middle Miocene Hanjiang Formation is a littoral and shallow marine facies; the lower part consists of light gray and grayish green sandstone and mudstone with alternations of mudstone and limestone; the upper part is composed of dark gray silty mudstone with intercalations of asphalt-bearing shale and sandstone. Southward in the Shenhuansha Uplift area, the formation becomes a fossiliferous reef limestone, containing glauconite and foraminifera. The drilled thickness is 90–709 m.

The Upper Miocene Yuehai Formation is a shallow sea facies, consisting mainly of light gray and gray silty mudstone, and argillaceous siltstone with alternations of different thicknesses of gray fine-grained conglomerates with gray sandstone and thin layers of limestone. The drilled thickness is 51–382 m.

The Pliocene Wanshan Formation is a set of shallow marine facies composed of nonconsolidated sediment. It consists of gray clay and silty clay with light gray sand layers and contains glauconite, fossil foraminifera, and bioclastics. The drilled thickness is 208–644 m.

2.4.3.2 Tectonic Units and Oil Fields

The tectonic framework of the Zhujiang Mouth Basin (Fig. 2.8) is that of north–south zonation and east–west block faulting. The basin is arranged from north to south as a series of E–N–E trending uplifts and depressions. A series of graben faults that formed during the Middle Yanshan Orogeny trends in a west to east direction. Within the basin, about eight tectonic units are subdivided, including the Northern Fault Terrace Belt, Zhu#1 Depression, Zhu#2 Depression, Zhu#3 Depression, Central fold belt, Weitan Uplift, Shenhu'ansha Uplift, and the South Slope. The Zhu#5 Well in the western part of the Zhu#2 Depression of the northeastern basin began oil production in August 1979, thereby becoming an important prospect for further oil and gas exploration and development of this basin (Zeng and Guo 1981; Jiang et al. 1994).

References

Division of Earth Science of Academia Sinica, Xinjiang Petroleum Administration. (1989). *Origin and evolution of the Junggar Basin and petroleum generation* (pp. 1–173). Beijing: Science Press. (in Chinese).

Hu B. L., Jiang D. X., Yang H. Q., Fu H., & Sun F. (1991). *Petroleum source rocks in East Xinjiang* (pp. 1–140). Lanzhou: Gansu Science and Technology Press (in Chinese).

Huang X. Z., Shao H. S., & Gu S. S. (1993). *Petroleum generation and in search of oil and gas fields in Qaidam Basin* (pp. 1–516). Lanzhou: Gansu Science and Technology Press (in Chinese).

Jiang Z. X., Zeng L., & Li M. X. (1994). *Tertiary of petroliferous provinces of China (VIII) Petroliferous province in northern shelf of South China Sea* (pp. 1–145). Beijing: Petroleum Industry Press (in Chinese).

Lanzhou Institute of Geology, Academia Sinica. (1960). *Formation and distribution of non-marine petroleum fields in Northwest China* (pp. 1–340). Beijing: Science Press (in Chinese).

Li M. C. (2004). *Petroleum and natural gas migration* (3rd edn.) (pp. 1–350). Beijing: Petroleum Industry Press (in Chinese).

Ma B. L., & Wen C. Q. (1991). *Sedimentary rock formation, evolution and petroleum in Tarim Basin* (pp. 1–195). Beijing: Science Press (in Chinese).

Ouyang S., Wang Z., Zhan J. Z., & Zhou Y. X. (2003). *Palynology of the Carboniferous and Permian strata of northern Xinjiang, northwestern China* (pp. 1–700). Hefei: University of Science and Technology of China Press (in Chinese with English summary).

Research Institute of Exploration and Development, Qinghai Petroleum Administration, Nanjing Institute of Geology and Palaeontology, Academia Sinica. (1985). *A research on tertiary palynology from the Qaidam Basin, Qinghai Province* (pp. 1–297). Beijing: Petroleum Industry Press (in Chinese with English abstract).

Research Institute of Petroleum Exploration and Development, Ministry of Petroleum Chemistry Industry, Nanjing Institute of Geology and Palaeontology, Chinese Academy of Sciences. (1978b). *On the Paleogene Dinoflagellates and Acritarchs from the coastal region of Bohai* (pp. 1–190). Beijing: Science Press (in Chinese with English abstract).

Sung, T. C. (1958). Tertiary spore and pollen complexes from the red beds of Chiuchuan, Kansu and their geological and botanical significance. *Acta Palaeontologica Sinica, 6*(2), 159–167. (in Chinese with English summary).

Wang, C. G., Ling, S. R., & Long, D. J. (1993). Jurassic petroleum forecasting prospect of the Turpan-Hami Basin. *Petroleum Geology, 9*(1), 1–16. (in Chinese).

Xu Y. C., Shen P., Liu W. H., Chen J. F., & Tao M. X. (1994). *Natural gas origin theory and application* (pp. 1–414). Beijing: Science Press (in Chinese).

Xu Y. C., Fu J. M., & Zheng J. J. (2000). *Origin of natural gas and geoscience foundation of large-middle gas-field formation* (pp. 1–228). Beijing: Science Press (in Chinese).

Zeng D. Q., & Guo B. (1981). *Tertiary of North continental shelf of South China Sea* (pp. 1–258). Guangzhou: Guangdong Science and Technology Press (in Chinese).

Zhong X. C., Zhao C. B., Yang S. Z., & Shen H. (2003). *Jurassic System in the North of China (II) Palaeoenvironment and oil-gas source* (pp. 1–201). Beijing: Petroleum Industry Press. (in Chinese with English abstract).

Zhou Q. J., & Zheng J. J. (1990). *Tectonics of the Tarim Basin* (pp. 1–144). Beijing: Science Press (in Chinese).

Zhou Z. Y. (2001). *Stratigraphy of the Tarim Basin* (pp. 1–359). Beijing: Science Press (in Chinese).

Zhou Z. Y., & Chen P. J. (1990). *Biostratigraphy and geological evolution of the Tarim Basin* (pp. 1–366). Beijing: Science Press (in Chinese).

Zhu X. (1986). *Tectonics of the petroliferous basins in China* (pp. 1–132). Beijing: Petroleum Industry Press (in Chinese).

Fossil Spores and Pollen in Crude Oils

3

Abstract

Crude oils from both inland and coastal shelf basins were analyzed for their spores and pollen content. The inland petroliferous basins include the Tarim, Junggar, Turpan, Qaidam, and Jiuquan basins in West China. The coastal shelf petroliferous basins include the Liaohe, Beibu Gulf, and Zhujiang Mouth basins in the East and South China Seas. Pollen and spores from crude oils were extracted using a Soxhlet apparatus. Those in potential source rocks were extracted using standard palynological processing techniques. The fossils were then compared by age and assemblage to others around the world to ensure that correlation between potential source rocks and crude oils in China would be accurate. Abundant fossil spores and pollen in crude oils, such as 183 species from the Tarim Basin, 211 species from the Junggar Basin, and 103 species from the Qaidam Basin, were identified. These spores and pollen can provide important information for assessing petroleum source rocks.

Keywords

Spores and pollen from crude oils · Carboniferous-Permian · Triassic · Jurassic · Cretaceous · Paleogene

3.1 Materials and Methods

3.1.1 Materials

The materials used for study in this special volume were taken from both inland and coastal shelf basins. The inland petroliferous basins include the Tarim, Junggar, Turpan, Qaidam, and Jiuquan basins in West China. The coastal shelf petroliferous basins include the Liaohe, Beibu Gulf, and Zhujiang Mouth basins in the East and South China Seas. Crude oil samples, 176 in number, were analyzed.

The oil fields in the Tarim Basin include the Yakela, Lunnan, Donghetang, Yingmaili, Yiqikelike, Kelatu, Kekeya, and Maigati (Markit) fields. The oils fields in the Junggar Basin include the Karamay, Baikouquan, Xiazijie, Beisantai, Huonan, Qigu, and Dushanzi fields. Those in the

Turpan Basin include the Qiktim and Shengjinkou fields. The oil fields in the Qaidam Basin include the Lenghu, Yuka, Youquanzi, Xianshuiquan, and Dafengshan fields. The oil fields in the Jiuquan Basin include the Yaerxia, Laojunmiao, and Baiyanghe fields. In the Liaohe Basin, the Xinglongtai oil field was analyzed. In the Beibu Gulf Basin, the Weizhou oil field was included. Finally, the Zhuhai oil field was analyzed from the Zhujiang Mouth Basin. More or less well-preserved fossil spores and pollen were found in crude oils from most of these oil fields. The most abundant were in crude oils from the Tarim Basin.

In addition, 164 rock samples collected from the potential petroleum source rocks were prepared for correlation. The spores and pollen in the crude oils were compared with those in their potential source rocks.

The success rate of finding pollen and spores following Soxhlet extraction varies. Some samples have none. Most samples have about 30–50 fossils. The most we have ever found in 5 L of oil was 204 fossils.

Abbreviations in the following tables are standard. These are as follows: C, Carboniferous; P, Permian; T, Triassic; J, Jurassic; K, Cretaceous; N, Neogene; and E, Paleogene.

3.1.2 Methods

The processing technique that extracted fossil spores and polled from crude oils used a Soxhlet apparatus. The procedure is summarized from Jiang et al. (1974) and Jiang (1990) as follows (Fig. 3.1):

(1) Crude oil samples were diluted with benzene or gasoline for filtration. Three to five liters of crude oil was used for each sample.
(2) The diluted crude oil samples were heated to 70–75 °C and filtered through filter papers to remove the petroleum.
(3) The filter papers together with any remaining mineral slags were extracted in a Soxhlet apparatus using benzene, ether, ketone, benzene-alcohol, and alcohol to remove the oily, waxy, and bituminous matter. Other organic solvents may be useful also.
(4) The residues were concentrated, water washed by centrifuging, and treated using 10 % hydrochloric acid to remove calcareous matter and 40 % hydrofluoric acid to remove siliceous matter.
(5) The insoluble organic matter (kerogen) and mineral matter were separated by heavy liquid (CdI–KI) flotation; the specific gravities of the heavy liquids were 2.0–2.2 (for CdI) and 2.2–2.4 (for KI).
(6) The insoluble organic matter was preserved in 50 % glycerin and 1 % carbolic acid. The insolubles included the fossil spores, pollen, fungi, and algae. These were mounted in glycerin jelly for study under light microscopy.

In contrast to the crude oil samples, rock samples were prepared by standard methods using 5 % potassium hydroxide, 10 % hydrochloric acid, and 40 % hydrofluoric acid to extract their fossil pollen and spores. Then, heavy liquids were used for concentrating microfossils having different densities. The fossils in the rocks were analyzed for correlation between crude oil fossils versus their potential rock sources.

3.2 Tarim Basin

3.2.1 North Tarim Upheaval

The North Tarim Upheaval is located in the north part of the Tarim Basin. Within this upheaval, the Yakela structural high was prospected in 1984. Ordovician dolomite was drilled in well SC-2 near a spring that gushed petroleum in Yakela. Since then, Ordovician, Carboniferous, Triassic, Jurassic, Cretaceous, and Tertiary petroleum pools have been found in this petroliferous region.

One hundred and thirty-six species of fossil spores and pollen referred to 66 genera were extracted from crude oil samples collected in six

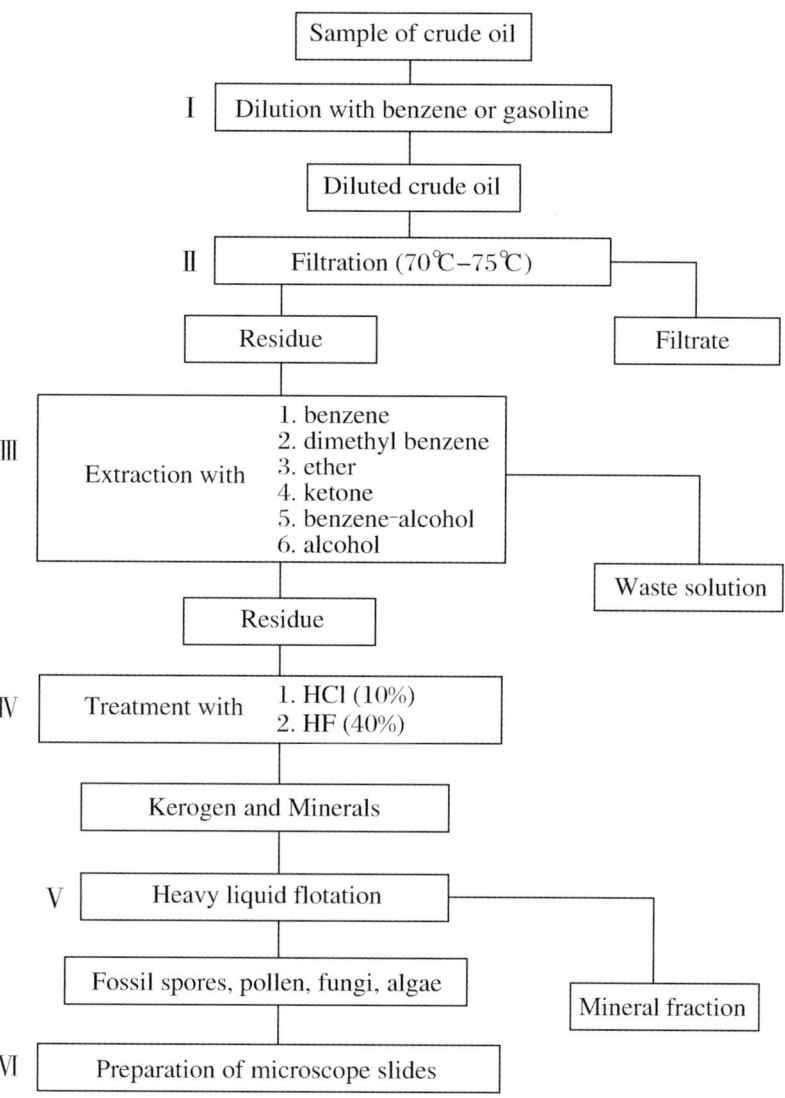

Fig. 3.1 Method of extraction of spores, pollen, and algae from crude oil samples

reservoirs within the Yakela, Lunnan, Donghetang, and Yingmaili oil fields in the North Tarim Upheaval. Spores and pollen from these fields are listed in Table 3.1, and illustrated on Plates I–VIII, XXXVII–XXXIX.

The spores and pollen of the North Tarim Upheaval (Table 3.1) are Mesozoic and Late Paleozoic species. Most are Triassic and Jurassic index species as well as common species. A few range down into the Carboniferous. Among the spores and pollen, *Punctatisporites minutus* and *Calamospora pedata* were found in the Middle to Upper Carboniferous (Pennsylvanian) formation in Illinois, USA (Kosanke 1950). *P. minutus* was found in the Lower Triassic Jiucaiyuan Formation, Middle Triassic Karamay Formation, and Upper Triassic Haojiagou Formation in the Dalongkou area, Jimsar, Xinjiang, China. *C. pedata* was found in the Upper Permian Guodikeng Formation in the Dalongkou area, Xinjiang, China (Institute of Geology, Chinese Academy of Geological Sciences, and Institute of Geology, Xinjiang Bureau of Geology and Mineral Resources 1986). *Granulatisporites adnatoides* and *Triquitrites desperatus* were found in the Upper Carboniferous

Table 3.1 Spores and pollen in crude oils from the North Tarim Upheaval

Spores and pollen	Number of specimens	Geological age
Spores		
Deltoidospora perpusilla (Bolchovitina) Pocock, 1970	7	J
Cyathidites australis Couper, 1953	8	J-K
C. minor Couper, 1953	14	J-K
Gleicheniidites senonicus Ross, 1949	6	J-K
G. rouseii Pocock, 1970	5	J
G. nilssonii Pocock, 1970	4	J
G. conflexus (Chlonova) Xu et Zhang, 1980	3	J
Concavisporites toralis (Leschik) Nilsson, 1958	5	J
Dictyophyllidites harrisii Coupet, 1958	11	J
Undulatisporites concavus Kedves, 1971	7	J
U. pflugii Pocock, 1970	7	J
Punctatisporites minutus Kosanke, 1950	3	C-T
P. triassicus Schulz, 1964	12	T
P. microtumulosus Playford et Dettmann, 1965	4	T
P. ambiguus Leschik, 1955	2	T
Calamospora pedata Kosanke, 1950	2	C-P
C. nathorstii (Halle) Klaus, 1960	4	T-J
C. tener (Leschik) Mädler, 1964	2	T
Cibotiumspora paradoxa (Maljavkina) Chang, 1965	8	J
C. humilis Zhang, 1984	5	T
Retusotriletes mesozoicus Klaus, 1960	3	T
R. arcticus Qu et Wang, 1986	4	T
Granulatisporites jurassicus Pocock, 1970	5	J
G. adnatoides (Potonié et Kremp) Smith et Butterworth, 1967	5	C-P
G. minor de Jersey, 1959	3	J
Verrucosisporites contactus Clarke, 1965	5	T
V. remyanus Mädler, 1964	4	T
Osmundacidites wellmanii Couper, 1953	11	J-K
O. alpinus Klaus, 1960	6	T
Manumia verrucata Pocock, 1970	6	J
M. irregularis Pocock, 1970	4	J
Lophotriletes corrugatus Ouyang et Li, 1980	4	T
Apiculatisporis variabilis Pocock, 1970	5	J
A. globosus (Leschik) Playford et Dettmann, 1965	7	T
A. parvispinosus (Leschik) Qu, 1980	3	T
A. spiniger (Leschik) Qu, 1980	4	T
Conbaculatisporites mesozoicus Klaus, 1960	2	T
Lycopodiacidites kuepperi Klaus, 1960	5	T

(continued)

Table 3.1 (continued)

Spores and pollen	Number of specimens	Geological age
L. rhaeticus Schulz, 1967	4	T
Lycopodiumsporites paniculatoides Tralau, 1968	4	J
L. subrotundum (Kara-Murza) Pocock, 1970	2	J
Tigrisporites halleinis klaus, 1960	3	T
Zebrasporites kahleri Klaus, 1960	2	T
Retispora florida Staplin, 1960	3	C
Triquitrites subrotundus Ouyang et Li, 1980	2	C
T. desperatus Potonié et Kremp, 1956	2	C
Lycospora pusilla (Ibrahim) Somers, 1973	2	C
Camarozonosporites rudis (Leschik) Klaus, 1960	3	T
Duplexisporites gyratus Playford et Dettmann, 1965	5	T-J
D. amplectiformis (Kara-Murza) Playford et Dettmann, 1965	7	T-J
D. anagrammensis (Kara-Murza) Playford et Dettmann, 1965	4	T-J
D. scanicus (Nilsson) Playford et Dettmann, 1965	5	J
D. problematicus (Couper) Playford et Dettmann, 1965	6	J-K
Multinodisporites junctus Ouyang et Li, 1980	3	T
Limatulasporites dalongkouensis Qu et Wang, 1986	4	T
L. parvus Qu et Wang, 1986	2	T
Lundbladispora playfordi Balme, 1963	2	T
L. subornata Ouyang et Li, 1980	2	T
L. nejburgii Schulz, 1964	3	T
L. plicata Bai, 1983	3	T
Schizosporis bilobatus (Faddeeva) Qu, 1980	2	T
Torispora securis (Balme) Alpern, Doubinger et Horst, 1965	3	C-T
Punctatosporites minutus Ibrahim, 1933	5	C
Marattisporites scabratus Couper, 1958	3	J-K
Aratrisporites coryliseminis Klaus, 1960	6	T
A. scabratus Klaus, 1960	5	T
A. paenulatus Playford et Dettmann, 1965	4	T
A. fischeri (Klaus) Playford et Detttmann, 1965	5	T
A. granulatus (Klaus) Playford et Dettmann, 1965	5	T
A. strigosus Playford, 1965	15	T
A. tenuispinosus Playford, 1965	4	T
A. paraspinosus Klaus, 1965	2	T
A. parvispinosus Leschik, 1955	3	T
Pollen		
Cordaitina uralensis (Luber) Samoilovich, 1953	6	C-P
Enzonalasporites tenuis Leschik, 1955	3	T
E. vigens Leschik, 1955	5	T

(continued)

Table 3.1 (continued)

Spores and pollen	Number of specimens	Geological age
Lueckisporites triassicus Clarke, 1965	4	T
L. virkkiae Potonié et Klaus, 1954	4	P-T
L. tattooensis Jansonius, 1962	2	P-T
Taeniaesporites pellucidus (Doubin) Balme, 1970	4	P-T
T. albertae Jansonius, 1962	3	P-T
T. rhaeticus Schulz, 1967	5	T
T. divisus Qu, 1982	3	T
Protohaploxypinus samoilovichii (Jansonius) Hart, 1964	3	P-T
P. microcorpus (Schaarschmidt) Clarke, 1965	3	P-T
Striatoabietites duivenii (Jansonius) Hart, 1964	4	P-T
S. richteri (Klaus) Hart, 1964	2	P-T
Colpectopollis pseudostriatus (Kopytova) Qu et Wang, 1986	9	T
C. scitulus (Qu et Pu) Qu et Wang, 1986	6	T
Chordasporites singulichorda Klaus, 1960	5	T
C. orientalis Ouyang et Li, 1980	4	T
C. impensus Quyang et Li, 1980	2	T
Vitreisporites itunensis Pocock, 1970	3	J
Pityosporites parvisaccatus de Jersey, 1959	4	J
Vesicaspora wilsonii Schemel, 1951	4	C
Granosaccus ornatus (Pautsch) Pautsch, 1973	7	T
Klausipollenites schaubergeri (Potonié et Klaus) Jansonius, 1962	2	P-T
Alisporites parvus de Jersey, 1962	5	T
A. australis de Jersey, 1962	3	T
A. aequalis Mädler, 1964	4	T
A. fusiformis Ouyang et Li, 1980	4	T
A. lowoodensis de Jersey, 1983	3	J
A. bilateralis Rouse, 1959	5	J-K
Piceites latens Bolchovitina, 1956	6	J-K
P. expositus Bolchovitina, 1956	4	J
P. pseudorotundiformis (Maljavkina) Pocock, 1970	5	J
P. podocarpoides Bolchovitina, 1956	2	J
Pinuspollenites normalis Qu et Wang, 1986	4	T
P. divulgatus (Bolchovitina) Qu, 1980	8	T-J
Cedripites priscus Balme, 1970	4	P-T
C. parvisaccus Quyang et Li, 1980	7	T
C. minor Pocock, 1970	3	J
Platysaccus undulatus Ouyang et Li, 1980	5	T
Podocarpidites multicinus (Bolchovitina) Pocock, 1970	5	J
P. multesimus (Bolchovitina) Pocock, 1962	3	J-K

(continued)

Table 3.1 (continued)

Spores and pollen	Number of specimens	Geological age
P. unicus (Bolchovitina) Pocock, 1970	2	J
P. rousei Pocock, 1970	3	J
P. queenslandi (de Jersey) Qu, 1980	7	T
Quadraeculina limbata Maljavkina, 1949	3	J-K
Parcisporites rarus Ouyang et Li, 1980	3	T
P. solutus Leschik, 1955	2	T
Minutosaccus parcus Qu et Wang, 1986	3	T
Rugubivesiculites lepidus Bai, 1983	5	T
Ephedripites tertiarius Krutzsch, 1970	11	E-N
Bennettiteaepollenites lucifer (Thiergart) Potonié, 1958	5	J
Eucommiidites troedssonii Erdtman, 1948	3	J-K
Chasmatosporites magnolioides (Erdtman) Nilsson, 1958	2	T
Cycadopites nitidus (Balme) Pocock, 1970	22	T-K
C. typicus (Maljavkina) Pocock, 1970	8	J-K
C. minimus (Cookson) Pocock, 1970	14	J-K
C. subgranulosus (Couper) Clarke, 1965	9	T-J
C. carpentieri (Delcourt et Sprumont) Singh, 1964	3	J-K
Araucariacites australis Cookson, 1947	5	J-K
Classopollis qiyangensis Shang, 1981	3	J
Callialasporites minus (Tralau) Guy, 1971	3	J
C. dampieri (Balme) Sukh Dev, 1961	2	J-K
Total	638	

Westphalian C stage of the Ruhr region, Germany (Potonié and Kremp 1955, 1956). *G. adnatoides* was reported from Carboniferous coal seams in Great Britain (Smith and Butterworth 1967) and the Upper Permian Hsuanwei Formation in Fuyuan, Yunnan, China (Ouyang 1986). *T. desperatus* is considered to be a Carboniferous index species. *Triquitrites subrotundus* was found in the Upper Carboniferous Benxi Formation in Shuoxian, Shanxi, China (Ouyang and Li 1980a), and its known distribution is limited to the Carboniferous System. *Lycospora pusilla* and *Punctatosporites minutus* were reported from the Upper Carboniferous Westphalian B stage of the Ruhr region, Germany (Potonié and Kremp 1956). *L. pusilla* was reported from the Upper Carboniferous Westphalian A stage of Britain (Owens 1996) and the Upper Carboniferous Jingyuan Formation in the Zhongwei region, Ningxia, China (Gao 1994), both being Carboniferous index species. *Retispora florida* was found in the Upper Mississippian stage of the Lower Carboniferous System in Alberta, Canada (Staplin 1960). *Torispora securis* was found in the Upper Carboniferous Westphalian stage of Britain (Balme 1952); it is common in Middle to Upper Carboniferous rocks in Europe and is distributed from Middle Carboniferous to Triassic in China (Wang 1984; Ouyang 1986). *Cordaitina uralensis* was found in the Permian of the Ural region, Russia, and the Qipan-Duwa region of the Tarim Basin (Samoilovich 1953; Wang 1989); it is common in the Carboniferous to Permian deposits of Kazakhstan (Luber 1955). *Vesicaspora wilsonii* was found in the Middle to Upper Carboniferous (Pennsylvanian) of Iowa, USA (Schemel 1951); it is considered as a Carboniferous index species.

Lueckisporites virkkiae and *Klausipollenites schaubergeri* were found in Upper Permian rocks in Germany, Austria, and Britain (Potonié and Klaus 1954; Clarke 1965b). *L. virkkiae* was reported from the Upper Permian Pusige Formation in the Tarim Basin (Zhu 1997). *Striatoabietites richteri* was found in Upper Permian deposits of Germany and Britain (Klaus 1955; Clarke 1965b). *Protohaploxypinus samoilovichii, Stratoabietites duivenii, Lueckisporites tattooensis,* and *Taeniaesporites albertae* were found in the Triassic System of the Peace River area in Canada (Jansonius 1962) and in the Permian of Europe (Hart 1965). *Taeniaesporites pellucidus* and *Cedripites priscus* were found in the Permian to Triassic systems of Pakistan (Balme 1970), and these species were also reported from the Upper Permian Guodikeng Formation and the Lower Triassic Jiucaiyuan Formation in the Dalongkou area, Jimsar, Xinjiang, China (Institute of Geology, Chinese Academy of Geological Sciences, and Institute of Geology, Xinjiang Bureau of Geology and Mineral Resources 1986). *Protohaploxypinus microcorpus* was found in the Upper Permian Zechstein Series of Germany, Austria, and Britain (Clarke 1965b). *Punctatisporites ambiguus, Calamospora tener, Apiculatisporis globosus, A. spiniger A. parvispinosus, Camarozonosporites rudis, Aratrisporites parvispinosus, Enzonalasporites tenuis,* and *Parcisporites solutus* were found in the middle part of the Upper Triassic Keuper Stage in Basel, Switzerland (Leschik 1955). *Calamospora nathorst, Retusotriletes mesozoicus, Osmundacidites alpinus, Conbaculatisporites mesozoicus, Lycopodiacidites kuepperi, Tigrisporites halleinis, Zebrasporites kahleri, Aratrisporites fischeri, A. coryliseminis, A. scabratus, A. granulatus, A. paraspinosus,* and *Chordasporites singulichorda* were found in the Triassic System of Austria (Klaus 1960). *Aratrisporites strigosus* and *A. tenuispinosus* were found in the Triassic of Tasmania (Playford 1965). *Punctatisporites triassicus* and *Lundbladispora nejburgii* were found in the middle part of the Lower Triassic Buntsandstein Series of Germany and were also reported from the Lower Triassic of Rumania (Venkatachala et al. 1968). *Verrucosisporites contactus* and *Lueckisporites triassicus* were found in the Upper Triassic Keuper Stage in Worcestershire, England (Clarke 1965a). *Verrucosisporites remyanus* and *Alisporites aequalis* were found in the Triassic of Germany (Mädler 1964). *Lycopodiacidites rhaeticus* and *Taeniaesporites rhaeticus* were found in the Upper Triassic Rhaetian Stage in Germany (Schulz 1967). *Alisporites australis, A. parvus,* and *Podocarpidites queenslandi* were found in the Middle to Upper Triassic Ipswich Formation of the Ipswich coalfield in Queensland, Australia (De Jersey 1962). *Lundbladispora playfordi* was reported from the Lower Triassic Scythian Stage of Western Australia (Balme 1963). *Lophotriletes corrugatus, Multinodisporites junctus, Lundbladispora subornata, Chordasporites orientalis, C. impensus, Alisporites fusiformis, Cedripites parvisaccus, Platysaccus undulatus,* and *Parcisporites rarus* were first found in the Lower Triassic Kayitou Formation in the Fuyuan region, Yunnan, China (Ouyang and Li 1980b). *Taeniasporites divisus* was reported from the Lower Triassic Liujiagou Formation in Jiaocheng, Shanxi, China (Qu 1982). *Cibotiumspora humilis* was reported from the Upper Triassic Series in Weiyuan, Sichuan, China (Zhang 1984). *Limatulasporites dalongkouensis* and *L. parvus* were reported from the Lower Triassic Jiucaiyuan Formation and Shaofanggou Formation in the Dalongkou area, Jimsar, Xinjiang, China. *Retusotriletes arcticus* was reported from the Lower Triassic Shaofanggou Formation and the Middle Triassic Karamay Formation in Dalongkou, China. *Minutosaccus parvus* was reported from the Karamay Formation in Dalongkou, China. *Pinuspollenites normalis* was recorded from the Karamay Formation and the Upper Triassic Huangshanjie Formation in Dalongkou, China. *Colpectopollis pseudostriatus* and *C. scitulus* were reported from the Middle Triassic Karamay Formation and the Upper Triassic Haojiagou Formation in Dalongkou, China (Institute of Geology, Chinese Academy of Geological Sciences, and Institute of Geology, Xinjiang Bureau of Geology and Mineral Resources 1986). These spores and pollen may be considered as Triassic index species.

Dictyophyllidites harrisii and *Duplexisporites problematicus* were found in the Middle Jurassic Bajocian Stage of Yorkshire, England (Couper 1958). *Lycopodiumsporites paniculatoides* and *Callialasporites minus* were found in the Middle Jurassic Bajocian and Bathonian Stages of southern Sweden (Tralau 1968; Guy 1971). *Bennettiteaepollenites lucifer* was found in the Middle Jurassic Dogger Series of Germany (Potonié 1958). *Gleicheniidites rouseii, G. nilsonii, Undulatisporites pflugii, Granulatisporites jurassicus, Apiculatisporis variabilis, Vitreisporites itunensis, Cedripites minor,* and *Podocarpidites rousei* were found in the Jurassic System of western Canada (Pocock 1970). *Granulatisporites minor* and *Pityosporites parvisaccatus* were found in the Jurassic Walloon Formation of the Rosewood coalfield, Queensland, Australia (De Jersey 1959). *Alisporites lowoodensis* was reported from the Jurassic Marburg Sandstone in Queensland, Australia (De Jersey 1963). *Cibotiumspora paradoxa* was found in the Middle Jurassic Series in Emba, Russia, the Lower to Middle Jurassic Yima coal-bearing Formation in Mianchi, Henan, China, and the Lower and Middle Jurassic Series in the Tarim Basin (Maljavkina 1949; Zhang 1965; Liu 2003; Jiang et al. 2008). *Piceites expositus* and *Podocarpidites multicinus* were found in the Upper Jurassic Series and Lower Jurassic Series, respectively, in Yakutsk, Russia (Bolchovitina 1956). *Piceites pseudorotundiformis* was found in the Jurassic System of Russia and Canada (Pocock 1970). These spores and pollen may be regarded as Jurassic index species. In addition, *Cyathidites australis, C. minor, Osmundacidites wellmanii, Marattisporites scabratus,* and *Araucariacites australis* were found in the Jurassic System and the Lower Cretaceous Series of New Zealand and Britain (Couper 1953, 1958). *Quadraeculina limbata* was found in the Jurassic System and the Lower Cretaceous Series in Emba, Russia (Maljavkina 1949); it was also reported from the Jurassic System of Canada and China (Pocock 1970; Institute of Geology, Chinese Academy of Geological Sciences 1980). *Piceites latens* was found in the Lower Cretaceous Series in Yakutsk, Russia (Bolchovitina 1956); it is an important member of the Jurassic palynoflora of western Canada (Pocock 1970). *Podocarpidites multesimus* was found in the Lower Jurassic Series in Yakutsk, Russia (Bolchovitina 1956), and the Lower Cretaceous Series of western Canada (Pocock 1962). These species of spores and pollen are common in the Jurassic System of Europe, America, and China; they may also be distributed in the Lower Cretaceous Series. *Cycadopites subgranulosus* was found in the Lower Jurassic Lias Stage in Brora, Scotland (Couper 1958), and was reported from the Upper Triassic Keuper Stage in Worcestershire, England (Clarke 1965a). These spores and pollen may be considered as Jurassic common species.

3.2.2 Kuqa Depression

The Kuqa Depression is a petroleum-bearing region in the northeastern part of the Tarim Basin. The Yiqikelike oil field, discovered in 1958, was the first commercial oil field in the basin. The Kangcun oil seepage at the Kuqatawu structure and the Jilishen oil seepage at the Dongqiulitake structure are well known in Xinjiang.

Thirty-four species of fossil spores and pollen referred to 23 genera were extracted from crude oil samples collected in the Kuqa Depression from Middle Jurassic, Upper Jurassic, and Lower Cretaceous reservoir beds of the Yiqikelike oil field and the Miocene oil seep at Kangcun and Jilishen. The species are listed in Table 3.2 and illustrated on Plate IX.

With the exception of the Tertiary-aged *Ephedripites, Quercoidites, Chenopodipollis, Artemisiaepollenites,* and *Pinuspollenites*, most of the species in the Kuqa Depression (Table 3.2) are Jurassic. However, some are Jurassic to Cretaceous, and a few are strictly Cretaceous species.

Among the spores and pollen, *Deltoidospora gradata* and *Cibotiumspora paradoxa* were found in the Middle Jurassic Series in Emba, Russia (Maljavkina 1949). *Podocarpidites multicinus* was found in the Lower Jurassic Series of Russia and was reported from the Jurassic

Table 3.2 Spores and pollen in crude oils from the Kuqa Depression

Spores and pollen	Number of specimens	Geological age
Spores		
Deltoidospora perpusilla (Bolchovitina) Pocock, 1970	5	J
D. gradata (Maljavkina) Pocock, 1970	3	J
Cyathidites australis Couper, 1953	5	J-K
C. minor Couper, 1953	7	J-K
Concavisporites toralis (Leschik) Nilsson, 1958	4	J
Dictyophyllidites harrisii Couper, 1958	7	J
Hymenophyllumsporites deltoidus Rouse, 1957	3	K
Cibotiumspora paradoxa (Maljavkina) Chang, 1965	5	J
Osmundacidites wellmanii Couper, 1953	6	J-K
Duplexisporites scanicus (Nilsson) Playford et Dettmann, 1965	4	J
Polypodiaceaesporites haardti (Potonié et Venitz) Thiergart, 1937	5	N
Pollen		
Alisporites grandis (Cookson) Dettmann, 1963	3	J-K
A. bilateralis Rouse, 1959	5	J-K
Protopicea exilioides (Bolchovitina) Pocock, 1970	4	J
Abietineaepollenites microalatus Potonié, 1951	4	J-K
A. dunrobinensis Couper, 1958	3	J
A. minimus Couper, 1958	6	J-K
Pinuspollenites labdacus (Potonié) Raatz, 1938	11	N
Piceaepollenites alatus Potonié, 1931	9	N
Cedripites minor Pocock, 1970	6	J
C. cretaceus Pocock, 1962	4	K
Podocarpidites multesimus (Bolchovitina) Pocock, 1962	4	J-K
P. multicinus (Bolchovitina) Pocock, 1970	3	J
Parvisaccites enigmatus Couper, 1958	4	J
Ephedripites tertiarius Krutzsch, 1970	15	E-N
Cycadopites nitidus (Balme) Pocock, 1970	7	T-K
C. typicus (Maljavkina) Pocock, 1970	5	J-K
C. minimus (Cookson) Pocock, 1970	8	J-K
C. subgranulosus (Couper) Clarke, 1965	4	T-J
Classopollis annulatus (Verbitzkaya) Li, 1974	5	J-K
Quercoidites henrici Potonié, Thomson et Thiergart, 1950	3	N
Q. microhenrici (Potonié) Potonié, 1950	6	N
Chenopodipollis multiporatus (Pflug et Thomson) Zhou, 1981	14	N
Artemisiaepollenites sellularis Nagy, 1969	12	N
Total	199	

System of Canada (Bolchovitina 1956; Pocock 1970). *Concavisporites toralis* and *Duplexisporites scanicus* were found in the Upper Triassic Rhaetian Stage and the Lower Jurassic Lias Stage of Sweden (Nilsson 1958). *Cycadopites subgranulosus* and *Abietineaepollenites dunrobinensis* were found in the Lower Jurassic Lias Stage of Britain. *Dictyophyllidites harrisii* and *Parvisaccites enigmatus* were found in the Middle Jurassic Bajocian and Bathonian stages of England. *Abietineaepollenites microalatus* and *A. minimus* were found in the Lower Cretaceous Wealden Stage of England; these are widely distributed in the Jurassic System and the Lower Cretaceous Series (Couper 1958). *Cedripites minor* was found in the Jurassic System of Canada (Pocock 1970). *Hymenophyllumsporites deltoidus* and *Cedripites cretaceus* were found in the Cretaceous System of Canada (Pocock 1962); the latter was reported from the Lower Cretaceous Shushanhe Formation and Baxigai Formation of the Tarim Basin (Jiang et al. 1988). The other spores and pollen, such as *Cyathidites minor*, *Cycadopites nitidus*, *Podocarpidites multesimus*, are common Jurassic species of Europe, America, and China.

3.2.3 Southwest Tarim Depression

3.2.3.1 Kashi Sag

The Kashi Sag is in the northern part of the Southwest Tarim Depression. The Kelatu oil field was discovered there in 1958. The reservoir rock was red sandstones of Neogene age.

Forty-six species of fossil spores and pollen referred to 31 genera were extracted from crude oil samples collected in the Neogene reservoir bed at Kelatu from the Kashi Sag. Species of fossil spores and pollen are listed in Table 3.3 and illustrated on Plate X.

With the exception of the Tertiary *Piceaepollenites alatus*, *Pinuspollenites labdacus*, *Ephedripites tertiarius*, *Caryapollenites simplex*, *Chenopodipollis multiplex*, and *Artemisiaepollenites sellularis*, most of the spores and pollen from the Kashi Sag of the Tarim Basin (Table 3.3) are Jurassic index species. However, some are distributed in Jurassic to Cretaceous age rocks, and a few range from Triassic to Jurassic or Triassic to Cretaceous in age. Among the spores and pollen, *Dictyophyllidites harrisii*, *Todisporites major*, *Leptolepidites major*, *Klukisporites variegatus*, and *Pteruchipellenites thomasii* were found in the Middle Jurassic Bajocian Stage in Yorkshire, England (Couper 1958). *Gleicheniidites rouseii*, *Murospora jurassica*, *M. minor*, *Chasmatosporites canadensis*, *Cerebropollenites carlylensis*, *Vitreisporites jurassicus*, *V. shouldicei*, *V. jansonii*, *Podocarpidites florinii*, *P. langii*, and *Cedripites minor* were found in the Jurassic System of western Canada (Pocock 1970). *Deltoidospora lineata*, *Paleoconiferus asaccatus*, *Podocarpidites multicinus*, *Piceites expositus*, and *Protopiea exilioides* were found in the Jurassic System in Yakutsk, Russia (Bolchovitina 1956). *Chasmatosporites major*, *C. elegans*, and *Protopinus scanicus* were found in the Lower Jurassic Series of Sweden (Nilsson 1958). *Granulatisporites minor* and *Alisporites lowoodensis* were found in the Jurassic System in Queensland, Australia (De Jersey 1959, 1963). *Cibotiumspora paradoxa* was found in the Middle Jurassic Series of Russia and the Lower to Middle Jurassic Series of China (Maljavkina 1949; Zhang 1965; Liu 2003; Jiang et al. 2008). These spores and pollen may be considered as Jurassic index species; they are widely distributed in the Jurassic System of Europe, Asia, and North America. *Cyathidites australis*, *C. minor*, *Osmundacidites wellmanii*, *Podocarpidites multesimus*, *Cycadopites nitidus*, *C. typicus*, *C. minimus*, *Classopollis classoides*, and *C. annulatus* are also widely distributed in the Jurassic System, but they are also found in the Lower Cretaceous Series in the Tarim Basin (Jiang et al. 1988, 2006, 2007). These species may be considered as Jurassic common species.

3.2.3.2 Yecheng Sag

The Yecheng Sag is a small depression in the middle part of the larger southwest Tarim Depression. This sag, which adjoins the Kashi Sag to the north, the Hetian Sag to the south, and the Maigaiti Clinoform (a clinothem zone) on the east, might become the main petroleum-bearing

Table 3.3 Spores and pollen in crude oils from the Kashi Sag

Spores and pollen	Number of specimens	Geological age
Spores		
Deltoidospora lineata (Bolchovitina) Pocock, 1970	2	J
Cyathidites australis Couper, 1953	5	J-K
C. minor Couper, 1953	12	J-K
Gleicheniidites rouseii Pocock, 1970	6	J
Dictyophyllidites harrisii Couper, 1958	7	J
Cibotiumspora paradoxa (Maljavkina) Chang, 1965	4	J
Todisporites major Couper, 1958	8	J
Granulatisporites minor de Jersey, 1959	3	J
Leptolepidites major Couper, 1958	3	J
Osmundacidites wellmanii Couper, 1953	3	J-K
Klukisporites variegatus Couper, 1958	7	J
Murospora jurassica Pocock, 1970	5	J
M. minor Pocock, 1970	4	J
Pollen		
Paleoconiferus asaccatus Bolchovitina, 1956	6	J
Vitreisporites jurassicus Pocock, 1970	7	J
V. shouldicei Pocock, 1970	4	J
V. jansonii Pocock, 1970	5	J
Pteruchipollenites thomasii Couper, 1958	4	J
Alisporites lowoodensis de Jersey, 1963	4	J
Pinuspollenites labdacus (Potonié) Raatz, 1938	11	N
Protopinus scanicus Nilsson, 1958	3	J
Piceaepollenites alatus Potonié, 1931	9	N
Protopicea exilioides (Bolchovitina) Pocock, 1970	4	J
Piceites latens Bolchovitina, 1956	5	J-K
P. expositus Bolchovitina, 1956	6	J
Cedripites minor Pocock, 1970	8	J
Podocarpidites florinii Pocock, 1970	5	J
P. langii Pocock, 1970	5	J
P. multicinus (Bolchovitina) Pocck, 1970	4	J
P. multesimus (Bolchovitina) Pocock, 1962	3	J-K
Quadraeculina limbata Maljavkina, 1949	3	J-K
Ephedripites tertiarius Krutzsch, 1970	12	E-N
Chasmatosporites major Nilsson, 1958	5	J
C. elegans Nilsson, 1958	3	J
C. minor Nilsson, 1958	3	J
C. canadensis Pocock, 1970	2	J
Cycadopites nitidus (Balme) Pocock, 1970	8	T-K

(continued)

Table 3.3 (continued)

Spores and pollen	Number of specimens	Geological age
C. typicus (Maljavkina) Pocock, 1970	3	J-K
C. subgranulosus (Couper) Clarke, 1965	7	T-J
C. minimus (Cookson) Pocock, 1970	5	J-K
Cerebropollenites carlylensis Pocock, 1970	3	J
Classopollis classoides (Pflug) Pocock et Jansonius, 1961	4	J-K
C. annulatus (Verbitzkaya) Li, 1974	8	J-K
Caryapollenites simplex (Potonié) Raatz, 1937	9	N
Chenopodipollis multiplex (Weyland et Pflug) Krutzsch, 1966	17	N
Artemisiaepollenites sellularis Nagy, 1969	21	N
Total	275	

region of the Tarim Basin. The Kekeya oil field was discovered in the southern part of this sag in 1977. Its reservoir rock is Neogene red sandstones.

Fifty-three species of fossil spores and pollen referred to 36 genera were extracted from crude oil samples collected from the Yecheng Sag in the Neogene reservoir of the Kekeya oil field of the Tarim Basin. These are listed in Table 3.4 and illustrated on Plates XI–XII, XL.

With the exception of the Tertiary *Piceacepollenites alatus, Pinuspollenites labdacus, Ephedripites tertiarius, Quercoidites microhenrici, Chenopodipollis multiplex,* and *Artemisiaepollenites sellularis,* most of the spores and pollen in the Yecheng Sag of the Tarim Basin (Table 3.4) are Jurassic index species. However, some are distributed from Jurassic to Cretaceous, and a few from Triassic to Jurassic or Triassic to Cretaceous. Among the spores and pollen, *Dictyophyllidites harrisii, Dictyophyllum rugosum, Leptolepidites major, Pteruchipollenites thomasii,* and *Parvisaccites enigmatus* were found in the Middle Jurassic Series of England (Couper 1958). *Leptolepidites verrucatus* and *Podocarpidites major* were found in the Jurassic System of New Zealand (Couper 1953). *Gleicheniidites rouseii, Murospora minor, Platysaccus lopsinensis, Podocarpidites wapellensis,* and *Cedripites minor* were found in the Jurassic System of Canada (Pocock 1970). *Bennettiteaepollenites lucifer* was found in the Middle Jurassic Series of Germany (Potonié 1958). *Chasmatosporites major, C. minor,* and *C. elegans* were found in the Lower Jurassic Series of Sweden (Nilsson 1958). *Paleoconiferus asaccatus, Podocarpidites multicinus, Protopicea exilioides,* and *Piceites expositus* were found in the Jurassic System in Yakutsk, Russia (Bolchovitina 1956). *Deltoidospora gradata* was found in the Middle Jurassic Series in Emba, Russia and the Jurassic System of western Canada (Maljavkina 1949; Pocock 1970). *Cibotiumspora paradoxa* was found in the Jurassic System of Russia and China (Maljavkina 1949; Zhang 1965). These spores and pollen may be considered as Jurassic index species. In addition, *Apiculatisporis ovalis* was found in the Lower Jurassic Series of Sweden and the Upper Triassic Rhaetian Stage of Antarctica and North Sea Basin (Nilsson 1958; Norris 1965; Lund 1977). *Lycopodiumsporites clavatoides, Caytonipollenites pallidus,* and *Eucommiidites troedssonii* were found in the Jurassic System and the Lower Cretaceous Series of Britain (Couper 1958). *Quadraeculina limbata* was found in the Jurassic System and the Lower Cretaceous Series of Russia (Maljavkina 1949). *Gleicheniidites senonicus* was found in the Jurassic System and the Cretaceous System of Europe. *Cyathidites australis, C. minor, Osmundacidites wellmanii, Cycadopites nitidus,* and *C. typicus* are widely distributed in the Jurassic System and the Lower Cretaceous Series of Europe, Asia, North

Table 3.4 Spores and pollen in crude oils from the Yecheng Sag

Spores and pollen	Number of specimens	Geological age
Spores		
Deltoidospora perpusilla (Bolchovitina) Pocock, 1970	8	J
D. gradata (Maljavkina) Pocock, 1970	5	J
Cyathidites australis Couper, 1953	9	J-K
C. minor Couper, 1953	15	J-K
Gleicheniidites senonicus Ross, 1949	6	J-K
G. rouseii Pocock, 1970	11	J
Dictyophyllidites harrisii Couper, 1958	17	J
Dictyophyllum rugosum L. et Harris., 1944	5	J
Undulatisporites concavus Kedves, 1971	7	J
Cibotiumspora paradoxa (Maljavkina) Chang, 1965	8	J
C. jurienensis (Balme) Filatoff, 1975	5	J
Leptolepidites major Couper, 1958	7	J
L. verrucatus Couper, 1953	5	J
Osmundacidites wellmanii Couper, 1953	12	J-K
Apiculatisporis ovalis (Nilsson) Norris, 1965	11	T-J
Lycopodiumsporites clavatoides Couper, 1858	5	J-K
Biretisporites potoniaei Delcourt et Sprumont, 1955	4	J-K
Murospora minor Pocock, 1970	5	J
Pollen		
Paleoconiferus asaccatus Bolchovitina, 1956	7	J
Caytonipollenites pallidus (Reissinger) Couper, 1958	7	J-K
Pteruchipollenites thomasii Couper, 1958	6	J
Alisporites bilateralis Rouse, 1959	8	J-K
Abietineaepollenites minimus Couper, 1958	5	J-K
Pinuspollenites labdacus (Potonié) Eaatz, 1938	19	N
Piceaepollenites alatus Potonié, 1931	11	N
Protopicea exilioides (Bolchovitina) Pocock, 1970	5	J
Piceites expositus Bolchovitina, 1956	7	J
P. podocarpoides Bolchovitina, 1956	7	J-K
P. latens Bolchovitina, 1956	9	J-K
Cedripites minor Pocock, 1970	14	J
Platysaccus lopsinensis (Maljavkina) Pocock, 1970	7	J
Podocarpidites major Couper, 1953	4	J
P. wapellensis Pocock, 1970	4	J
P. multesimus (Bolchovitina) Pocock, 1962	6	J-K
P. multicinus (Bolchovitina) Pocock, 1970	8	J
Parvisaccites enigmatus Couper, 1958	7	J
Quadraeculina limbata Maljavkina, 1949	5	J-K

(continued)

Table 3.4 (continued)

Spores and pollen	Number of specimens	Geological age
Bennettiteaepollenites lucifer (Thiergart) Potonié, 1958	7	J
Eucommiidites troedssonii Erdtman, 1948	12	J-K
Ephedripites tertiarius Krutzsch, 1970	21	E-N
Chasmatosporites major Nilsson, 1958	8	J
C. minor Nilsson, 1958	11	J
C. elegans Nilsson, 1958	9	J
Cycadopites nitidus (Balme) Pocock, 1970	21	J-K
C. typicus (Maljavkina) Pocock, 1970	9	T-K
C. minimus (Cookson) Pocock, 1970	23	J-K
C. subgranulosus (Couper) Clarke, 1965	9	T-J
C. carpentieri (Delcourt et Sprumant) Singh, 1964	5	J-K
Classopollis classoides (Pflug) Pocock et Jansonius, 1961	9	J-K
C. annulatus (Verbitzkaya) Li, 1974	14	J-K
Quercoidites microhenrici (Potonié) Potonié, 1950	12	N
Chenopodipollis multiplex (Weyland et Pflug) Krutzsch, 1966	27	N
Artemisiaepollenites sellularis Nagy, 1969	25	N
Total	513	

America, and Australia. These species may be considered as Jurassic common species.

Besides fossil spores and pollen, some fossil fungi and algae were found in crude oils of the North Tarim Upheaval and southwest Tarim Depression. Twenty-seven species (or form species) of fungi, algae, and acritarchs referred to 16 genera were extracted from the crude oil samples from the North Tarim Upheaval, the Yecheng Sag, and the Maigaiti Clinoform (Table 3.5).

Most of the fungi, algae, and acritarchs from the Tarim Basin have been reported from the Paleogene Shahejie Formation of the coastal region of Bohai, China. The chlorophyte *Campenia* was found in the Jurassic System of Germany (Research Institute of Petroleum Exploration and Development, Ministry of Petroleum Chemistry Industry, and Nanjing Institute of Geology and Palaeontology, Chinese Academy of Sciences 1978b). The fossil fungus *Glomus* belonging to Endogonaceae of Zygomycetes was found in the Middle to Upper Carboniferous (Pennsylvanian) of North America (Wagner and Taylor 1982).

3.3 Junggar Basin

3.3.1 East Junggar Depression

3.3.1.1 Beisantai Oil Field

The Beisantai oil field is located in an actively uplifting area in the southwestern part of the East Junggar Depression. It is a prospective target for petroleum accumulation. This important oil field was discovered in 1980; the reservoir rock was found to be Carboniferous igneous rock (basalt).

Ninety-six species of fossil spores and pollen referred to 52 genera were extracted from crude oil samples collected in this igneous rock reservoir from the Beisantai oil field. These are listed in Table 3.6 and illustrated on Plates XIII–XVI, and XLI.

The spores and pollen from the Beisantai oil field are mostly distributed in the Carboniferous through the Triassic systems. A few extend into the Jurassic System. Among the spores and pollen, *Calamospora breviradiata, Punctatisporites minutus, Laevigatosporites medius, Latosporites latus,* and *Wilsonia vesicata* were found in the

Table 3.5 Fungi, algae, and acritarchs in crude oils from the Tarim Basin

Category		Name of fossils	Distribution of fossils		
			North Tarim Upheaval	Yecheng Sag	Maigaiti Clinoform
Fungi		*Multicellaesporites ovatus* Sheffy et Dilcher, 1971		+	
		Multicellaesporites pachydermus Ke et Shi, 1978		+	
		Multicellaesporites margaritus Ke et Shi, 1978			+
		Glomus sp.			+
Algae	Pyrrhophyta	*Rhombodella* cf. *baculata* Jiabo, 1978		+	
		Rhombodella cf. *variabilis* Jiabo, 1978		+	
		Rhombodella sp.		+	
		Tenua cf. *bellula* Jiabo, 1978	+	+	
		Tenua sp.	+		
		Conicoidium cf. *granorugosum* Jiabo, 1978	+		
		Conicoidium sp.	+		
		Prominangularia cf. *dongyingensis* Jiabo, 1978		+	
		Dinogymnium granulatum Jiabo, 1978		+	
	Chlorophyta	*Campenia* cf. *circellata* Jiabo, 1978	+		+
		Hungarodiscus cf. *punctatus* Jiabo, 1978	+		
		Hungarodiscus sp.	+		
Acritarchs		*Granodiscus* cf. *granulatus* Madler, 1963			+
		Verrucosphaera cf. *verrucosa* Jiabo, 1978	+		
		Dictyotidium cf. *asperatum* Jiabo, 1978	+		
		Dictyotidium sp.	+		
		Granoreticella sp.	+		
		Porusphaera sp.	+		+
		Heliospermopsis sp.	+		
		Tectocorpidium cf. *rimosum* Jiabo, 1978	+		
		Acritarcha indet. Type 1	+		
		Acritarcha indet. Type 2	+		
		Acritarcha indet. Type 3	+		

Table 3.6 Spores and pollen in crude oils from the Beisantai oil field

Spores and pollen	Number of specimens	Geological age
Spores		
Leiotriletes ornatus Ishchenko, 1956	4	C-T
Dictyophyllidites mortoni (de Jersey) Playford et Dettmann, 1965	5	T-J
D. discretus Ouyang, 1986	4	P
Calamospora breviradiata Kosanke, 1950	4	C-T
C. nathorstii (Halle) Klaus, 1960	6	T-J
C. densirugosus Hou et Wang, 1986	4	P
Punctatisporites minutus Kosanke, 1950	5	C-P
P. triassicus Schulz, 1964	5	T
P. cathayensis Ouyang, 1962	4	P
P. parasolidus Ouyang, 1964	6	P
P. gansuensis Geng, 1985	3	C
P. densus Geng, 1985	4	C
Granulatisporites granulatus Ibrahim, 1933	3	C-P
G. parvus (Ibrahim) Potonié et Kremp, 1955	4	C-T
Cyclogranisporites leopoldii (Kremp) Potonié et Kremp, 1955	6	C-T
Verrucosisporites verrucosus Ibrahim, 1933	3	C-P
V. morulae Klaus, 1960	4	T
Lophotriletes gibbosus (Ibrahim) Potonié et Kremp, 1954	3	C
L. delicatus Ouyang et Li, 1980	3	T
L. humilus Hou et Wang, 1986	6	P
Apiculatisporis decorus Singh, 1964	3	P
A. globosus (Leschik) Playford et Dettmann, 1965	3	T-J
A. xiaolongkouensis Hou et Wang, 1986	7	P
Acanthotriletes ciliatus (Knox) Potonié et Kremp, 1955	3	C-P
A. microspinosus (Ibrahim) Potonié et Kremp, 1955	4	C-P
Discisporites psilatus de Jersey, 1964	5	T
Duplexisporites gyratus Playford et Dettmann, 1965	3	T-J
D. sp.	4	P
Multinodisporites junctus Ouyang et Li, 1980	3	T
Endosporites ornatus Wilson et Coe, 1940	3	C-P
E. multirugulatus Hou et Wang, 1986	7	P
Crassispora orientalis Ouyang et Li, 1980	6	C-T
C. sp	2	P
Densosporites paranulatus Ouyang, 1986	4	P
Limatulasporites limatulus (Playford) Helby et Foster, 1979	7	P-T
L. fossulatus (Balme) Helby et Foster, 1979	6	P-T
L. dalongkouensis Qu et Wang, 1986	9	T

(continued)

Table 3.6 (continued)

Spores and pollen	Number of specimens	Geological age
L. parvus Qu et Wang, 1986	11	T
Kraeuselisporites argutus Hou et Wang, 1986	6	P
K. spinulosus Hou et Wang, 1986	7	P
Lundbladispora iphilegna Foster, 1979	3	P-T
L. subornata Ouyang et Li, 1980	4	T
Laevigatosporites desmoinesensis (Wilson et Coe) Schopf, Wilson et Bentall, 1944	2	C-P
L. medius Kosanke, 1950	3	C-P
Latosporites latus (Kosanke) Potonié et Kremp, 1954	3	C-P
Torispora securis (Balme) Alpern, Doubinger et Horst, 1965	4	C-T
T. medius Ouyang, 1962	3	P
Punctatosporites minutus Ibrahim, 1933	5	C
Aratrisporites parvispinosus Leschik, 1955	4	T
A. coryliseminis Klaus, 1960	3	T
A. fischeri (Klaus) Playford et Dettmann, 1965	4	T
A. strigosus Playford, 1965	5	T
Wilsonia vesicata Kosanke, 1950	2	C
Pollen		
Cordaitina uralensis (Luber) Samoilovich, 1953	8	C-P
C. rotata (Luber) Samoilovich, 1953	5	P
C. duralimita Hou et Wang, 1986	4	P
Florinites cf. *antiquus* Schopf, 1944	3	C-P
F. millotti Butterworth et Williams, 1954	4	T
F. minutus Hou et Wang, 1986	3	P
Vestigisporites dalongkouensis Hou et Wang, 1986	5	P
Chordasporites orientalis Ouyang et Li, 1980	4	T
C. impensus Ouyang et Li, 1980	3	T
Gardenasporites xinjiangensis Hou et Wang, 1986	6	P
Lueckisporites virkkiae (Potonié et Klaus) Clarke, 1965	4	P
L. prolixus (Luber) Ouyang, 1962	3	P
Taeniaesporites kraeuseli Leschik, 1955	3	T
T. pellucidus (Goubin) Balme, 1970	4	P-T
Striatopodocarpites tojmensis (Sedova) Hart, 1964	12	P
S. crassus Singh, 1964	4	P
Striatites sp.	3	P
Protohaploxypinus samoilovichiae (Jansonius) Hart, 1964	3	P-T
P. dvinensis (Sedova) Hart, 1964	4	P
Striatoabietites richteri (Klaus) Hart, 1964	3	P-T
Limitisporites rectus Leschik, 1956	3	P

(continued)

Table 3.6 (continued)

Spores and pollen	Number of specimens	Geological age
L. rhombicorpus Zhou, 1979	4	P
Vittatina cryptosaccata Wang, 2003	3	P
Labiisporites granulatus Leschik, 1956	3	P
Vesicaspora xinjiangensis Hou et Wang, 1986	7	P
Vitreisporites signatus Leschik, 1955	4	T
Pityosporites evolutus Ouyang, 2003	7	C-P
Klausipollenites schaubergeri (Potonié et Klaus) Jansonius, 1962	6	P-T
Cedripites parvisaccus Ouyang et Li, 1980	3	T
Platysaccus papilionis Potonié et Kaus, 1954	4	P-T
P. cf. undulatus Ouyang et Li, 1980	3	T
Podocarpidites queenslandi (de Jersey) Qu, 1980	3	T
P. transversus Qu et Wang, 1986	3	T
Parcisporites rarus Ouyang et Li, 1980	3	T
Minutosaccus parcus Qu et Wang, 1986	3	T
Alisporites australis de Jersey, 1962	12	T
A. parvus de Jersey, 1962	8	T
A. auritus Ouyang et Li, 1980	4	T
A. fusiformis Ouyang et Li, 1980	3	T
Triangulisaccus primitivus Hou et Wang, 1986	3	P
Entylissa spinosa (Samoilovich) Du, 1986	6	P
Cycadopites subgranulosus (Couper) Clarke, 1965	5	T-J
C. granulatus Ouyang, 1986	4	P
Total	426	

Middle to Upper Carboniferous Pennsylvanian Stage in Illinois, USA (Kosanke 1950). *Granulatisporites granulatus, G. parvus, Cyclogranisporites leopoldii, Verrucosisporits verrucosus, Lophotriletes gibbosus, Acanthotriletes microspinosus, A. ciliatus,* and *Punctatosporites minutus* were found in the Middle Carboniferous Westphalian Stage in the Rhur region of Germany (Potonié and Kremp 1955, 1956). *Torispora securis* was found in the Middle Carboniferous Westphalian Stage of Britain (Balme 1952). *Laevigatosporites desmoinesensis* was found in the Middle Carboniferous Des Moines Member in Iowa, USA (Wilson and Coe 1940). *Leiotriletes ornatus* was found in the Lower Carboniferous Visean Stage and Namurian Stage in the Donetsk Basin, Russia (Ishchenko 1956). *Punctatisporites gansuensis* was found in the Upper Carboniferous Taiyuan Formation in Huanxian, Ganxu, China, and *Punctatisporites densus* was found in the Middle Carboniferous Yanghugou Formation in Ningxia, Nei Monggol (Inner Mongolia) and Gansu, China (Geng 1985a, b). *Crassispora orientalis* was found in the Upper Carboniferous Benxi Formation in Shuoxian, Shanxi, China, and the Lower Triassic Kayitou Formation in Fuyuan, Yunnan, China (Ouyang and Li 1980a, b). *Endosporites ornatus* and *Florinites antiguus* were found in the Middle Carboniferous Series in Iowa, USA (Wilson and Coe 1940; Schopf et al. 1944). *Cordaitina uralensis* was found in the Permian System in Solikamsk of the Ural region, Russia (Samoilovich 1953), and the

Carboniferous to Permian System of Kazakhstan (Luber 1955). *Cordaitina rotata* was found in the Permian System in Siberia, Russia (Samoilovich 1953). *Pityosporites evolutus* was found in the Upper Carboniferous Chepaizi Formation and the Permian Jingjingzigou Formation, Lucaogou Formation, and Guodikeng Formation in northern Xinjiang, China (Ouyang et al. 2003). *Lueckisporites virkkiae*, *Klausipollenites schaubergeri*, and *Platysaccus papilionis* were found in the Upper Permian Zechstein Series of Germany, Austria, and Britain (Potonié and Klaus 1954; Clarke 1965b). *Striatoabietites richteri* was found in the Upper Permian Zechstein Series of Germany (Klaus 1955; Hart 1965). *Striatopodocarpites tojmensis* and *Protohaploxypinus dvinensis* were found in the Permian Kazanian Series in the North Dvina Basin, Russia (Hart 1965). *Apiculatisporis decorus* and *Striatopodocarpites crassus* were found in the Permian System in Mosul, Iraq (Singh 1964b). *Protohaploxypinus samoilovichii* was found in the Lower Triassic Series in the Peace River area of Canada, and in the Lower Permian Series in South America and the Tarim Basin of China (Jansonius 1962; Hart 1965; Wang 1989). *Punctatisporites cathayensis* and *Torispora medius* were found in the Upper Permian Lungtan Formation in Changxing, Zhejiang, China (Ouyang 1962). *Lueckisporites prolixus* was found in the Lungtan Formation in Changxing and the Permian System of Russia (Ouyang 1962). *Punctatisporites parasolidus* was found in the Lower Permian Lower Shihhotze Series in Hokü, Shanxi, China (Ouyang 1964). *Limitisporites rectus* and *Labiisporites granulatus* were found in the Upper Permian Zechstein Series of Germany (Leschik 1956). *Limatulasporites fossulatus* and *Taeniaesporites pellucidus* were found in the Permian to Triassic System of West Pakistan and Xinjiang, China (Balme 1970; Institute of Geology, Chinese Academy of Geological Sciences, and Institute of Geology, Xinjiang Bureau of Geology and Mineral Resources 1986). *Entylissa spinosa* was found in the Permian System in the Ural region, Russia and the Lower Permian Shansi Formation in Pingliang, Gansu, China (Samoilovich 1953; Du 1986). *Dictyophyllidites discretus* and *Densosporites paranulatus* were found in the Upper Permian Hsuanwei Formation in Fuyuan, Yunnan, China (Ouyang 1986). *Aratrisporites parvispinosus*, *Taeniaesporites kraeuseli*, and *Vitreisporites signatus* were found in the Upper Triassic Keuper Stage in Basel, Switzerland (Leschik 1955). *Calamospora nathorstii*, *Verrucosisporites morulae*, *Aratrisporites coryliseminis*, and *A. fischeri* were first found in the Triassic System of Austria (Klaus 1960). *Limatulasporites limatulus* and *Aratrisporites strigosus* were found in the Triassic System of Tasmania (Playford 1965). *Punctatisporites triassicus* was found in the Lower Triassic Series of Germany and Romania (Venkatachala et al. 1968). *Discisporites psilatus*, *Alisporites australis*, *A. parvus*, and *Podocarpidites queenslandi* were found in the Triassic System in Queensland, Australia (De Jersey 1962). *Dictyophyllidites mortoni*, *Duplexisporites gyratus*, and *Apiculatisporis globosus* were found in the Upper Triassic Rhaetian Stage to Lower Jurassic Liassic Stage of South Australia (Playford and Dettmann 1965). *Multinodisporites junctus*, *Lundbladispora subornata*, *Chordasporites orientalis*, *C. impensus*, *Alisporites auritus*, *A. fusiformis*, and *Parcisporites rarus* were found in the Lower Triassic Kayitou Formation in Fuyuan, Yunnan, China (Ouyang and Li 1980b). *Cycadopites subgranulosus* was found in the Upper Triassic Keuper Stage in Worcestershire, England, and the Lower Jurassic Liassic Stage in Yorkshire, England (Couper 1958; Clarke 1965a). *Cycadopites granulatus* was found in the upper member of the Upper Permian Hsuanwei Formation in Fuyuan, Yunnan, China (Ouyang 1986). *Calamospora densirugosus*, *Lophotriletes humilus*, *Apiculatisporis xiaolongkouensis*, *Kraeuselisporites argutus*, *K. spinulosus*, *Endosporites multirugulatus*, *Cordaitina duralimita*, *Vestigisporites dalongkouensis*, *Gardenasporites xinjiangensis*, *Vesicaspora xinjiangensis*, and *Triangulisaccus primitivus* were found in the Upper Permian Wutonggou Formation in Dalongkou, Jimsar, Xinjiang. *Limatulasporites dalongkouensis*, *L. parvus*, *Podocarpidites transversus*, and *Minutosaccus parcus* were found in the Lower Triassic Jiucaiyuan Formation and Middle Triassic Karamay Formation in Dalongkou, Jimsar, Xinjiang (Institute of Geology, Chinese Academy of Geological Sciences, and Institute of Geology, Xinjiang Bureau of Geology and Mineral Resources 1986). *Vittatina cryptosaccata* was found in the Upper Permian Lucaogou Formation in Dalongkou, Jimsar, Xinjiang and the Upper Permian Pindiquan Formation in the eastern Junggar Basin (Ouyang et al. 2003).

3.3.1.2 Huonan Oil Field

The Huonan oil field is located in the Huoshaoshan down-faulted trough in the middle part of the East Junggar Depression. It is an anticlinal trap oil pool. The reservoir bed of the Huonan oil field belongs to the Upper Permian Pindiquan Formation which corresponds with the Upper Permian Lucaogou and Hongyanchi formations.

Sixty species of fossil spores and pollen referred to 36 genera were extracted from crude oil samples collected in the Upper Permian reservoir bed of the Huonan oil field. These are listed in Table 3.7 and illustrated on Plates XVII–XX and XLI–XLII. In addition, one species of the fossil megaspore, *Triangulatisporites junggarensis*, was also extracted and identified (Plate XX, Fig. 12. and Plate XLII, Fig. 12).

Most of the spores and pollen of the Huonon oil field are Permian species. However, some are Carboniferous and others Triassic. Among the spores and pollen, *Leiotriletes adnatus* was found in the Upper Carboniferous Series in Illinois, USA (Kosanke 1950), and it was reported from the Upper Permian Series in Yunnan and Xinjiang, China (Ouyang 1986; Ouyang et al. 2003). *Leiotriletes microtriangulus* was found in the Upper Carboniferous Series in Zonguldak, Turkey (Artüz 1957) and the Carboniferous System in northern Xinjiang, China (Ouyang et al. 2003). *Verrucosisporites verrucosus, Lophotriletes pseudoculeatus, Acanthotriletes microspinosus, A. ciliatus,* and *Punctatosporites minutus* were found in the Middle Carboniferous Westphalian Stage in Ruhr region, Germany (Potonié and Kremp 1955, 1956). *Torispora securis* was found in the Middle Carboniferous Westphalian Stage of Britain (Balme 1952). *Convolutispora triangularis* and *Crassispora orientalis* were found in the Upper Carboniferous Benxi Formation in Shuoxian, Shanxi, China (Ouyang and Li 1980a). *Lophotriletes novicus* and *Apiculatisporis decorus* were found in the Permian System in Mosul, Iraq (Singh 1964b); the former was found in the Lower Triassic Kayitou Formation in Fuyuan, Yunnan, China (Ouyang and Li 1980b) and the Lower Triassic Shaofanggou Formation in Jimsar, Xinjiang, China (Institute of Geology, Chinese Academy of Geological Sciences, and Institute of Geology, Xinjiang Bureau of Geology and Mineral Resources 1986). *Punctatisporites triassicus* was first found in the Lower Triassic Series of Germany and Romania (Venkatachala et al. 1968). *Leiotriletes exiguus, Lophotriletes corrugatus,* and *L. delicatus* were first found in the Lower Triassic Kayitou Formation in Fuyuan, Yunnan, China (Ouyang and Li 1980b). *Converrucosisporites mictus* and *Triqutrites attenuatus* were first found in the Upper Permian Hsuanwei Formation in Fuyuan, Yunnan, China (Ouyang 1986). *Calamospora densirugosus, Lophotriletes humilus, Apiculatisporis xiaolongkouensis, A. setaceformis, Kraeuselisporites argutus,* and *Tuberculatosporites homotubercularis* were found in the Upper Permian Wutonggou Formation in Jimsar, Xinjiang, China. *Lophotriletes flavus* was found in the Lower Triassic Shaofanggou Formation in Jimsar, Xinjiang (Institute of Geology, Chinese Academy of Geological Sciences, and Institute of Geology, Xinjiang Bureau of Geology and Mineral Resources 1986). *Apiculatisporis spiniger* was first found in the Upper Triassic Keuper Stage in Basel, Switzerland (Leschik 1955). *Dictyophyllidites mortoni* was found in the Triassic System of the Ipswich coalfield in Queensland, Australia, the Upper Triassic Series to Lower Jurassic Series in southern Australia, and the Triassic System of Tasmania (de Jersey 1962; Playford 1965; Playford and Dettmann 1965); it has also been reported from the Upper Permian Series in Yunnan, China (Ouyang 1986). *Florinites antiquus* was found in the Middle Carboniferous Series in Iowa, USA (Schopf et al. 1944).

Table 3.7 Spores and pollen in crude oils from the Huonan oil field

Spores and pollen	Number of Specimens	Geological age
Spores		
Leiotriletes adnatus (Kosanke) Potonié et Kremp, 1955	5	C-P
L. microtriangulus Artuz, 1957	3	C
L. exiguus Ouyang et Li, 1980	4	P-T
Dictyophyllidites mortoni (de Jersey) Playford et Dettmann, 1965	3	T-J
Calamospora densirugosus Hou et Wang, 1986	4	P
Punctatisporites triassicus Schulz, 1964	3	T
Verrucosisporites verrucosus Ibrahim, 1933	2	C-P
Converrucosisporites mictus Ouyang, 1986	3	P
Lophotriletes novicus Singh, 1964	2	P-T
L. pseudaculeatus Potonié et Kremp, 1955	5	C-T
L. corrugatus Ouyang et Li, 1980	3	P-T
L. delicatus Ouyang et Li, 1980	3	T
L. humilus Hou et Wang, 1986	4	P
L. flavus Qu et Wang, 1986	3	T
Apiculatisporis decorus Singh, 1964	3	P
A. xiaolongkouensis Hou et Wang, 1986	5	P
A. setaceformis Hou et Wang, 1986	4	P
A. spiniger (Leschik) Qu, 1980	4	T
Acanthotriletes microspinosus (Ibrahim) Potonié et Kremp, 1955	5	C-P
A. ciliatus (Knox) Potonié et Kremp, 1955	4	C-P
Convolutispora triangularis Ouyang et Li, 1980	2	C
C. grandicornis (Luber) Ouyang, 2003	3	C
Triquitrites attenuatus Ouyang, 1986	3	P
Endosporites ornatus Wilson et Coe, 1940	2	C-P
E. multirugulatus Hou et Wang, 1986	4	P
Crassispora orientalis Ouyang et Li, 1980	7	C-T
Kraeuselisporites argutus Hou et Wang, 1986	5	P
Punctatosporites minutus Ibrahim, 1933	4	C
Tuberculatosporites homotubercularis Hou et Wang, 1986	5	P
Torispora securis (Balme) Alpern, Doubinger et Horst, 1965	3	C-T
Pollen		
Cordaitina uralensis (Luber) Samoilovich, 1953	8	C-P
C. rotata (Luber) Samoilovich, 1953	6	P
C. radialis Ouyang et Li, 1980	5	P-T
Florinites cf. *antiquus* Schopf, 1944	1	C-P
Gardenasporites xinjiangensis Hou et Wang, 1986	4	P
G. latisectus Hou et Wang, 1986	5	P
Lueckisporites virkkiae (Potonié et Klaus) Clarke, 1965	6	P

(continued)

Table 3.7 (continued)

Spores and pollen	Number of Specimens	Geological age
Striatoabietites duivenii (Jansonius) Hart, 1964	2	P-T
Limitisporites rectus Leschik, 1956	5	P
L. rhombicorpus Zhou, 1979	4	P
Hamiapollenites bullaeformis (Samoilovich) Jansonius, 1962	3	P
H. obliquus Zhan, 2003	3	P
Vittatina subsaccata Samoilovich, 1953	4	P
V. cryptosaccata Wang, 2003	5	P
V. specialis Wang, 2003	3	P
Vesicaspora cf. *salebra* Gao, 1987	3	C-P
Granosaccus ornatus (Pautsch) Pautsch, 1973	4	T
Klausipollenites decipiens Jansonius, 1962	2	P-T
K. caperatus Ouyang, 2003	3	C-P
Pteruchipollenites reticorpus Ouyang et Li, 1980	3	P-T
Verrucorpipollis archaicus Ouyang, 2003	3	C-P
Cedripites priscus Balme, 1970	4	P-T
C. parvisaccus Ouyang et Li, 1980	4	T
C. cf. *parvisaccus* Ouyang et Li, 1980	2	T
Triangulisaccus primitivus Hou et Wang, 1986	2	P
Entylissa spinosus (Samoilovich) Du, 1986	11	P
Cycadopites granulatus Ouyang, 1986	5	P
C. caperatus (Luber et Valts) Hart, 1965	5	P
Welwitschipollenites clarus (Qu et Wang) Ouyang, 2003	3	P-T
Ricciisporites tuberculatus Lundblad, 1954	4	T-J
Total	232	

Cordaitina uralensis and *Vittatina subsaccata* were found in the Permian System and the Lower Permian Kungurian Stage in Solikamsk, in the Ural region, Russia; the former was found in the Carboniferous to Permian System of Kazakhstan (Samoilovich 1953; Luber 1955). *Cordaitina rotata* and *Hamiapollenites bullaeformis* were found in the Permian System in Siberia and the Ural region, Russia (Samoilovich 1953). *Lueckisporites virkkiae* was found in the Upper Permian Zechstein Series of Germany, Austria, and Britain (Potonié and Klaus 1954; Clarke 1965b). *Limitisporites rectus* was found in the Upper Permian Zechstein Series of Germany (Leschik 1956). *Striatoabietites duivenii* and *Klausipollenites decipiens* were found in the Triassic System of western Canada and the Permian System of northern Xinjiang, China (Jansonius 1962; Hart 1964; Ouyang et al. 2003). *Cedripites priscus* was found in the Permian to Triassic System of Pakistan (Balme 1970). *Limitisporites rhombicorpus* was found in the Permian System in northern Shandong, China (Zhou 1987). *Vesicaspora salebra* was found in the Carboniferous System in Jingyuan, Gansu, China (Gao 1987). *Cordaitina radialis, Pteruchipollenites reticorpus,* and *Cedripites parvisaccus* were found in the Lower Triassic Kayitou Formation in Fuyuan, Yunnan, China (Ouyang and Li 1980b). *Entylissa spinosa* was found in the Permian System in the Ural region of Russia and the Lower Permian Shansi Formation in Pingliang, Gansu, China (Samoilovich 1953; Du 1986). *Cycadopites caperatus* was found in the

Permian System in the Kusnets Basin, Russia (Hart 1965) and the Permian System in northern Xinjiang, China (Ouyang et al. 2003). *Cycadopites granulatus* was found in the Upper Permian Hsuanwei Formation in Fuyuan, Yunnan, China (Ouyang 1986). *Endosporites multirugulatus, Gardenasporites xinjiangensis, G. latisectus,* and *Triangulisaccus primitivus* were found in the Upper Permian Wutonggou Formation in Jimsar, Xinjiang, China. *Vittatina cryptosaccata* and *V. specialis* were found in the Upper Permian Pindiquan Formation in the eastern part of the Junggar Basin. *Hamiapollenites obliquus, Klausipollenites caperatus,* and *Verrucorpipollis archaicus* were found in the Carboniferous to Permian System in the northern part of Xinjiang, China. *Welwitschipollenites clarus* was found in the Upper Permian Lucaogou, Wutonggou, and Guodikeng formations in northern Xinjiang and in the Lower Triassic Jiucaiyuan and Shaofanggou formations in Jimsar, Xinjiang, China (Institute of Geology, Chinese Academy of Geological Sciences, and Institute of Geology, Xinjiang Bureau of Geology and Mineral Resources 1986; Ouyang et al. 2003). *Ricciisporites tuberculatus* was found in the Lower Jurassic Liassic Stage in Scania, Sweden (Lundblad 1954), but it also ranges into the Upper Triassic Rhaetian Stage.

3.3.2 South Junggar Depression

3.3.2.1 Qigu Oil Field

The Qigu oil field is located to the southwest of Manas County in the west part of the South Junggar Depression. The reservoir bed belongs to the Middle Jurassic Xishanyao and Toutunhe formations.

Thirty-eight species of fossil spores and pollen referred to 21 genera were extracted from crude oil samples collected in the Middle Jurassic reservoir bed of the Qigu oil field of the Junggar Basin. These are listed in Table 3.8 and illustrated on Plates XXI–XXII and XLIII.

Most of the spores and pollen in the Qigu oil field of the Junggar Basin (Table 3.8) are Jurassic species. Some may extend to the Cretaceous, and a few range from the Triassic. Among the spores and pollen, *Gleicheniidites rouseii, G. nilssonii, Granulatisporites jurassicus, Concavissimisporites delcourtii, C. southeyensis, C. subgranulosus, Verrucosisporites variabilis, Converrucosisporites minor, Apiculatisporis variabilis, Ceratosporites jurassicus, C. varispinosus* and *Cedripites minor* were found in the Jurassic System of western Canada (Pocock 1970). *Dictyophyllidites harrisii* and *Pteruchipollenites thomasii* were found in the Middle Jurassic Bajocian Stage in Yorkshire, England (Couper 1958). *Granulatisporites minor* was found in the Jurassic System in the Rosewood coalfield of Australia (De Jersey 1959). *Cibotiumspora paradoxa* was found in the Middle Jurassic Series in Emba, Russia, and the Lower to Middle Jurassic in Mianchi, Henan, China (Maljavkina 1949; Zhang 1965). *Cibotiumspora juriensis* and *Pinuspollenites globosaccus* were found in the Jurassic System in the Perth Basin, Western Australia (Filatoff 1975). *Apiculatisporis ovalis* was found in the Lower Jurassic Lias Stage in Schonen, Sweden, and the Upper Triassic Rhaetian Stage of Antarctica and North Sea Basin (Nilsson 1958; Norris 1965; Lund 1977). *Paleoconiferus asaccatus, Protopicea vilujensis,* and *P. exilioides* were found in the Jurassic System in Yakutsk of Russia (Bolchovitina 1956). *Gleicheniidites conflexus* was found in the Middle Jurassic Yan'an Formation in Shaanxi and Nei Monggol, China, and the Lower to Middle Jurassic Series in Turpan, Xinjiang, China (Institute of Geology, Chinese Academy of Geological Sciences 1980; Jiang and Wang 2002a; Sun 1989). *Converrucosisporites elegans* was found in the Lower Jurassic Ziliujing Formation in Beipei, Sichuan, China (Chengdu Institute of Geology and Mineral Resources and Ministry of Geology and Mineral Resources 1983). Except for *Apiculatisporis ovalis*, these spores and pollen may be considered as Jurassic index species. In addition, *Cyathidites australis* and *C. minor* were found in the Jurassic System and the Lower Cretaceous Series of New Zealand and Britain (Couper 1953, 1958). *Podocarpidites multesimus* was found in the Lower Jurassic Series in Yakutsk, Russia, (Bolchovitina 1956)

Table 3.8 Spores and pollen in crude oils from the Qigu oil field

Spores and pollen	Number of specimens	Geological age
Spores		
Cyathidites australis Couper, 1953	6	J-K
C. minor Couper, 1953	11	J-k
Gleicheniidites rouseii Pocock, 1970	4	J
G. nilssonii Pocock, 1970	5	J
G. conflexus (Chlonova) Xu et Zhang, 1980	4	J
Dictyophyllidites harrisii Couper, 1958	8	J
Cibotiumspora paradoxa (Maljavkina) Chang, 1965	5	J
C. jurienensis (Balme) Filatoff, 1975	4	J
Granulatisporites minor de Jersey, 1959	3	J
G. jurassicus Pocock, 1970	6	J
Concavissimisporites delcourtii Pocock, 1970	4	J
C. southeyensis Pocock, 1970	5	J
C. subgranulosus (Couper) Pocock, 1970	4	J
Verrucosisporites variabilis Pocock, 1970	4	J
Converrucosisporites minor Pocock, 1970	5	J
C. venitus Batten, 1973	5	J-K
C. elegans Bai, 1983	4	J
Apiculatisporis ovalis (Nilsson) Norris, 1965	3	T-J
A. variabilis Pocock, 1970	4	J
Ceratosporites jurassicus Pocock, 1970	6	J
C. varispinosus Pocock, 1970	5	J
Rouseisporites cf. *reticulatus* Pocock, 1962	3	J-K
Pollen		
Paleoconiferus asaccatus Bolchovitina, 1956	7	J
Pteruchipollenites thomasii Couper, 1958	5	J
Alisporites bilateralis Rouse, 1959	4	J-K
Pinuspollenites globosaccus Filatoff, 1975	5	J
P. alatipollenites (Rouse) Liu, 2003	3	T-J
Protopicea exilioides (Bolchovitina) Pocock, 1970	5	J
P. vilujensis Bolchovitina, 1956	4	J
Piceites latens Bolchovitina, 1956	7	J-K
P. expositus Bolchovitina, 1956	2	J
P. podocarpoides Bolchovitina, 1956	3	J-K
Cedripites minor Pocock, 1970	6	J
Podocarpidites multesimus (Bolchovitina) Pocock, 1962	4	J-K
P. multicinus (Bolchovitina) Pocock, 1970	2	J
Parvisaccites sp.	2	J-K
Cycadopites nitidus (Balme) Pocock, 1970	4	T-K
C. carpentieri (Delcourt et Sprumont) Singh, 1964	3	J-K
Total	174	

and the Lower Cretaceous Series of Canada (Pocock 1962). *Piceites latens* was found in the Lower Cretaceous of Russia and the Jurassic System of Canada and China (Bolchovitina 1956; Pocock 1970; Huang 1995; Jiang and Wang 2002a, b). *Converrucosisporites venitus* was found in the Lower Cretaceous Series of Europe (Batten 1973) and the Lower Jurassic Ziliujing Formation and the Middle Jurassic lower Shaximiao Formation in Sichuan, China (Chengdu Institute of Geology and Mineral Resources and Ministry of Geology and Mineral Resources 1983). *Pinuspollenites alatipollenites* was found in the Upper Jurassic Series of Canada (Rouse 1959) and the Triassic System and Jurassic System in the Tarim Basin, China (Liu 2003). *Alisporites bilateralis* was found in the Upper Jurassic Series and the Lower Cretaceous Series of Canada (Rouse 1959; Pocock 1962; Norris 1967). *Cycadopites carpentieri* was found in the Middle Jurassic to Lower Cretaceous Series of Canada (Singh 1964a). These species may be considered as common Jurassic species.

3.3.2.2 Dushanzi Oil Field

The Dushanzi oil field is located to the west of Anjihai County in the western part of the South Junggar Depression. Both structural and lithological traps control the petroleum reservoirs. The Dushanzi oil field was found in the Tertiary sedimentary center of the basin; petroleum reservoirs are the Oligocene to Miocene Shawan Formation (E_3-N_{1s}) and the Miocene Taxihe Formation (N_1t).

Fifty-one species of fossil spores and pollen referred to 35 genera were extracted from crude oil samples collected in the Oligocene to Miocene reservoir beds of the Dushanzi oil field of the Junggar Basin. These are listed in Table 3.9 and illustrated on Plates XXIII–XXIV.

Most of the spores and pollen in the Dushanzi oil field of the Junggar Basin (Table 3.9) are Jurassic species; some are Tertiary in age. Among the spores and pollen, *Leiotriletes parvus* was found in the Lower Jurassic Series of Sweden (Nilsson 1958). *Dictyophyllidites harrisii, Marattisporites scabratus, Pteruchipollenites thomasii, Abietineaepollenites dunrobinensis,* and *Parvisaccites enigmatus* were found in the Lower and Middle Jurassic Series of Britain (Couper 1958). *Murospora jurassica, M. minor, Pseudowalchia ovais, P. landesii,* and *Cedripites minor* were found in the Lower and Middle Jurassic Series of western Canada (Pocock 1970). *Podocarpidites multicinus* and *P. multesimus* were found in the Lower Jurassic Series in Yakutsk, Russia. *Paleoconiferus asaccatus* was found in the Middle Jurassic Series in Yakutsk, and *Protopicea exilioides* was found in the Upper Jurassic Series in Yakutsk (Bolchovitina 1956). *Piceites pseudorotundiformis* was found in the Upper Jurassic Series in western Siberia, Russia, (Maljavkina 1949) and was reported from the Middle Jurassic Series of Canada (Pocock 1970). *Aratrisporites xiangxiensis* was found in the Lower Jurassic Xiangxi Formation in Xiangxi, Hubei, China (Li and Shang 1980). *Schizosporis microreticulatus* was found in the Middle Jurassic lower Shaximiao Formation in Shaximiao, Sichuan, China (Chengdu Institute of Geology and Mineral Resources, Ministry of Geology and Mineral Resources 1983). *Callialasporites radius* was found in the Middle Jurassic Yan'an Formation in Yan'an, Huangling, and Zhidan, Shaanxi, China (Institute of Geology, Chinese Academy of Geological Sciences 1980). These spores and pollen may be considered as Jurassic index species. In addition, *Cyathidites minor, Alisporites grandis, Araucariacites australis, Callialasporites dampieri, Eucommiidites troedssonii, Cycadopites nitidus,* and *C. typicus* are common in Jurassic to Cretaceous rocks. These species may be considered as Jurassic common species. The Jurassic spores and pollen were reported from the Lower to Middle Jurassic Series in the Junggar Basin (Zhang 1990; Liu 1993). *Ephedripites eocenipites* was found in the Eocene Series of the USA (Wodehouse 1933), which was reported from the Eocene Series in Yingcheng county, Hubei, China, and the Eocene to Oligocene Series in Jintan county, Jiangsu, China (Li et al. 1978; Song et al. 1981). *Salixipollenites discoloripites* was found in the Eocene Green River Formation of the USA (Wodehouse 1933) and also found in the Eocene Liushagang Formation in the northern part of the South China Sea, China (Hou et al.

Table 3.9 Spores and pollen in crude oils from the Dushanzi oil field

Spores and pollen	Number of Specimens	Geological age
Pteridophyta		
Leiotriletes parvus Nilsson, 1958	3	J
L. adriensis (Potonié et Gelletich) Krutzsch, 1959	3	E-N
Cyathidites minor Couper, 1953	5	J-K
Dictyophyllidites harrisii Couper, 1958	7	J
Murospora jurassica Pocock, 1970	2	J
M. minor Pocock, 1970	2	J
Marattisporites scabratus Couper, 1958	3	J
Aratrisporites xiangxiensis Li et Shang, 1980	2	J
Schizosporis microreticulatus Bai, 1983	2	J
Gymnospermae		
Paleoconiferus asaccatus Bolchovitina, 1956	5	J
Pseudowalchia ovalis Pocock, 1970	4	J
P. landesii Pocock, 1970	4	J
Pteruchipollenites thomasii Couper, 1958	4	J
Alisporites grandis (Cookson) Dettmann, 1963	2	J-K
Abietineaepollenites dunrobinensis Couper, 1958	3	J
A. minimus Couper, 1958	4	J-K
A. cembraeformis (Zaklinskaya) Ke et Shi, 1978	5	E-N
Pinuspollenites labdacus maximus (Potonié) Potonié, 1958	3	E-N
P. labdacus minor (Potonié) Potonié, 1958	5	E-N
Piceaepollenites alatus (Potonié) Potonié, 1958	6	E-N
Protopicea exilioides (Bolchovitina) Pocock, 1970	5	J
Piceites pseudorotundiformis (Maljavkina) Pocock, 1970	3	J
Cedripites minor Pocock, 1970	3	J
Podocarpidites multicinus (Bolchovitina) Pocock, 1970	5	J
P. multesimus (Bolchovitina) Pocock, 1962	3	J-K
Parvisaccites enigmatus Couper, 1958	4	J
Tsugaepollenites viridifluminipites (Wodehouse) Potonié, 1958	5	E-N
Taxodiaceaepollenites hiatus (Potonié) Kremp, 1949	5	E-N
Araucariacites australis Cookson, 1947	3	J-K
Callialasporites dampieri (Balme) Sukh Dev, 1961	3	J-K
C. radius Xu et Zhang, 1980	4	J
Ephedripites eocenipites (Wodehouse) Krutzsch, 1961	5	E
E. fusiformis (Shakhmundes) Krutzsch, 1970	4	E-N
E. tertiarius Krutzsch, 1970	7	E-N
E. scabridus Song et Zheng, 1981	4	E-N
Eucommiidites troedssonii Erdtman, 1948	4	J-K
Cycadopites nitidus (Balme) Pocock, 1970	6	T-K

(continued)

Table 3.9 (continued)

Spores and pollen	Number of Specimens	Geological age
C. typicus (Maljavkina) Pocock, 1970	4	J-K
C. minimus (Cookson) Pocock, 1970	3	J-K
Angiospermae		
Salixipollenites discoloripites (Wodehouse) Srivastava, 1966	3	E
Caryapollenites simplex (Potonié) Raatz, 1937	2	E-N
Juglanspollenites verus Raatz, 1937	3	E-N
Quercoidites henrici (Potonié) Potonié, Thomson et Thiergart, 1950	4	E-N
Q. microhenrici (Potonié) Potonié, 1950	6	E-N
Q. asper (Thomson et Pflug) Sung et Zheng, 1976	5	E-N
Chenopodipollis psilatoides (Trevisan) Kedves, 1981	7	N
C. minor Song, 1985	8	N
Liquidambarpollenites stigmosus (Potonié) Raatz, 1937	4	E-N
Rhoipites pseudocingulum (Potonié) Potonié, 1951	3	E
Artemisiaepollenites sellularis Nagy, 1969	11	N
Tubulifloridites macroechinatus (Trevisan) Song et Zhu, 1985	9	E-N
Total	219	

1981). *Leiotriletes adriensis* was found in the Tertiary System of Europe (Thomson and Pflug 1953; Krutzsch 1959), and in the Eocene and Oligocene Series in Jiangsu, China (Song et al. 1981). *Abietineaepollenites cembraeformis* was found in the Lower Oligocene Series in the Irtysh Basin, Russia (Zaklinskaja 1957), in the Oligocene Series in the coastal region of Bohai, China, and in the Oligocene to Pliocene Series in the Qaidam Basin, China (Research Institute of Petroleum Exploration and Development, Ministry of Petroleum Chemistry Industry, and Nanjing Institute of Geology and Palaeontology, Chinese Academy of Sciences 1978a; Research Institute of Exploration and Development, Qinghai Petroleum Administration, Nanjing Institute of Geology and Palaeontology, Academia Sinica 1985). *Abietineaepollenites minimus, Pinuspollenites labdacus maximus, P. labdacus minor, Piceaepollenites alatus, Tsugaepollenites viridifluminipites, Caryapollenites simplex, Juglanspollenites verus, Quercoidites henrici, Q. microhenrici,* and *Liquidambarpollenites stigmosus* were found in the Miocene Series in Europe (Potonié 1958, 1960). *Rhoipites pseudocingulum* was found in the Eocene Series in Europe (Potonié 1960).

Taxodiaceaepollenites hiatus was found in the Miocene Series of Germany and the Tertiary System of Europe (Kremp 1949; Potonié et al. 1950). *Chenopodipollis psilatoides* was found in the Neogene System of Egypt (Kedves 1981). *Chenopodipollis minor* was found in the Neogene System in the Longjing Structural Area of the East China Sea, China (Song et al. 1985). *Artemisiaepollenites sellularis* was first found in the Miocene Series of Hungary (Nagy 1969). These species of pollen are widely distributed in the Tertiary System in China.

3.4 Turpan Basin

3.4.1 Qiktim Oil Field

The Qiktim oil field is located in the eastern section of the Huoyanshan anticlinal zone in the Turpan Depression. It is an anticlinal trap oil pool. The reservoir bed of the Qiktim oil field is the Middle Jurassic Toudunhe Formation.

Sixty-eight species of fossil spores and pollen referred to 40 genera were extracted from crude oil samples collected in the Middle Jurassic

3.4 Turpan Basin

reservoir bed of the Qiktim oil field in the Turpan Basin. These are listed in Table 3.10 and illustrated on Plates XXV–XXVI and XLIV.

Most of the spores and pollen from the Qiktim oil field of the Turpan Basin (Table 3.10) are Jurassic species. However, some extend into the

Table 3.10 Spores and pollen in crude oils from the Qiktim oil field

Spores and pollen	Number of specimens	Geological age
Spores		
Leiotriletes medius Nilsson, 1958	4	J
Deltoidospora gradata (Maljavkina) Pocock, 1970	4	J
D. lineata (Bolchovitina) Pocock, 1970	3	J
D. magna (de Jersey) Norris, 1963	4	J
Cyathidites australis Couper, 1953	9	J-K
C. minor Couper, 1953	19	J-K
Dictyophyllidites harrisii Couper, 1958	17	J
Cibotiumspora paradoxa (Maljavkina) Chang, 1965	4	J
C. jurienensis (Balme) Filatoff, 1975	3	J
Gleicheniidites rouseii Pocock, 1970	5	J
G. conflexus (Chlonova) Xu et Zhang, 1980	4	J
Undulatisporites pflugii Pocock, 1970	3	J
U. concavus Kedves, 1971	4	J
Divisisporites undulatus Huang, 1995	5	J
Tripartina variabilis Maljavkina, 1949	3	J
Granulatisporites jurassicus Pocock, 1970	2	J
G. minor de Jersey, 1959	3	J
Osmundacidites wellmanii Couper, 1953	4	J-K
Todisporites minor Couper, 1958	7	J
Verrucosisporites variabilis Pocock, 1970	4	J
Converrucosisporites minor Pocock, 1970	3	J
C. venitus Batten, 1973	2	J-K
Leptolepidites major Couper, 1958	4	J
Apiculatisporis ovalis (Nilsson) Norris, 1965	5	T-J
A. variabilis Pocock, 1970	3	J
Marattisporites scabratus Couper, 1958	3	J-K
Pollen		
Protoconiferus minor Pocock, 1970	3	J
Paleoconiferus asaccatus Bolchovitina, 1956	3	J
Pseudowalchia ovalis Pocock, 1970	4	J
P. landesii Pocock, 1970	3	J
Pteruchipollenites thomasii Couper, 1958	7	J
Alisporites grandis (Cookson) Dettmann, 1963	12	J-K
A. bilateralis Rouse, 1959	14	J-K

(continued)

Table 3.10 (continued)

Spores and pollen	Number of specimens	Geological age
Abietineaepollenites dunrobinensis Couper, 1958	11	J
A. minimus Couper, 1958	6	J-K
Pityosporites similis Balme, 1957	4	J
P. divulgatus (Bolchovitina) Pocock, 1970	3	J-K
Piceaepollenites complanatiformus (Bolchovitina) Xu et Zhang, 1980	8	J-K
P. omoriciformis (Bolchovitina) Xu et Zhang, 1980	4	J
Protopicea exilioides (Bolchovitina) Pocock, 1970	8	J
Pseudopicea variabiliformis (Maljavkina) Bolchovitina, 1956	2	T-J
Piceites expositus Bolchovitina, 1956	5	J
P. latens Bolchovitina, 1956	2	J-K
P. podocarpoides Bolchovitina, 1956	3	J-K
Cedripites minor Pocock, 1970	14	J
Caytonipollenites pallidus (Reissinger) Couper, 1958	3	J-K
Vitreisporites jurassicus Pocock, 1970	14	J
V. jansonii Pocock, 1970	2	J
Platysaccus lopsinensis (Maljavkina) Pocock, 1970	4	J
Podocarpidites major Couper, 1953	12	J
P. unicus (Bolchovitina) Pocock, 1970	12	J-K
P. multicinus (Bolchovitina) Pocock, 1970	14	J
P. multesimus (Bolchovitina) Pocock, 1962	12	J-K
P. rousei Pocock, 1970	6	J
P. wapellaensis Pocock, 1970	3	J
Parvisaccites enigmatus Couper, 1958	16	J
Inaperturopollenites dettmannii Pocock, 1970	5	J
Callialasporites dampieri (Balme) Sukh Dev, 1961	11	J-K
C. radius Xu et Zhang, 1980	4	J
Classopollis classoides (Pflug) Pocock et Jansonius, 1961	24	J-K
C. annulatus (Verbitzkaya) Li, 1974	37	J-K
C. qiyangensis Shang, 1981	6	J
Eucommiidites troedssonii Erdtman, 1948	5	J-K
Chasmatosporites elegans Nilsson, 1958	8	J
Cycadopites subgranulosus (Couper) Clarke, 1965	2	T-J
C. nitidus (Balme) Pocock, 1970	6	T-K
C. typicus (Maljavkina) Pocock, 1970	3	J-K
C. minimus (Cookson) Pocock, 1970	3	J-K
Total	459	

Cretaceous, and a few trace back into the Triassic. Among the spores and pollen, *Deltoidospora lineata*, *Paleoconiferus asaccatus*, *Piceaepollenites omoriciformis*, *Protopicea exilioides*, *Piceites expositus*, and *Podocarpidites multicinus* were found in the Jurassic System in Yakutsk, Russia (Bolchovitina 1956). *Leiotriletes medius*, *Apiculatisporis ovalis*, and *Chasmatosporites elegans* were found in the Lower Jurassic Lias Stage of Sweden (Nilsson 1958). *Apiculatisporis ovalis* may range into the Upper Triassic Series (Norris 1965; Lund 1977). *Cycadopites subgranulosus* and *Abietineaepollenites dunrobinensis* were found in the Lower Jurassic Lias Series of Britain. *Dictyophyllidites harrisii*, *Todisporites minor*, *Leptolepidites major*, *Pteruchipollenites thomasii*, and *Parvisaccites enigmatus* were found in the Middle Jurassic Dogger Series of Britain (Couper 1958). *Cycadopites subgranulosus* was reported from the Upper Triassic Series in Worcestershire, Britain (Clarke 1965a). *Cibotiumspora paradoxa*, *Tripartina variabilis*, and *Pseudopicea variabiliformis* were found in the Middle Jurassic Series in Emba, Russia (Maljavkina 1949), *Cibotiumspora paradoxa* was reported from the Lower to Middle Jurassic Yima coal-bearing Formation in Mianchi, Henan, China (Zhang 1965), and the Lower and Middle Jurassic Series in the Tarim Basin, China (Liu 2003; Jiang et al. 2008). *Cibotiumspora jurienensis* was found in the Jurassic System in the Perth Basin, Western Australia (Filatoff 1975). *Gleicheniidites rouseii*, *Undulatisporites pflugii*, *Granulatisporites jurassicus*, *Verrucosisporites variabilis*, *Converrucosisporites minor*, *Apiculatisporis variabilis*, *Protoconiferus minor*, *Pseudowalchia ovalis*, *P. landesii*, *Cedripites minor*, *Vitreisporites jurassicus*, *V. jansonii*, *Podocarpidites rousei*, *P. wapellaensis*, and *Inaperturopollenites dettmannii* were found in the Jurassic System of western Canada (Pocock 1970). *Podocarpidites major* was found in the Jurassic System of New Zealand (Couper 1953). *Pityosporites similis* was found in the Jurassic System of West Australia (Balme 1957). *Granulatisporites minor* was found in the Jurassic System in the Rosewood coalfield, Australia (De Jersey 1959). *Classopollis qiyangensis* was found in the Lower Jurassic Guanyintan Formation in Hunan, China (Shang 1981). *Gleicheniidites conflexus* and *Callialasporites radius* were found in the Middle Jurassic Yan'an Formation in Yan'an, Ganquan, and Zhidan, Shaanxi, China (Institute of Geology, Chinese Academy of Geological Sciences 1980). *Divisisporites undulatus* was found in the Lower to Middle Jurassic Series in the Dananhu coalfield in the Turpan-Hami Basin, China (Huang 1995). *Cyathidites australis*, *C. minor*, *Osmundacidites wellmanii*, *Marattisporites scabratus*, *Alisporites bilateralis*, *Abietineaepollenites minimus*, *Pityosporites divulgatus*, *Piceaepollenites complanatiformus*, *Piceites latens*, *P. podocarpoides*, *Caytonipollenites pallidus*, *Podocarpidites multesimus*, *Callialasporites dampieri*, *Classopollis classoides*, *C. annulatus*, *Eucommiidites troedssonii*, *Cycadopites typicus*, and *C. minimus* are common in the Jurassic rocks, and they may extend into the Cretaceous (Couper 1958; Pocock 1970; Institute of Geology, Chinese Academy of Geological Sciences 1980; Jiang and Wang 2002a). In addition, *Pseudopicea variabiliformis* was found in the Lower Triassic Ehuobulake Formation, the Middle Triassic Karamay Formation, the Upper Triassic Huangshanjie and Taliqike formations, the Lower Jurassic Ahe and Yangxia formations, and the Middle Jurassic Kezilenuer and Qiakemake formations in Kuqa, Xinjiang, China (Liu 2003). These spores and pollen are widely distributed in the Lower and Middle Jurassic Series in the Turpan Basin (Sun 1989; Huang 1995; Wang et al. 1998).

3.4.2 Shengjinkou Oil Field

The Shengjinkou oil field is located in the western section of the Huoyanshan anticlinal zone in the Turpan Depression. It is an anticlinal trap oil pool. The reservoir bed of the Shengjinkou oil field is the Middle Jurassic Toudunhe Formation.

Twenty-three species of fossil spores and pollen referred to 20 genera were extracted from crude oil samples collected in the Middle Jurassic reservoir bed of the Shengjinkou oil field of the Turpan Basin (Table 3.11).

Table 3.11 Spores and pollen in crude oils from the Shengjinkou oil field

Spores and pollen	Number of specimens	Geological age
Spores		
Leiotriletes medius Nilsson, 1958	3	J
Deltoidospora lineata (Bolchovitina) Pocock, 1970	2	J
Cyathidites minor Couper, 1953	7	J-K
Dictyophyllidites harrisii Couper, 1958	5	J
Cibotiumspora paradoxa (Maljavkina) Chang, 1965	2	J
Gleicheniidites rouseii Pocock, 1970	3	J
Granulatisporites jurassicus Pocock, 1970	2	J
Osmundacidites wellmanii Couper, 1953	3	J-K
Todisporites minor Couper, 1958	4	J
Leptolepidites major Couper, 1958	2	J
Apiculatisporis variabilis Pocock, 1970	3	J
Pollen		
Paleoconiferus asaccatus Bolchovitina, 1956	2	J
Alisporites bilateralis Rouse, 1959	2	J-K
Pityosporites divulgatus (Bolchovitina) Pocock, 1970	3	J-K
Protopicea exilioides (Bolchovitina) Pocock, 1970	5	J
Piceites expositus Bolchovitina, 1956	3	J
P. latens Bolchovitina, 1956	2	J-K
Cedripites minor Pocock, 1970	3	J
Podocarpidites multicinus (Bolchovitina) Pocock, 1970	3	J
P. multesimus (Bolchovitina) Pocock, 1962	3	J-K
Classopollis classoides (Pflug) Pocock et Jansonius, 1961	7	J-K
C. annulatus (Verbitzkaya) Li, 1974	8	J-K
Cycadopites nitidus (Bolchovitina) Pocock, 1970	4	T-K
Total	81	

Most of the spores and pollen from the Shengjinkou oil field of the Turpan Basin (Table 3.11) are Jurassic species. Some extend into the Cretaceous, and one species ranges down into the Triassic. Among the spores and pollen, *Deltoidospora lineata*, *Podocarpidites multicinus*, and *P. multesimus* were found in the Lower Jurassic Series in Yakutsk, Russia. *Paleoconiferus asaccatus* was found in the Middle Jurassic Series, and *Protopicea exilioides* and *Piceites expositus* were found in the Upper Jurassic Series, and *Pityosporites divulgatus* and *Piceites latens* were found in the Lower Cretaceous Series in Yakutsk, Russia (Bolchovitina 1956). Except for *Podocarpidites multesimus*, these species were found in the Jurassic System of Canada (Pocock 1970). *Dictyophyllidites harrisii*, *Todisporites minor*, and *Leptolepidites major* were found in the Middle Jurassic Bajocian Stage in Yorkshire, England (Couper 1958). *Leiotriletes medius* was found in the Lower Jurassic Series of Sweden (Nilsson 1958). *Cibotiumspora paradoxa* was found in the Middle Jurassic series in Emba, Russia, (Maljavkina 1949) and reported from the Lower and Middle Jurassic Series of China (Zhang 1965; Liu 2003; Jiang et al. 2008). *Gleicheniidites rouseii*, *Granulatisporites jurassicus*, *Apiculatisporis*

3.5 Qaidam Basin

3.5.1 North Border Block-fault Zone

3.5.1.1 Lenghu Oil Field

The Lenghu oil field is located in the Lenghu structural fault zone in the northwestern part of the North Border Block-fault Zone. It is a fault-block trap oil pool. The reservoir rocks of the Lenghu oil field are Middle Jurassic sandstones.

Forty-four species of fossil spores and pollen referred to 25 genera were extracted from crude oil samples collected in the Middle Jurassic reservoir bed of the Lenghu oil field of the Qaidam Basin. These are listed in Table 3.12 and illustrated on Plates XXVII–XXVIII and XLV.

Most of the spores and pollen from the Lenghu oil field of the Qaidam Basin (Table 3.12) are Jurassic species. Some range down to the Triassic and others range up into the Cretaceous. Among the spores and pollen, *Dictyophyllidites harrisii, Leptolepidites major, Lycopodiacidites rugulatus*, and *Parvisaccites enigmatus* were found in the Middle Jurassic Series of Britain. *Lycopodiacidites rugulatus* was found in the Lower Jurassic Series of Germany (Couper 1958; Schulz 1967). *Duplexisporites scanicus, Protopinus scanicus, Chasmatosporites major*, and *C. elegans* were found in the Lower Jurassic Lias Stage of Sweden (Nilsson 1958). *Protopicea vilujensis, P. minutereticulata, Protopodocarpus mollis, Podocarpidites multicinus*, and *P. multesimus* were found in the Lower Jurassic Series in Yakutsk in the Vilyui Basin, Russia. *Paleoconiferus asaccatus* was found in the Middle Jurassic Series in Yakutsk. *Protopicea exilioides, Piceaepollenites omoriciformis, Piceites expositus*, and *Podocarpidites paulus* were found in the Upper Jurassic Series in

(Text on previous columns continues:) *variabilis*, and *Cedripites minor* were found in the Jurassic System of Canada (Pocock 1970). The other spores and pollen are widely distributed in the Jurassic to Cretaceous sediments.

Table 3.12 Spores and pollen in crude oils from the Lenghu oil field

Spores and pollen	Number of specimens	Geological age
Spores		
Cyathidites australis Couper, 1953	4	J-K
C. minor Couper, 1953	5	J-K
Dictyophyllidites harrisii Couper, 1958	4	J
Gleicheniidites rouseii Pocock, 1970	4	J
G. conflexus (Chlonova) Xu et Zhang, 1980	7	J
Osmundacidites wellmanii Couper, 1953	8	J-K
O. elegans (Verbitzkaya) Xu et Zhang, 1980	5	J
Leptolepidites major Couper, 1958	4	J
Apiculatisporis variabilis Pocock, 1970	9	J
Acanthotriletes midwayensis Pocock, 1970	5	J
Lycopodiacidites rugulatus (Couper) Schulz, 1967	4	J
Duplexisporites gyratus Playford et Dettmann, 1965	6	T-J
D. amplectiformis (Kara-Murza) Playford et Dettmann, 1965	3	T-J
D. scanicus (Nilsson) Playford et Dettmann, 1965	9	J
D. anagrammensis (Kara-Murza) Playford et Dettmann, 1965	3	T-J

(continued)

Table 3.12 (continued)

Spores and pollen	Number of specimens	Geological age
Pollen		
Paleoconiferus asaccatus Bolchovitina, 1956	4	J
Pityosporites similis Balme, 1957	3	J
Protopinus scanicus Nilsson, 1958	5	J
Protopicea vilujensis Bolchovitina, 1956	5	J
P. minutereticulata Bolchovitina, 1956	4	J
P. exilioides (Bolchovitina) Pocock, 1970	6	J
Piceaepollenites omoriciformis (Bolchovitina) Xu et Zhang, 1980	3	J
Piceites expositus Bolchovitina, 1956	5	J
P. latens Bolchovitina, 1956	4	J-K
Cedripites minor Pocock, 1970	5	J
Protopodocarpus mollis Bolchovitina, 1956	5	J
Podocarpidites multicinus (Bolchovitina) Pocock, 1970	6	J
P. multesimus (Bolchovitina) Pocock, 1962	3	J-K
P. wapellaensis Pocock, 1970	3	J
P. langii Pocock, 1970	4	J
P. paulus (Bolchovitina) Xu et Zhang, 1980	6	J
Parvisaccites enigmatus Couper, 1958	5	J
Quadraeculina limbata Maljavkina, 1949	6	J-K
Callialasporites dampieri (Balme) Sukh Dev, 1961	2	J-K
C. radius Xu et Zhang, 1980	5	J
Cerebropollenites carlylensis Pocock, 1970	3	J
Classopollis annulatus (Verbitzkaya) Li, 1974	6	J-K
Chasmatosporites major Nilsson, 1958	3	J
C. elegans Nilsson, 1958	4	J
Cycadopites subgranulosus (Couper) Clarke, 1965	5	T-J
C. carpentieri (Delcourt et Sprumont) Singh, 1964	4	J-K
C. nitidus (Balme) Pocock, 1970	6	T-K
C. typicus (Maljavkina) Pocock, 1970	3	J-K
C. minimus (Cookson) Pocock, 1970	5	J-K
Total	208	

Yakutsk, Russia (Bolchovitina 1956). *Gleicheniidites rouseii, Apiculatisporis variabilis, Acanthotriletes midwayensis, Cedripites minor, Podocarpidites wapellaensis, P. langii,* and *Cerebropollenites carlylensis* were found in the Jurassic System of western Canada (Pocock 1970). *Pityosporites similis* was found in the Jurassic System of Australia and Canada (Balme 1957; Pocock 1970). *Gleicheniidites conflexus,* *Osmundacidites elegans,* and *Callialasporites radius* were found in the Middle Jurassic Yan'an Formation in Shaanxi, China (Institute of Geology, Chinese Academy of Geological Sciences 1980). These spores and pollen are widely distributed in the Jurassic sediments in Europe, Asia, North America, and Australia; they are considered as Jurassic index species. *Duplexisporites gyratus, D. amplectiformis,* and

D. anagrammensis were found in the Upper Triassic Rhaetian Stage to Lower Jurassic Lias Stage of South Australia (Playford and Dettmann 1965). The other spores and pollen such as *Cyathidites australis*, *C. minor*, *Osmundacidites wellmanii*, *Quadraeculina limbata*, *Classopollis annulatus*, *Cycadopites subgranulosus*, *C. nitidus*, and *C. typicus* are all Jurassic common species.

3.5.1.2 Yuka Oil Field

The Yuka oil field is located in the Yuka fault sag in the east part of the North Border Block-fault Zone. It is an anticlinal trap oil pool. The reservoir rock of the Yuka oil field is Upper Jurassic sandstones.

Twenty species of fossil spores and pollen referred to 16 genera were extracted from crude oil samples collected in the Upper Jurassic reservoir bed of the Yuka oil field of the Qaidam Basin (Table 3.13).

Among the spores and pollen in the Yuka oil field of the Qaidam Basin (Table 3.13), *Dictyophyllidites harrisii* and *Densoisporites perinatus* were found in the Middle Jurassic Bajocian Stage in Yorkshire, England; the latter may be distributed in the Lower Cretaceous Series (Couper 1958). *Chasmatosporites major* was found in the Lower Jurassic Lias Stage of Sweden (Nilsson 1958). *Paleoconiferus asaccatus*, *Protopicea exilioides*, *Piceites expositus*, *Podocarpidites multicinus*, and *P. paulus* were found in the Jurassic System in Yakutsk, Russia (Bolchovitina 1956). *Cedripites minor* and *Cerebropollenites carlylensis* were found in the Jurassic System of Canada (Pocock 1970). *Pityosporites similis* was found in the Jurassic System of Australia (Balme 1957). *Gleicheniidites*

Table 3.13 Spores and pollen in crude oils from the Yuka oil field

Spores and pollen	Number of specimens	Geological age
Spores		
Cyathidites autstralis Couper, 1953	3	J-K
C. minor Couper, 1953	4	J-K
Dictyophyllidites harrisii Couper, 1958	2	J
Gleicheniidites conflexus (Chlonova) Xu et Zhang, 1980	2	J
Duplexisporites gyratus Playford et Dettmann, 1965	1	T-J
Densoisporites perinatus Couper, 1958	2	J-K
Pollen		
Paleoconiferus asaccatus Bolchovitina, 1956	3	J
Pityosporites similis Balme, 1957	2	J
Protopicea exilioides (Bolchovitina) Pocock, 1970	3	J
Piceites expositus Bolchovitina, 1956	2	J
Cedripites minor Pocock, 1970	4	J
Podocarpidites multicinus (Bolchovitina) Pocock, 1970	3	J
P. paulus (Bolchovitina) Xu et Zhang, 1980	3	J
Callialasporites radius Xu et Zhang, 1980	1	J
Cerebropollenites carlylensis Pocock, 1970	1	J
Classopollis annulatus (Verbitzkaya) Li, 1974	4	J-K
Chasmatosporites major Nilsson, 1958	2	J
Cycadopites subgranulosus (Couper) Clarke, 1965	2	T-J
C. nitidus (Balme) Pocock, 1970	3	T-K
C. minimus (Cookson) Pocock, 1970	4	J-K
Total	51	

conflexus and *Callialasporites radius* were found in the Middle Jurassic Series in Shaanxi, China (Institute of Geology, Chinese Academy of Geological Sciences 1980). Except for *Densoisporites perinatus*, these species are distributed within the limits of Jurassic System. They may be considered as Jurassic index species. The other spores and pollen such as *Cyathidites minor*, *Duplexisporites gyratus*, *Classopollis annulatus*, *Cycadopites subgranulosus*, and *C. nitidus* may be considered as common Jurassic species.

3.5.2 Mangnai Depression

3.5.2.1 Youquanzi Oil Field

The Youquanzi oil field is located in the Mangnai Sag in the middle part of the Mangnai Depression. It is an anticlinal trap oil pool. The reservoir bed of the Youquanzi oil field is the Pliocene Upper Youshashan Formation.

Fifty-nine species of fossil spores and pollen referred to 26 genera were extracted from crude oil samples collected in the Pliocene reservoir bed of the Youquanzi oil field of the Qaidam Basin. They are listed in Table 3.14 and illustrated on Plates XXIX–XXXI and XLVI–XLVII.

Most of the spores and pollen of the Youquanzi oil field of the Qaidam Basin (Table 3.14) are Tertiary species; a few range down into the Cretaceous. Among the spores and pollen, *Deltoidospora regularis* and *Lygodiumsporites pseudomaximus* were found in the Tertiary System in middle Europe (Thomson and Pflug 1953; Krutzsch 1959), and the former was found in the Cretaceous System and the Oligocene Series in Jiangsu, China, and the latter was found in the Paleogene System in Jiangsu, China (Song et al. 1981). *Polypodiaceaesporites haardti* was found in the Miocene Series of Germany while *Polypodiisporites favus* was found in the Oligocene Series and the Miocene Series of Germany (Potonié 1956); these species of spores were found in the Eocene to Oligocene Series in the coastal region of Bohai, China, and the Upper Oligocene to Miocene Series in the Yangtze-Han River Plain, China (Research Institute of Petroleum Exploration and Development, Ministry of Petroleum Chemistry Industry, Nanjing Institute of Geology and Palaeontology, Chinese Academy of Sciences 1978a; Li et al. 1978). *Ephedripites eocenipites* and *Momipites coryloides* were found in the Eocene Series of the USA (Wodehouse 1933) and the Eocene Series and the Oligocene Series of China (Li et al. 1978; Song et al. 1981). *Piceaepollenites alatus*, *Pinuspollenites labdacus maximus*, *P. labdacus minor*, *Tsugaepollenites igniculus major*, and *T. igniculus minor* were found in the Miocene Series of Germany (Potonié 1958). *Podocarpidites nageiaformis* and *Abietineaepollenites cembraeformis* were found in the Lower Oligocene Series in the Irtysh Basin, Russia. *Pinuspollenites banksianaeformis* was found in the Cretaceous System to the Paleocene Series in the Irtysh Basin (Zaklinskaja 1957). *Piceaepollenites planoides* was found in the Tertiary System of Europe (Krutzsch 1971) and the Oligocene Series of the South China Sea (Hou et al. 1981). *Abietineaepollenites microalatus major* was found in the Miocene Series of Europe (Potonié 1958) and the Paleocene to Eocene Funing Formation, the Oligocene Dainan Formation, and the Miocene to Pliocene Yancheng Formation in Jiangsu, China (Song et al. 1981). *Keteleeriaepollenites dubius* was found in the Upper Cretaceous Series of Russia and the Middle to Upper Tertiary System of middle Europe (Krutzsch 1971), which was reported from the Paleogene Shahejie Formation in the coastal region of Bohai, China, and the Neogene Hailongjing, Yuquan, and Santan formations in the Shelf Basin of the East China Sea (Song et al. 1985). *Cedripites deodariformis* was found in the Oligocene Series of Kazakhstan and was reported from the Eocene to Oligocene Series in Liaoning, Shandong, Hebei, Henan, Jiangsu, China. *Podocarpidites paranageiaformis* was found in the Oligocene Series in Liaoning and Shandong, China. *Cedripites microsaccoides* was found in the Paleocene to Eocene Series in Jiangsu, China (Research Institute of Petroleum Exploration and Development, Ministry of Petroleum Chemistry Industry, Nanjing Institute of Geology and Palaeontology, Chinese Academy of Sciences 1978a; Song et al. 1981).

Table 3.14 Spores and pollen in crude oils from the Youquanzi oil field

Spores and pollen	Number of specimens	Geological age
Pteridophyta		
Deltoidospora regularis (Pflug) Song et Zheng, 1981	5	K-N
Granulatisporites pteridiumoides Zhang, 1985	3	N
Lygodiumsporites pseudomaximus (Thomson et Pflug) Song et Zhang, 1981	3	E
Polypodiaceaesporites haardti (Potonié et Venitz) Thiergart, 1938	7	E-N
Polypodiisporites favus (Potonié) Potonié, 1956	5	E-N
Echinosporis qaidamensis Zhang, 1985	3	N
Gymnospermae		
Podocarpidites nageiaformis (Zaklinskaja) Krutzsch, 1971	5	E
P. paranageiaformis Ke et Shi, 1978	4	E
P. verrucorpus Wu, 1985	8	E-N
P. qigequanensis Wu, 1985	6	E-N
Abietineaepollenites microalatus major (Potonié) Potonié, 1958	5	E-N
A. cembraeformis (Zaklinskaja) Ke et Shi, 1978	5	E
A. lenghuensis Zhu, 1985	4	E-N
Pinuspollenites labdacus maximus (Potonié) Potonié, 1958	6	E-N
P. labdacus minor (Potonié) Potonié, 1958	8	E-N
P. banksianaeformis (Zaklinskaja) Ke et Shi, 1978	3	K-E
P. mangnaiensis Zhu, 1985	5	E-N
Piceaepollenites alatus (Potonié) Potonié, 1958	7	E-N
P. tobolicus (Panova) Ke et Shi, 1978	5	E
P. planoides (Krutzsch) Sun et Li, 1981	6	E-N
P. quadracorpus Zhu et Xi, 1985	5	E-N
Cedripites deodariformis (Zauer) Krutzsch, 1971	8	E
C. pachydermus (Zauer) Krutzsch, 1971	7	E-N
C. microsaccoides Song et Zheng, 1981	5	E
Abiespollenites lenghuensis Zhu, 1985	5	E-N
Keteleeriaepollenites dubius (Chlonova) Li, 1985	7	K-N
K. mangnaiensis Zhu, 1985	5	E-N
Tsugaepollenites igniculus major (Potonié) Potonié, 1958	4	E-N
T. igniculus minor (Potonié) Potonié, 1958	6	E-N
T. viridifluminipites (Wodehouse) Potonié, 1958	3	E-N
T. spinulosus (Krutzsch) Ke et Shi, 1978	4	E-N
Ephedripites eocenipites (Wodehouse) Krutzsch, 1961	8	E
E. fusiformis (Shakhmundes) Krutzsch, 1970	6	E-N
E. claricristatus (Shakhmundes) Krutzsch, 1970	4	E-N
E. neogenicus Zhu et Wu, 1985	5	N
E. mangnaiensis Zhu et Wu, 1985	3	E-N

(continued)

Table 3.14 (continued)

Spores and pollen	Number of specimens	Geological age
E. dafengshanensis Zhu et Wu, 1985	4	E-N
E. ganchaigouensis Zhu et Wu, 1985	4	E-N
Angiospermae		
Salixipollenites elegans Sung et Tsao, 1976	3	E
S. minor Song et Zhu, 1985	4	E-N
Betulaepollenites lenghuensis Song et Zhu, 1985	3	N
Momipites coryloides Wodehouse, 1933	3	E
Quercoidites henrici (Potonié) Potonié, Thomson et Thiergart, 1950	8	E-N
Q. microhenrici (Potonié) Potonié, 1950	12	E-N
Q. asper (Thomson et Pflug) Sung et Zheng, 1978	4	K-N
Cupuliferoipollenites pusillus (Potonié) Potonié, 1951	5	E
Meliaceoidites rhomboiporus Wang, 1980	5	E-N
M. ovatus Zhu et Xi, 1985	3	E-N
Nitrariadites pachypolarus Zhu et Xi, 1985	4	E-N
N. subrotundus Zhu et Xi, 1985	5	E-N
N. communis Zhu et Xi, 1985	6	E-N
Sparganiaceaepollenites sparganioides (Meyer) Krutzsch, 1970	5	E-N
Chenopodipollis multiplex (Weyland et Pflug) Krutzsch, 1966	5	E-N
C. multiporatus (Pflug et Thomson) Zhou, 1981	6	K-N
C. minor Song, 1985	4	E-N
Tubulifloridites macroechinatus (Trevisan) Song et Zhu, 1985	7	E-N
T. minispinulosus Song et Zhu, 1985	5	E-N
Artemisiaepollenites sellularis Nagy, 1969	6	N
A. communis Song et Zhu, 1985	9	N
Total	308	

Cupuliferoipollenites pusillus was found in the Eocene Series and *Quercoidites henrici* was found in the Miocene Series of Germany (Potonié 1960). *Salixipollenites elegans* was found in the Paleocene Series in Fushun, Liaoning, China (Sung and Tsao 1976). *Artemisiaepollenites sellularis* was found in the Miocene Series of Hungary (Nagy 1969). *Chenopodipollis minor* was found in the Miocene to Pliocene Series in the Shelf Basin of the East China Sea (Song et al. 1985). *Granulatisporites pteridiumoides, Echinosporis qaidamensis, Podocarpidites verrucorpus, P. qigequanensis, Abietineaepollenites lenghuensis, Pinuspollenites mangnaiensis, Piceaepollenites quadracorpus, Abiespollenites lenghuensis, Keteleeriaepollenites mangnaiensis, Ephedripites neogenicus, E. mangnaiensis, E. dafengshanensis, E. ganchaigouensis, Salixipollenites minor, Betulaepollenites lenghuensis, Meliaceoidites ovatus, Nitrariadites pachypolarus, N. subrotundus, N. communis, Tubulifloridites minispinulosus,* and *Artemisiaepollenites communis* are new species found in the Tertiary System in the Qaidam Basin, China (Research Institute of Exploration and Development, Qinghai Petroleum Administration, Nanjing Institute of Geology and Palaeontology, Academia Sinica 1985).

3.5.2.2 Xianshuiquan Oil Field

The Xianshuiquan oil field is located in the Altun Clinoform in the northwestern part of the Mangnai Depression. It is a nose-trap oil pool. The reservoir bed of the Xianshuiquan oil field is the Pliocene Upper Youshashan Formation.

Twenty-three species of fossil spores and pollen referred to 14 genera were extracted from crude oil samples collected in the Pliocene reservoir bed of the Xianshuiquan oil field of the Qaidam Basin. These are listed in Table 3.15 and illustrated on Plate XXXII.

Among the spores and pollen from the Xianshuiquan oil field of the Qaidam Basin (Table 3.15), *Deltoidospora regularis* was found in the Tertiary System of Europe and the Cretaceous System and the Oligocene Series in Jiangsu, China (Thomson and Pflug 1953; Song et al. 1981). *Polypodiaceaesporites haardti* was found in the Miocene Series of Germany and the Upper Oligocene to Miocene Series in the Yangtze-Han River Plain, China (Potonié 1956; Li et al. 1978). *Momipites coryloides* was found in the Eocene Green River Formation in Colorado and Utah, USA (Wodehouse 1933). *Pinuspollenites labdacus, Piceaepollenites alatus,* and *Tsugaepollenites igniculus* were found in the Miocene Series of Germany (Potonié 1958).

Table 3.15 Spores and pollen in crude oils from the Xianshuiquan oil field

Spores and pollen	Number of specimens	Geological age
Pteridophyta		
Deltoidospora regularis (Pflug) Song et Zheng, 1981	4	K-N
Granulatisporites pteridiumoides Zhang, 1985	6	N
Polypodiaceaesporites haardti (Potonié et Venitz) Thiergart, 1938	5	E-N
Gymnospermae		
Abietineaepollenites lenghuensis Zhu, 1985	3	E-N
Pinuspollenites labdacus maximus (Potonié) Potonié, 1958	3	E-N
P. labdacus minor (Potonié) Potonié, 1958	4	E-N
P. mangnaiensis Zhu, 1985	5	E-N
Piceaepollenites alatus (Potonié) Potonié, 1958	4	E-N
P. quadracorpus Zhu et Xi, 1985	6	E-N
Cedripites deodariformis (Zauer) Krutzsch, 1971	3	E
C. ovatus Ke et Shi, 1978	4	E
Keteleeriaepollenites megasaccus Zhu, 1985	2	E-N
Tsugaepollenites igniculus major (Potonié) Potonié, 1958	3	E-N
T. spinulosus (Krutzsch) Ke et Shi, 1978	3	E-N
Ephedripites fusiformis (Shakhmundes) Krutzsch, 1970	5	E-N
E. neogenicus Zhu et Wu, 1985	7	N
E. mangnaiensis Zhu et Wu, 1985	3	E-N
E. dafengshanensis Zhu et Wu, 1985	4	E-N
Angiospermae		
Momipites coryloides Wodehouse, 1933	2	E
Nitrariadites subrotundus Zhu et Xi, 1985	5	E-N
N. communis Zhu et Xi, 1985	7	E-N
Chenopodipollis microporatus (Nakoman) Liu, 1981	3	K-N
Artemisiaepollenites sellularis Nagy, 1969	5	N
Total	96	

Cedripites ovatus and *Tsugaepollenites spinulosus* were found in the Paleogene System in the coastal region of Bohai, China (Research Institute of Petroleum Exploration and Development, Ministry of Petroleum Chemistry Industry, Nanjing Institute of Geology and Palaeontology, Chinese Academy of Sciences 1978a). *Chenopodipollis microporatus* was found in the Upper Cretaceous Taizhou Formation in Jiangsu, China, and the Tertiary System in the Qaidam Basin, China (Song et al. 1981). *Artemisiaepollenites sellularis* was found in the Miocene Series of Hungary (Nagy 1969). *Granulatisporites pteridiumoides, Abietineaepollenites lenghuensis, Pinuspollenites mangnaiensis, Piceaepollenites quadracorpus, Keteleeriaepollenites megasaccus, Ephedripites neogenicus, E. mangnaiensis, E. dafengshanensis, Nitrariadites subrotundus*, and *N. communis* are new species found in the Tertiary System in the Qaidam Basin, China (Research Institute of Exploration and Development, Qinghai Petroleum Administration, Nanjing Institute of Geology and Palaeontology, Academia Sinica 1985).

3.6 West Jiuquan Basin

3.6.1 Laojunmiao Anticlinal Zone

3.6.1.1 Yaerxia Oil Field

The Yaerxia oil field is located in the northwestern part of the Laojunmiao Anticlinal Zone. It is an anticlinal trap oil pool. The petroleum reservoir rocks of the Yaerxia oil field include Silurian metamorphic rocks, Lower Cretaceous sandstone, and a Neogene sandstone.

Cycadopites minimus and *Psophosphaera tenuis* were extracted from crude oil samples from the Silurian metamorphic rock reservoir. *Cibotiumspora juncta, Schizaeoisporites zizyphinus, Cycadopites minimus, Bennettiteaepollenites* sp., and *Psophosphaera tenuis* were extracted from crude oil samples from the Lower Cretaceous sandstone reservoir. *Schizaeoisporites zizyphinus, Cycadopites minimus, Psophosphaera tenuis* and angiospermous pollen *Lilium, Chenopodium, Nymphaea, Tamarix, Artemisia, Bidens, Graminidites*, and *Cyperaceaepollis* were extracted from crude oil samples from the Neogene sandstone reservoir (Plate XXXIII).

With the exception of the angiosperm pollen from the Neogene reservoir, these spores and pollen are all found in the Lower Cretaceous of the Jiuquan Basin (Hsü et al. 1974).

3.6.1.2 Laojunmiao Oil Field

The Laojunmiao oil field is located in the middle part of the Laojunmiao Anticlinal Zone. It is an anticlinal trap oil pool. The reservoir bed of the Laojunmiao oil field is the Neogene Baiyanghe Series.

Twenty-five species of fossil spores and pollen referred to 23 genera were extracted from crude oil samples collected in the Neogene reservoir bed of the Laojunmiao oil field of the West Jiuquan Basin. These are listed in Table 3.16 and illustrated on Plate XXXIII.

With the exception of the angiosperm pollen and the three gymnosperms *Pinus, Picea*, and *Ephedra*, the spores and pollen from the Laojunmiao oil field of the West Jiuquan Basin (Table 3.16) are all distributed in the Lower Cretaceous Series in the Jiuquan and Huahai basins in Gansu, China (Hsü et al. 1974; Chiang and Young 1978).

3.6.2 Baiyanghe Monoclinal Zone

The Baiyanghe oil field is located in the middle part of the Baiyanghe Monoclinal Zone. It is a nose-fault trap oil pool. Both the Paleogene Huoshaogou and the Neogene Baiyanghe series are petroleum reservoirs in the Baiyanghe oil field.

The spores of the pteridophyte *Schizaeoisporites zizyphinus*, the gymnosperms *Cycadopites minimus, Pinus* sp., *Ephedra* sp., and the angiosperms *Eutrema* sp., and *Nitraria* sp., were extracted from crude oil samples collected in the Paleogene petroleum reservoir. The pteridophyte spores *Cibotiumspora juncta, Schizaeoisporites zizyphinus*, and the angiosperms *Potamogeton, Graminidites, Chenopodium, Eutrema, Solidago*,

Table 3.16 Spores and pollen in crude oils from the Laojunmiao oil field

Spores and pollen	Number of specimens	Geological age
Pteridophyta		
Cibotiumspora juncta (Kara-Murza) Xu et Zhang, 1980	5	J-K
Osmundacidites wellmanii Couper, 1953	3	J-K
Schizaeoisporites zizyphinus Hsü, Chiang et Young, 1974	6	K
S. gansuensis Chiang et Young, 1978	3	K
Cingulatisporites ruginosus Hsü, Chiang et Young, 1974	3	K
Gymnospermae		
Cycadopites minimus (Cookson) Pocock, 1970	5	J-K
Bennettiteaepollenites sp.	3	J-K
Psophosphaera grandis Bolchovitina, 1953	4	K
P. tenuis Naumova ex Bolchovitina, 1953	3	K
Classopollis annulatus (Verbitzkaya) Li, 1974	4	J-K
Pinus sp.	7	N
Picea sp.	6	N
Ephedra sp.	5	N
Angiospermae		
Potamogeton sp.	5	N
Graminidites sp.	7	N
Cyperaceaepollis sp.	6	N
Lilium sp.	8	N
Ulmus sp.	5	N
Chenopodium sp.	12	N
Nymphaea sp.	6	N
Eutrema sp.	8	N
Lens sp.	5	N
Tamarix sp.	7	N
Artemisia sp.	14	N
Bidens sp.	4	N
Total	144	

Achillea, and *Artemisia* were extracted from crude oil samples from the Neogene petroleum reservoir (Plate XXXIII).

Of the previous spores and pollen, *Cibotiumspora juncta*, *Schizaeoisporites zizyphinus*, and *Cycadopites minimus* are widely distributed in the Lower Cretaceous Series of the Jiuquan Basin (Hsü et al. 1974); the others are distributed in the Tertiary System of the region (Sung 1958).

3.7 Liaohe Basin of Bohai Gulf

The Liaohe Basin is a Tertiary petroliferous basin in the northern part of the Bohai Gulf. The Xinglongtai oil field is located in the West Sag of the Liaohe Basin. The Paleogene Shahejie and Dongying formations are petroleum reservoirs in the Xinglongtai oil field.

Twenty-one species of fossil spores and pollen referred to 16 genera were extracted from crude oil samples collected in the Paleogene petroleum reservoir of the Xinglongtai oil field of the Liaohe Basin. These are listed in Table 3.17 and illustrated on Plate XXXIV. In addition, some fossil algae such as *Campenia irregularis*, *Dictyotidium reticulatum*, *Dictyotidium microreticulatum*, *Palaeostomocystis minor*, and *Bohaidina laevigata minor* were extracted from crude oil samples from the petroleum reservoir in the Shahejie Formation.

Most of the spores and pollen from the Xinglongtai oil field of the Liaohe Basin (Table 3.17) are Paleogene species; a few range down into the Cretaceous. Among the spores and pollen, *Pterisisporites undulatus* was found in the Upper Cretaceous Taizhou Formation, the Paleocene to Eocene Funing Formation, and the Oligocene Sanduo Formation in Jiangsu, China (Song et al. 1981). *Plicifera decora* was found in the Cretaceous System of Russia (Bolchovitina 1953) and the Paleogene System in the coastal region of Bohai, China (Research Institute of Petroleum Exploration and Development, Ministry of Petroleum Chemistry Industry, Nanjing Institute of Geology and Palaeontology, Chinese Academy of Sciences 1978a). *Pinuspollenites strobipites* was found in the Eocene Series of the USA (Wodehouse 1933). *Pinuspollenites*

Table 3.17 Spores and pollen in crude oils from the Xinglongtai oil field

Spores and pollen	Number of specimens	Geological age
Pteridophyta		
Pterisisporites undulatus Sung et Zheng, 1976	7	K-E
Plicifera decora (Chlonova) Bolchovitina, 1953	5	K-E
Osmundacidites sp.	4	E
Gymnospermae		
Pinuspollenites strobipites (Wodehouse) Sun et Li, 1981	4	E
P. labdacus minor (Potonié) Potonié, 1958	6	E-N
Abietineaepollenites cembraeformis (Zaklinskaya) Ke et Shi, 1978	7	E-N
A. microsibiricus (Zaklinskaya) Ke et Shi, 1978	8	E
Cedripites pachydermus (Zauer) Krutzsch, 1971	5	E-N
C. diversus Ke et Shi, 1978	6	E
Keteleeriaepollenites dubius (Chlonova) Li, 1985	5	K-N
Ephedripites cheganicus (Shakhmundes) Ke et Shi, 1978	9	E
Angiospermae		
Quercoidites asper (Thomson et Pflug) Sung et Zheng, 1978	5	K-N
Q. microhenrici (Potonié) Potonié, 1950	6	E-N
Chenopodipollis multiporatus (Pflug et Thomson) Zhou, 1981	12	K-N
C. microporatus (Nakoman) Liu, 1981	14	K-N
Compositoipollenites sp.	9	E
Artemisiaepollenites sp.	12	E
Sparganiaceaepollenites sparganioides (Meyer) Krutzsch, 1970	11	E-N
Liliacidites sp.	9	E
Cyperaceaepollis sp.	11	E
Graminidites sp.	8	E
Total	163	

labdacus minor was found in the Miocene Series of Germany (Potonié 1958) and the Eocene to Pliocene Series in China. *Abietineaepollenites cembraeformis* and *A. microsibiricus* were found in the Oligocene Series in the Irtysh Basin, Russia (Zaklinskaja 1957). *Cedripites pachydermus* was found in the Tertiary System of Russia and middle Europe (Krutzsch 1971). *Keteleeriaepollenites dubius* was found in the Upper Cretaceous Series of Russia and the Tertiary System of middle Europe (Krutzsch 1971). *Quercoidites microhenrici* was found in the Tertiary System of Europe (Potonié 1960). *Quercoidites asper* was found in the Upper Cretaceous Taizhou Formation, the Paleocene to Eocene Funing Formation, the Oligocene Sanduo Formation, and the Miocene to Pliocene Yancheng Formation in Jiangsu, China (Song et al. 1981), as well as the Eocene and Oligocene series in the Yangtze-Han River Plain, China (Li et al. 1978). *Cedripites diversus* and *Ephedripites cheganicus* were found in the Paleogene Shahejie Formation in the coastal region of Bohai, China. *Chenopodipollis multiporatus, C. microporatus, Compositoipollenites* sp., *Artemisiaepollenites* sp., *Sparganiaceaepollenites sparganioides, Liliacidites* sp., *Cyperaceaepollis* sp., and *Graminidites* sp. were reported from the Paleogene Shahejie Formation and Dongying Formation in the coastal region of Bohai, China (Research Institute of Petroleum Exploration and Development, Ministry of Petroleum Chemistry Industry, Nanjing Institute of Geology and Palaeontology, Chinese Academy of Sciences 1978a). The fossil algae including *Campenia irregularis, Dictyotidium reticulatum,* and *Palaeostomocystis minor* were found in the Paleogene Shahejie Formation in the coastal region of Bohai (Research Institute of Petroleum Exploration and Development, Ministry of Petroleum Chemistry Industry, Nanjing Institute of Geology and Palaeontology, Chinese Academy of Sciences 1978b).

3.8 Shelf Basins of South China Sea

3.8.1 Beibu Gulf Basin

The Beibu Gulf Basin lies in the Beibu Gulf Depression of the western North Shelf of the South China Sea. It is a Tertiary petroleum-bearing basin. In the northwestern part of the basin, the famous Weizhou oil field is located in the southwest Weizhou Sag. The reservoir bed of the Weizhou oil field is the Eocene to Oligocene Liushagang Formation.

Thirty-six species of fossil spores and pollen referred to 27 genera were extracted from crude oil samples collected in the Eocene to Oligocene reservoir bed of the Weizhou oil field of the Beibu Gulf Basin. These are listed in Table 3.18 and illustrated on Plates XXXV and XLVIII.

The spores and pollen found in the Weizhou oil field for the Beibu Gulf Basin (Table 3.18) are widely distributed in the Tertiary sediments in Europe, Asia, North America, or Australia. Most of them are Paleogene species. Among the spores and pollen, *Leiotriletes adriensis, Polypodiaceaesporites haardti,* and *Polypodiisporites afavus* were found in the Tertiary System of Europe (Thomson and Pflug 1953; Potonié 1956; Krutzsch 1959). *Polypodiaceaesporites ovatus* was found in the Paleocene Fort Union Formation in Montana, USA (Wilson and Webster 1946). *Pinuspollenites strobipites, Cedripites eocenicus, Salixipollenites discoloripites,* and *Momipites coryloides* were found in the Eocene Green River Formation, USA (Wodehouse 1933). *Cupuliferoipollenites pusillus* was found in the Eocene Series of Germany (Potonié 1960). *Podocarpidites andiniformis* was found in the Upper Cretaceous Danian Series to Paleogene System in the Irtysh Basin, Russia, and *Pinuspollenites minutus* was found in the Oligocene Chegan Formation in the Irtysh Basin (Zaklinskaya 1957). *Abietineaepollenites microalatus minor,*

Table 3.18 Spores and pollen in crude oils from the Weizhou oil field

Spores and pollen	Number of specimens	Geological age
Pteridophyta		
Leiotriletes adriensis (Potonié et Gelletich) Krutzsch, 1959	5	E-N
Osmundacidites primarius (Wolff) Sun et Li, 1981	4	E-N
Crassoretitriletes nanhaiensis Zhang et Li, 1981	7	E-N
Polypodiaceaesporites haardti (Potonié et Venitz) Thiergart, 1938	4	E-N
P. ovatus (Wilson et Webster) Sun et Zhang, 1981	5	E-N
Polypodiisporites afavus (Krutzsch) Sun et Li, 1981	6	E-N
Gymnospermae		
Podocarpidites andiniformis (Zaklinskaya) Takahashi, 1971	3	E
Abietineaepollenites microalatus minor (Potonié) Potonié, 1958	4	E-N
Pinuspollenites labdacus minor (Potonié) Potonié, 1958	5	E-N
P. minutus (Zaklinskaya) Song et Zheng, 1978	7	E
P. strobipites (Wodehouse) Sun et Li, 1981	9	E
Piceaepollenites alatus (Potonié) Potonié, 1958	6	E
P. planoides (Krutzsch) Sun et Li, 1981	8	E-N
Cedripites eocenicus Wodehouse, 1933	7	E
C. cedroides (Thomson et Pflug) Sun et Li, 1981	5	E
Angiospermae		
Salixipollenites discoloripites (Wodehouse) Srivastava, 1966	8	E
Caryapollenites simplex (Potonié) Raatz, 1937	4	E-N
Juglanspollenites verus Raatz, 1937	4	E-N
Momipites coryloides Wodehouse, 1933	5	E
Quercoidites microhenrici (Potonié) Potonié, 1950	7	E-N
Q. minor He et Sun, 1977	6	E
Cupuliferoipollenites pusillus (Potonié) Potonié, 1951	8	E
Ulmipollenites undulosus Wolff, 1934	4	E-N
U. granopollenites (Rouse) Sun et Li, 1981	3	E
Corylopsis princeps Lubomirova, 1961	5	E
Liquidambarpollenites stigmosus (Potonié) Raatz, 1937	3	E-N
L. minutus Ke et Shi, 1978	4	E
Ilexpollenites membranous Sun, Kong et Li, 1980	3	E-N
Myrtaceidites parvus Cookson et Pike, 1954	5	E-N
Tricolpites tenuicolpus Sun, Kong et Li, 1980	7	E
Tricolporopollenites minutus He et Sun, 1977	5	E
Multiporopollenites punctatus Ke et Shi, 1978	6	E
Retimultiporopollenites liushaensis Li et Sun, 1981	8	E
Trilobapollis leptus Sun, Kong et Li, 1980	5	E
T. ellipticus Sun, Kong et Li, 1980	4	E
Verrutricolporites pachydermus Sun, Kong et Li, 1980	3	E
Total	192	

Pinuspollenites labdacus minor, Piceaepollenites alatus, Caryapollenites simplex, Juglanspollenites verus, Quercoidites microhenrici, and *Liquidambarpollenites stigmosus* were first found in the Miocene Series of Europe (Potonié 1958, 1960). *Osmundacidites primarius* and *Ulmipollenites undulosus* were found in the Pliocene Series of Germany (Wolff 1934). *Myrtaceidites parvus* was found in the Tertiary System of Australia (Potonié 1960). *Quercoidites minor* was found in the Eocene Qingjiang Formation in Jiangxi, China (He and Sun 1977). *Liquidambarpollenites minutus* and *Multiporopollenites punctatus* were found in the Paleogene System in the coastal region of Bohai, China (Research Institute of Petroleum Exploration and Development, Ministry of Petroleum Chemistry Industry, Nanjing Institute of Geology and Palaeontology, Chinese Academy of Sciences 1978a). These spores and pollen are widely distributed in the Paleogene System in the northern part of the South China Sea (Sun et al. 1981, 1982; Hou et al. 1981; Zhang 1981). In addition, *Crassoretitriletes nanhaiensis, Ilexpollenites membranous,* and *Trilobapollis ellipticus* are new species found in the Eocene to Oligocene Liushagang Formation and the Oligocene Weizhou Formation; *Retimultiporopollenites liushaensis* and *Tricolpites tenuicolpus* are new species found in the Liushagang Formation; and *Trilobapollis leptus* and *Verrutricolporites pachydermus* are new species found in the Weizhou Formation in the Beibu Gulf Depression and the Leizhou Peninsula of the North Shelf of the South China Sea (Sun et al. 1980; Hou et al. 1981; Zhang 1981).

3.8.2 Zhujiang Mouth Basin

The Zhujiang Mouth Basin lies in the Zhujiang Mouth Depression of the eastern North Shelf of the South China Sea. Similar to the Beibu Gulf Basin, it is a high value, petroleum-bearing basin of Tertiary age. The Zhuhai oil field was discovered in the northeastern part of the basin in 1979. The reservoir bed of the Zhuhai oil field is the Oligocene Zhuhai Formation.

Thirty species of fossil spores and pollen referred to 23 genera were extracted from crude oil samples collected in the Oligocene reservoir bed of the Zhuhai oil field of the Zhujiang Mouth Basin. These are listed in Table 3.19 and illustrated on Plate XXXVI.

The spores and pollen of the Zhuhai oil field of the Zhujiang Mouth Basin (Table 3.19) are Tertiary index species or common species in Europe, Asia, and North America. Among the spores and pollen, *Leiotriletes adriensis, Polypodiaceaesporites haardti, Polypodiisporites favus, P. afavus,* and *Polypodiaceoisporites vitiosus* were found in the Tertiary System of Europe (Thomson and Pflug 1953; Potonié 1956; Krutzsch 1959). *Polypodiaceaesporites gracilis* was found in the Paleocene Series in Montana, USA (Wilson and Webster 1946) and the Eocene to Oligocene Series in the coastal region of Bohai, China (Research Institute of Petroleum Exploration and Development, Ministry of Petroleum Chemistry Industry, Nanjing Institute of Geology and Palaeontology, Chinese Academy of Sciences 1978a). *Pinuspollenites strobipites, Cedripites eocenicus, Salixipollenites discoloripites,* and *Momipites coryloides* were found in the Eocene Series of the USA (Wodehouse 1933). *Alnipollenites verus* and *Cupuliferoipollenites pusillus* were found in the Eocene Series of Germany (Potonié 1960). *Pinuspollenites minutus* was found in the Oligocene Series in the Irtysh Basin, Russia (Zaklinskaya 1957). *Abietineaepollenites microalatus minor, Pinuspollenites labdacus minor, Piceaepollenites alatus, Caryapollenites simplex, Juglanspollenites verus, Quercoidites microhenrici, Liquidambarpollenites stigmosus,* and *Ilexpollenites margaritatus* were found in the Miocene Series of Europe (Potonié 1958, 1960). *Cedripites cedroides* and *Monocolpopollenites tranguillus* were found in the Eocene Series of Germany (Thomson and Pflug 1953). *Osmundacidites primarius* and *Ulmipollenites undulosus* were found in the Pliocene Series of Germany (Wolff 1934). *Salixipollenites hian* and *Tricolporopollenites minutus* were found in the Paleocene Series and Eocene Series in Jiangxi, China (He and Sun 1977). The spores and pollen are widely

Table 3.19 Spores and pollen in crude oils from the Zhuhai oil field

Spores and pollen	Number of specimens	Geological age
Pteridophyta		
Leiotriletes adriensis (Potonié et Gelletich) Krutzsch, 1959	3	E-N
Osmundacidites primarius (Wolff) Sun et Li, 1981	5	E-N
Polypodiaceaesporites haardti (Potonié et Venitz) Thiergart, 1938	4	E-N
P. gracilis (Wilson et Webster) Ke et Shi, 1978	5	E
Polypodiisporites favus (Potonié) Potonié, 1956	3	E-N
P. afavus (Krutzsch) Sun et Li, 1981	2	E-N
Polypodiaceoisporites vitiosus Krutzsch, 1959	4	E-N
Gymnospermae		
Abietineaepollenites microalatus minor (Potonié) Potonié, 1958	5	E-N
Pinuspollenites labdacus minor (Potonié) Potonié, 1958	6	E-N
P. minutus (Zaklinskaya) Song et Zheng, 1978	4	E
P. strobipites (Wodehouse) Sun et Li, 1981	6	E
Piceaepollenites alatus (Potonié) Potonié, 1958	4	E-N
Cedripites eocenicus Wodehouse, 1933	5	E
C. cedroides (Thomson et Pflug) Sun et Li, 1981	4	E
Angiospermae		
Salixipollenites discoloripites (Wodehouse) Srivastava, 1966	3	E
S. hians (Elsik) Sun et Li, 1981	4	E-N
Caryapollenites simplex (Potonié) Raatz, 1937	5	E-N
Juglanspollenites verus Raatz, 1937	4	E-N
Alnipollenites verus (Potonié) Potonié, 1934	3	E
Momipites coryloides Wodehouse, 1933	5	E
Quercoidites microhenrici (Potonié) Potonié, 1950	6	E-N
Cupuliferoipollenites pusillus (Potonié) Potonié, 1951	7	E
Ulmipollenites undulosus Wolff, 1934	3	E-N
Liquidambarpollenites stigmosus (Potonié) Raatz, 1937	4	E-N
Ilexpollenites margaritatus (Potonié) Raatz, 1937	5	E-N
I. membranous Sun, Kong et Li, 1980	4	E-N
Tricolpites tenuicoplus Sun, Kong et Li, 1980	5	E
Operculumpollis operculatus Sun, Kong et Li, 1980	4	E
Tricolporopollenites minutus He et Sun, 1977	4	E
Monocolpopollenites tranguillus (Potonié) Thomson et Pflug, 1953	5	E
Total	131	

distributed in the Paleogene System in the North Shelf of the South China Sea (Sun et al. 1981, 1982; Hou et al. 1981; Lei 1985). In addition, *Ilexpollenites membranous, Operculumpollis operculatus,* and *Tricolpites tenuicolpus* are new species found in the Eocene to Oligocene Liushagang Formation in the Beibu Gulf Depression and the Leizhou Peninsula of the North Shelf of the South China Sea (Sun et al. 1980; Hou et al. 1981).

References

Artüz, S. (1957). Die Sporae dispersae der turkischen Steinkohle von Zonguldak-Gebiet (Mit besonderer Beachtung der neuen Arten und Gener). *Istanbul Üniversitesi Fen Fakültesi mecmuasi, Series B, 22*(4), 239–263.

Balme, B. E. (1952). On some spore specimens from British Upper Carboniferous coals. *Geological Magazine, 89*, 175–184.

Balme, B. E. (1957). Spores and pollen grains from the Mesozoic of western Australia. *Commonwealth Scientific and Industrial Research Organization, Coal Research Section, Technical Communication, 25*, 1–50.

Balme, B. E. (1963). Plant microfossils from the Lower Triassic of western Australia. *Palaeontology, 6*, 12–40.

Balme, B. E. (1970). Palynology of Permian and Triassic strata in the Salt Range and Surghar Range, West Pakistan. In B. Kummel & C. Teichert (Eds.), *Stratigraphic boundary problems: Permian and Triassic of West Pakistan*. Kansas University Special Publication (Vol. 4, pp. 306–453).

Batten, D. J. (1973). Use of palynologic assemblage-types in Wealden correlation. *Palaeontology, 16*(1), 1–40.

Bolchovitina, N. A. (1953). Spores and pollen characteristic of Cretaceous deposits of central regions of USSR. *Transactions of the Institute of Geology, Academy of Sciences, USSR, 61*, 1–184. (in Russian).

Bolchovitina, N. A. (1956). Atlas of spores and pollen from the Jurassic and Lower Cretaceous deposits of the Vilyui depression. *Transactions of the Institute of Geology, Academy of Sciences, USSR, 2*, 1–188. (in Russian).

Chengdu Institute of Geology and Mineral Resources, & Ministry of Geology and Mineral Resources. (1983). *Paleontological Atlas of Southwest China. Volume of Microfossils* (pp. 1–791). Beijing: Geological Publishing House. (in Chinese).

Chiang, T. C., & Young, H. C. (1978). Early Cretaceous palynological assemblage of the Huahai Basin, western Kansu. *Journal of Lanchow University (Natural Sciences), 2*, 115–135. (in Chinese with English abstract).

Clarke, R. F. A. (1965a). Keuper miospores from Worcestershire, England. *Palaeontology, 8*(2), 294–321.

Clarke, R. F. A. (1965b). British Permian saccate and monosulcate miospores. *Palaeontology, 8*(2), 322–354.

Couper, R. A. (1953). Upper Mesozoic and Cainozoic spores and pollen grains from New Zealand. *New Zealand Geological Survey Palaeontological Bulletin, 22*, 1–77.

Couper, R. A. (1958). British Mesozoic microspores and pollen grains. A systematic and stratigraphic study. *Palaeontographica B, 103*, 75–179.

De Jersey, N. J. (1959). Jurassic spores and pollen grains from the Rosewood coalfield. *Queensland Government Mining Journal, 60*, 344–366.

De Jersey, N. J. (1962). Triassic spores and pollen grains from the Ipswich coalfield. *Geological Survey of Queensland Publication, 307*, 1–18.

De Jersey, N. J. (1963). Jurassic spores and pollen grains from the Marburg sandstone. *Geological Survey of Queensland Publication, 313*, 1–13.

Dettmann, M. E. (1963). Upper Mesozoic microfloras from south-eastern Australia. *Proceedings of the Royal Society of Victoria, 77*(1), 1–148.

Du, B. A. (1986). Sporo-pollen assemblage from Shansi Formation of Pingliang, Gansu and its geological age. *Acta Palaeontologica Sinica, 25*(3), 284–295. (in Chinese with English abstract).

Filatoff, J. (1975). Jurassic palynology of the Perth Basin, western Australia. *Palaeontographica B, 154*(1–4), 1–113.

Gao, L. D. (1987). Carboniferous Namurian spores and stratigraphic boundary in Jingyuan, Gansu of China. *Bulletin of the Institute of Geology, Chinese Academy of Geolological Science, 16*, 193–226. (in Chinese).

Gao, L. D. (1994). Carboniferous and Early Permian spore zones and boundary between Carboniferous and Permian in Ningxia of China. *Gansu Geology, 3*(1), 11–25. (in Chinese).

Geng, G. C. (1985a). Microfossil assemblages from Upper Carboniferous in Ordos Basin, Northwest China. *Acta Botanica Sinica, 27*(2), 208–216. (in Chinese with English abstract).

Geng, G. C. (1985b). Microfossil assemblage from the Late Middle Carboniferous in western Shaan-Gan-Ning Basin, Northwest China. *Acta Botanica Sinica, 27*(6), 652–660. (in Chinese with English abstract).

Guy, D. J. E. (1971). Palynological investigations in the Middle Jurassic of the Vilhelmsfalt boring, southern Sweden. *Publications of the Institut of the Mineralogy and Palaeontology and Geological, University of Lund, Sweden, 168*, 1–104.

Hart, G. F. (1964). A review of the classification and distribution of the Permian miospores: Disaccate Striatiti. *Comptes Rendus 5th Congrés International de la Stratigraphie et Géologie du Carbonifere, Paris, 1963*, 1171–1199.

Hart, G. F. (1965). *The systematics and distribution of Permian miospores* (pp. 1–252). Johannesburg: Witwatersrand University Press.

He, Y. M., & Sun, X. J. (1977). Eogene spores and pollen from Qingjiang Basin, Jiangxi. I. II. *Acta Botanica Sinica, 19*(72–82), 237–243. (in Chinese with English abstract).

Hou, Y. T., Li, Y. P., & Jin, Q. H. (1981). *Tertiary palaeontology of North Continental Shelf of South China Sea* (pp. 1–274). Guangzhou: Guangdong Science and Technology Press. (in Chinese).

Hsü, J., Chiang, T. C., & Young, H. C. (1974). Sporo-pollen assemblage and geological age of the Lower Xinminbu Formation of Chiuchüan, Kansu. *Acta Botanica Sinica, 16*(4), 365–379. (in Chinese with English abstract).

Huang, P. (1995). Early—Middle Jurassic sporopollen assemblages from Dananhu coalfield of Tuha Basin, Xinjiang and their stratigraphical significance. *Acta Palaeontologica Sinica, 34*(2), 171–193. (in Chinese with English abstract).

Institute of Geology, Chinese Academy of Geological Sciences. (1980). *Mesozoic stratigraphic paleontology of the Shaanxi-Gansu-Ningxia Basin* (pp. 1–230). Beijing: Geological Publishing House. (in Chinese).

Institute of Geology, Chinese Academy of Geological Sciences, & Institute of Geology, Xinjiang Bureau of Geology and Mineral Resources. (1986). *Permian and Triassic strata and fossil assemblages in the Dalongkou area of Jimsar* (pp. 1–262). Beijing: Geological Publishing House.

Ishchenko, A. M. (1956). Spores and pollen of the Lower Carboniferous deposits of the western extension of the Donets Basin and their stratigraphic importance. *Izd Akademiya Nauk Ukrainskoy SSR, Stratigrafii i paleontologii, Series, 11*, 1–187. (in Russian).

Jansonius, J. (1962). Palynology of Permian and Triassic sediments, Peace River area, western Canada. *Palaeontographica B, 110*, 35–98.

Jiang, D. X. (1990). Palynological evidence for identification of nonmarine petroleum source rocks, China. *Ore Geology Reviews, 5*, 553–575.

Jiang, D. X., & Wang, Y. D. (2002a). Middle Jurassic sporo-pollen assemblage form the Yanan Formation of Dongsheng, Nei Monggol, China. *Acta Botanica Sinica, 44*(2), 230–238.

Jiang, D. X., & Wang, Y. D. (2002b). Middle Jurassic palynoflora and its environmental significance of Dongsheng, Inner Mongolia. *Acta Sedimentologica Sinica, 20*(1), 47–54. (in Chinese with English abstract).

Jiang, D. X., Yang, H. Q., & Du, J. E. (1974). Method of extraction of spores/pollen from crude oils. *Journal of Botany, 1*(1), 31–32. (in Chinese).

Jiang, D. X., He, Z. S., & Dong, K. L. (1988). Early Cretaceous palynofloras from Tarim Basin, Xinjiang. *Acta Botanica Sinica, 30*(4), 430–440. (in Chinese with English abstract).

Jiang, D. X., Wang, Y. D., He, Z. S., Dong, K. L., Ni, Q., & Tian, N. (2006). Early Cretaceous palynofloras from the Kizilsu Group in the Tarim Basin, Xinjiang. *Acta Micropalaeontologica Sinica, 23*(4), 371–391. (in Chinese with English abstract).

Jiang, D. X., Wang, Y. D., He, Z. S., & Dong, K. L. (2007). Early Cretaceous Sporo-pollen assemblages from the Shushanhe Formation in Baicheng area of the Tarim Basin, Xinjiang. *Acta Micropalaeontologica Sinica, 24*(3), 247–260. (in Chinese with English abstract).

Jiang, D. X., Wang, Y. D., He, Z. S., & Dong, K. L. (2008). Middle Jurassic palynoflora from the Taerga Formation in the Tarim Basin, Xinjiang and its bearings on stratigraphy and palaeogeography. *Acta Micropalaeontologica Sinica, 25*(4), 333–344. (in Chinese with English abstract).

Kedves, M. (1981). Etudes palynologiques sur les sediments Préquaternaires de l'Egypte. Neogene I. *Grana, 20*, 119–130.

Klaus, W. (1955). Uber die Sporendiagnose des deutschen Zechsteinsalzes und des alpinen Salzgebirges. *Zeitschrift der Deutschen Geologischen Gesellschaft, 105*, 776–788.

Klaus, W. (1960). Sporen der Karnischen Stufe der ostalpinen Trias. *Jahrbuch, Geologisches Bundesanst, 5*, 107–184. (Austria).

Kosanke, R. M. (1950). Pennsylvanian spores of Illinois and their use in correlation. *Illinois Geological Survey Bulletin, 74*, 1–128.

Kremp, G. (1949). Pollenanalytische Untersuchung des Miozänen Braunkohlenlagers von Konin an der Warthe. *Palaeontographica, 90*, 53–93.

Krutzsch, W. (1959). Mikropalaeontologische (Sporenpalaeontologische) Untersuchungen in der Braunkohle des Geiseltales. *Beiheft zur Zeitschrift Geologie, 21–22*, 1–425.

Krutzsch, W. (1971). Atlas der mittel-und jungtertiaren dispersen Sporen und Pollen. Sowie der Mikroplanktonformen des nordlichen. Mitteleuropas. Lief. VI, Coniferenpollen. Volkseigener Betrieb Deutscher Verlag der Wissenschaften, Berlin, pp. 1–274.

Lei, Z. Q. (1985). Tertiary sporo-pollen assemblage of Zhujiangkou (Pearl River Mouth) Basin and its stratigraphical significance. *Acta Botanica Sinica, 27*(1), 94–105. (in Chinese with English abstract).

Leschik, G. (1955). Die Keuper Flora von Neuewelt bei Basel. II. Die Iso-und Mikrosporen. *Schweizerische Paläontologische Abhandlungen, 72*, 1–70.

Leschik, G. (1956). Sporen aus dem Salzton des Zechsteins von Neuhof (bei Fulda). *Palaeontographica B, 100*, 122–142.

Li, M. Y., Sung, T. C., & Li, Z. P. (1978). Some Cretaceous—Tertiary palynological assemblages from the Yangtze-Han River Plain. *Memoirs of Nanjing Institute of Geology and Palaeontogy. Academia Sinica, 9*, 1–44. (in Chinese with English abstract).

Li, W. B., & Shang, Y. K. (1980). Sporo-pollen assemblages from the Mesozoic coal series of western Hubei. *Acta Palaeontologica Sinica, 19*(3), 201–219. (in Chinese with English abstract).

Liu, Z. S. (1993). Jurassic Sporo-pollen assemblages from the Beishan Coal-field, Qitai, Xinjiang. *Acta Micropalaeontologica Sinica, 10*(1), 13–36. (in Chinese with English abstract).

Liu, Z. S. (2003). Triassic and Jurassic Sporo-pollen assemblages from the Kuqa Depression, Tarim Basin of Xinjiang, NW China. Palaeontologia Sinica, whole Number 190, New Series A, No. 14. Beijing: Science Press, pp. 1–244. (in Chinese with English summary).

Luber, A. A. (1955). Atlas of spores and pollen grains of the Paleozoic deposits of Kazakhstan. *Izd. Akademiya Nauk Kazakhstan SSR, Alma-Ata* 1–125. (in Russian).

Lund, J. J. (1977). Rhaetic to Lower Liassic Palynology of the onshore south-eastern North Sea Basin. *Geological Survey of Denmark, Publication, 2*(109), 6–128.

References

Lundblad, A. B. (1954). Liassic coal-mines of Skromberga (Province of Scania) Sweden. *Svenka Botanisk Tidsrift, 48*(2), 381–417.

Mädler, K. (1964). Bemerkenswerte Sporenformen aus dem Keuper und unterem Lias. *Fortschritte in der Geologie von Rheinland und Westfalen, 12*, 169–200.

Maljavkina, V. S. (1949). Identification of spores and pollen of the Jurassic and Cretaceous. *Trudy Vsesouuznyi Neftyanoi Nauchno-Issledovatel'skii Geologocheskii Institut, 33*, 1–138. (in Russian).

Nagy, E. (1969). Palynological elaborations of the Miocene layers of the Mecsek Mountains. *Annales Hungary Geological Institute, 52*(2), 235–648.

Nilsson, T. (1958). Über das Vorkommen eines mesozoischen Sapropelgesteins in Schonen. *Publications from the Institutes of Mineralogy, Palaeontology and Geology, University of Lund, Sweden, 54*, 5–112.

Norris, G. (1965). Triassic and Jurassic miospores and acritarchs from the Beacon and Ferrar Groups, Victoria Land, Antarctica. *New Zealand Journal of Geology and Geophysics, 8*, 236–277.

Norris, G. (1967). Spores and pollen from the Lower Colorado Group (Albian-? Cenomanian) of central Alberta. *Palaeontographica B, 120*, 72–115.

Ouyang, S. (1962). The microspore assemblage from the Lungtan Series of Changhsing, Chekiang. *Acta Palaeontologica Sinica, 10*(1), 76–119. (in Chinese with English summary).

Ouyang, S. (1964). A preliminary report on sporae dispersae from the Lower Shihhotze Series of Hokü district, NW Shansi. *Acta Palaeontologica Sinica, 12*(3), 486–519.

Ouyang, S. (1986). *Palynology of Upper Permian and Lower Triassic strata of Fuyuan district, eastern Yunnan. Palaeontologia Sinica* (Whole Number 169, New Series A, No 9, pp. 1–122). Beijing: Science Press. (in Chinese with English summary).

Ouyang, S., & Li, Z. P. (1980a). *Upper Carboniferous spores from Shuo Xian, northern Shanxi.* Paper for 5th International Palynological Conference, Nanjing Institute of Geology and Palaeontology, Academia Sinica, pp. 1–16.

Ouyang, S., & Li, Z. P. (1980b). Microflora from the Kayitou Formation in Fuyuan of Yunnan and its stratigraphic and paleobotanic significance. In Nanjing Institute of Geology and Palaeontology, Academia Sinica (Ed.), *Upper Permian coal-bearing strata and paleobiologic groups in western Guizhou and eastern Yunnan* (pp. 123–183). Beijing: Science Press. (in Chinese).

Ouyang, S., Wang, Z., Zhan, J. Z., & Zhou, Y. X. (2003). Palynology of the Carboniferous and Permian strata of northern Xinjiang, northwestern China (pp. 1–700). Hefei: University of Science and Technology of China Press. (in Chinese with English summary).

Owens, B. (1996). Upper Carboniferous spores and pollen. In J. Jansonius & D. C. McGregor (Eds.), *Palynology: principles and applications* (Vol. 2, pp. 597–606). Dallas, Texas: American Association of Stratigraphic Palynologists Foundation.

Playford, G. (1965). Plant microfossils from Triassic sediments near Poatina, Tasmania. *Journal of Geological Society of Australia, 12*(2), 173–210.

Playford, G., & Dettmann, M. E. (1965). Rhaeto-Liassic plant microfossils from the Leigh Creek coal measures, South Australia. *Senckenbergiana Lethaea, 46*, 127–181.

Pocock, S. A. J. (1962). Microfloral analysis and age determination of strata at the Jurassic-Cretaceous Boundary in the western Canada plains. *Palaeontographica B, 3*, 1–95.

Pocock, S. A. J. (1970). Palynology of the Jurassic sediments of western Canada. *Palaeontographica B, 130*, 12–136.

Potonié, R. (1956). Synopsis der Gattungen der Sporae dispersae. Teil I. Sporites. *Beihefte zum Geologischen Jahrbuch, 23*, 1–103.

Potonié, R. (1958). Synopsis der Gattungen der Sporae dispersae. Teil II. Sporites (Nachtrage), Saccites, Aletes, Praecolpates, Polyplicates, Monocolpates. *Beihefte zum Geologischen Jahrbuch, 31*, 1–114.

Potonié, R. (1960). Synopsis der Gattungen der Sporae dispersae. Teil III. Nachtrage Sporites. *Beihefte zum Geologischen Jahrbuch, 39*, 1–189.

Potonié, R., & Klaus, W. (1954). Einige Sporengattung des alpinen Salzgebirges. *Geologisches Jahrbuch, 68*, 517–546.

Potonié, R., & Kremp, G. (1955). Die Sporae dispersae des Ruhrkarbons. Teil I. *Palaeontographica B, 98*, 1–136.

Potonié, R., & Kremp, G. (1956). Die Sporae dispersae des Ruhrkarbons. Teil II. *Palaeontographica B, 99*, 85–191.

Potonié, R., Thomson, P. W., & Thiergart, F. (1950). Zur Nomenklatur der Neogenen Sporomorphae (Pollen und Sporen). *Geologisches Jahrbuch, 65*, 35–70.

Qu, L. F. (1982). The palynological assemblage from the Liujiagou Formation of Jiaocheng, Shanxi. *Bulletin of Geological Institute, Chinese Academy of Geological and Science, 4*, 83–93. (in Chinese with English abstract).

Research Institute of Exploration and Development, Qinghai Petroleum Administration, Nanjing Institute of Geology and Palaeontology, Academia Sinica. (1985). *A research on Tertiary palynology from the Qaidam Basin, Qinghai Province* (pp. 1–297). Beijing: Petroleum Industry Press. (in Chinese with English abstract).

Research Institute of Petroleum Exploration and Development, Ministry of Petroleum Chemistry Industry, Nanjing Institute of Geology and Palaeontology, Chinese Academy of Sciences. (1978a). *On the Paleogene Spores and Pollen from the Coastal Region of Bohai* (pp. 1–177). Beijing: Science Press. (in Chinese with English abstract).

Research Institute of Petroleum Exploration and Development, Ministry of Petroleum Chemistry Industry, Nanjing Institute of Geology and Palaeontology, Chinese Academy of Sciences. (1978b). *On the Paleogene Dinoflagellates and Acritarchs from the*

Coastal Region of Bohai (pp. 1–190). Beijing: Science Press. (in Chinese with English abstract).

Rouse, G. E. (1959). Plant microfossils from Kootenay coal-measures strata of British Columbia. *Micropaleontology, 5*(3), 303–324.

Samoilovich, S. R. (1953). Pollen and spores from the Permian deposits of the Cherdyn and Aktyubinsk areas, Cis-Urals. *Trudy VNIGRI, N. S., 75,* 5–57. (in Russian).

Schemel, M. P. (1951). Small spores of the Mystic coal of Iowa. *American Midland Nature, 46,* 743–750.

Schopf, J. M., Wilson, L. R., & Bentall, R. (1944). An annotated synopsis of Palaeozoic fossil spores and the definition of generic groups. *Illinois Geological Survey Report of Investigations, 91,* 1–77.

Schulz, E. (1967). Sporenpalaontologische Untersuchungen rätoliassischer Schichten im Zentralteil des Germanischen Beckens. *Paläontologische Abhandlungen B, 2*(3), 427–633.

Shang, Y. K. (1981). Early Jurassic sporo-pollen assemblages in southwestern Hunan, northeastern Guangxi. *Acta Palaeontologica Sinica, 20*(5), 428–440.

Singh, C. (1964a). Microflora of the Lower Cretaceous Mannville Group, east-central Alberta. *Research Council of Alberta Bulletin, 15,* 1–238.

Singh, H. P. (1964b). A miospore assemblage from the Permian of Iraq. *Palaeontology, 7,* 240–265.

Smith, A. H. V., & Butterworth, M. A. (1967). Miospores in the coal seams of the Carboniferous of Great Britain. *Special Papers in Palaeontology, 1,* 1–324.

Song, Z. C., Zheng, Y. H., Liu, J. L., Ye, P. Y., Wang, C. F., & Zhou, S. F. (1981). *Cretaceous—Tertiary palynological assemblages from Jiangsu* (pp. 1–268). Beijing: Geological Publishing House. (in Chinese with English abstract).

Song, Z. C., Guan, X. T., Zheng, Y. H., Li, Z. R., Wang, W. M., & Hu, Z. H. (1985). *A research on Cenozoic palynology of the Longjing structural area in the Shelf Basin of the East China Sea (Donghai) Region* (pp. 1–209). Hefei: Anhui Science and Technology Publishing House. (in Chinese with English summary).

Staplin, F. L. (1960). Upper Mississippian plant spores from the Golata Formation, Alberta, Canada. *Palaeontographica B, 107,* 1–40.

Sun, F. (1989). Early and Middle Jurassic sporo-pollen assemblages of Qiquanhu coalfield of Turpan, Xinjiang. *Acta Botanica Sinica, 31*(8), 638–646. (in Chinese with English abstract).

Sun, X. J., Kong, Z. C., & Li, M. X. (1980). Paleogene new pollen genera and species of South China Sea. *Acta Botanica Sinica, 22*(2), 191–197. (in Chinese with English abstract).

Sun, X. J., Kong, Z. C., Li, P., & Li, M. X. (1981). Oligocene palynoflora in the northern part of South China Sea. *Acta Phytotaxonomica Sinica, 19*(2), 186–194. (in Chinese with English abstract).

Sun, X. J., Kong, Z. C., Li, M. X., & Li, P. (1982). Palynoflora of the Liushagang Formation (Eocene–Early Oligocene) in the northern part of South China Sea. *Acta Phytotaxonomica Sinica, 20*(1), 63–72. (in Chinese with English abstract).

Sung, T. C. (1958). Tertiary spore and pollen complexes from the red beds of Chiuchuan, Kansu and their geological and botanical significance. *Acta Palaeontologica Sinica, 6*(2), 159–167. (in Chinese with English summary).

Sung, T. C., & Tsao, L. (1976). The Paleocene spores and pollen grains from the Fushun coalfield, Northeast China. *Acta Palaeontologica Sinica, 15*(2), 147–162. (in Chinese with English abstract).

Thomson, P. W., & Pflug, H. (1953). Pollen und Sporen des mitteleuropaischen Tertiars. *Palaeontographica B, 94,* 1–138.

Tralau, H. (1968). Botanical investigation into the fossil flora of Eriksdal in Fyledalen, Scania. II. The middle Jurassic microflora. *Sweden Geological Survey Bulletin, 633,* 1–132.

Venkatachala, B. S., Beju, D., & Kar, R. K. (1968). Palynological evidence on the presence of Lower Triassic in the Danubean (Moesian) Platform, Rumania. *Palaeobotanist, 16,* 29–37.

Wagner, C. A., & Taylor, T. N. (1982). Fungal Chlamydospores from the Pennsylvanian of North America. *Review of Palaeobotany and Palynology, 37,* 317–328.

Wang, H. (1984). Middle and Upper Carboniferous sporo-pollen assemblages from Hengshanbu Ningxia. *Acta Palaeontologica Sinica, 23*(1), 91–106.

Wang, H. (1989). Early Permian palynofloras from Qipan-Duwa areas, Tarim Basin, Xinjiang and their paleoenvironment. *Acta Palaeontologica Sinica, 28*(3), 402–414. (in Chinese with English abstract).

Wang, Y. D., Jiang, D. X., Yang, H. Q., & Sun, F. (1998). Middle Jurassic sporo-pollen assemblages from Turpan-Shanshan area, Xinjiang. *Acta Botanica Sinica, 40*(10), 969–976. (in Chinese with English abstract).

Wilson, L. R., & Coe, E. A. (1940). Description of some unassigned plant microfossils from the Des Moines Series of Iowa. *American Midland Nature, 23,* 182–186.

Wilson, L. R., & Webster, R. M. (1946). Plant microfossils from a Fort Union Coal of Montana. *American Journal of Botany, 33*(4), 271–278.

Wodehouse, R. P. (1933). Tertiary pollen II. The oil shales of the Eocene Green River Formation. *Torrey Botany Club, Bulletin, 60,* 479–524.

Wolff, H. (1934). Mikrofossilien des pliozaenen Humodils der Grube Freigericht bei Dettingen A. M. und Vergleich mit aelteren Schichten des Tertiaers sowie posttertiaeren Ablagerungen. *Arb. Inst. Palaeobot. U. Petrogr. Brennst. Preuss. Geol. L-A. Berlin, 5,* 55–86.

Zaklinskaja, E. D. (1957). Stratigraphic significance of pollen grains of Gymnosperms of the Cenozoic deposits of the Irtysh Basin and the northern Aral Basin. *Transactions of the Institute of Geology, Academy of Sciences, USSR, 6,* 1–184. (in Russian).

References

Zhang, L. J. (1965). Palynological assemblages from Yima coal-bearing rock formation and their significance in Mianchi, Henan Province. *Acta Palaeontologica Sinica, 13*(1), 160–196. (in Chinese with Russian abstract).

Zhang, L. J. (1984). *Late Triassic spores and pollen from central Sichuan. Palaeontologia Sinica* (whole Number 167, New Series A, No. 8, pp. 1–100). Beijing: Science Press. (in Chinese with English abstract).

Zhang, W. P. (1990). Jurassic sporo-pollen assemblages in Junggar Basin of Xinjiang. In Institute of Geology, Chinese Academy of Geological Sciences, and Research Institute of Petroleum Exploration and Development, Xinjiang Petroleum Administration (Ed.), *Permian to Tertiary strata and palynological assemblages in the North of Xinjiang* (pp. 57–121). Beijing: China Environmental Science Press. (in Chinese).

Zhang, Y. Y. (1981). Tertiary spores and pollen grains from the Leizhou Peninsula. *Acta Palaeontologica Sinica, 20*(5), 449–458. (in Chinese with English abstract).

Zhou, H. Y. (1987). Late Paleozoic sporo-pollen complexes of northern Shandong. In Institute of Geology of Shengli oil-field (Ed.), *Collected papers of stratigraphic paleontology in petroliferous regions of China* (pp. 1–17). Beijing: Petroleum Industry Press. (in Chinese).

Zhu, H. C. (1997). Permian spore and pollen assemblages from the Tarim Basin and their biostratigraphical significance. *Acta Palaeontologica Sinica, 36*, 38–64. (in Chinese with English summary).

Petroleum Sporo-pollen Assemblages and Petroleum Source Rocks

4

Abstract

Based on the investigations of sporo-pollen assemblages in individual petroleum reservoir beds, it is possible to predict the petroleum source rocks. Such investigation is applied to the basins in China selected for study. In the Tarim Basin, Triassic and Jurassic sequences should be the main petroleum source rocks, while Carboniferous sequences should have been the source of some petroleum, and the Permian and Cretaceous sequences might contain petroleum source rocks. Ten formations or groups are identifiable for the North Tarim petroliferous region, four formations for the Kuqa Depression, and three for the Kashi and Yecheng Sags. In the East Junggar Depression, four Permian formations are identified as the main source rocks, while five Triassic formations are secondary in importance. Carboniferous and Jurassic rocks also have some potential. In the South Junggar Depression, four Jurassic formations and one Oligocene formation should be the main source rocks. In the Qiktim oil field, only a Jurassic monotype sporo-pollen assemblage is found. In the Turpan Depression of the Turpan-Hami Basin, four Jurassic formations are implicated. In the Qaidam Basin, two Jurassic formations are implicated in the North Border Block-fault Zone. In the Mangnai Depression, Paleocene to Eocene and Oligocene to Miocene formations are identified as the main petroleum source rocks. In the West Jiuquan Basin, the black shales of the Lower Cretaceous Lower Xinminbu Formation should be the petroleum source rocks. In the Liaohe Basin, two Oligocene formations are implicated. In the Beibu Gulf Depression of the western North Shelf of South China Sea, Eocene and Oligocene formations are predicted as the source rocks. In the Zhujiang Mouth Depression of the eastern North Shelf of South China Sea, three formations of Eocene and Oligocene ages are implicated.

Keywords

Petroleum sporo-pollen assemblage · Petroleum source bed · Carrier bed · Reservoir bed

4.1 Definition and Classification of Petroleum Sporo-pollen Assemblages

4.1.1 Definition

A petroleum sporo-pollen assemblage is a random assemblage of fossil spores and pollen derived from three potential sources: the petroleum source bed, the carrier bed, and the reservoir bed itself. Evidence from all three sources can be found in crude oil collected from a reservoir (Jiang and Yang 1980, 1982).

The assemblage from the source bed is generally limited stratigraphically and is composed of spores and pollen of the original plants growing in that environment that is restricted in latitude. The assemblage in the carrier bed and in the reservoir can vary widely by age, environmental conditions, and latitude. So a stratigraphic sporo-pollen complex can provide a scientific basis for determination of stratigraphic ages; and a petroleum sporo-pollen assemblage can provide scientific basis for judging of petroleum source rocks. These can be significantly different.

Thus, the petroleum sporo-pollen assemblage of a reservoir bed is a gathering of dispersed spores and pollen of different geological ages or stratigraphic positions. This includes spores and pollen of the reservoir bed plus spores and pollen brought from source bed and carrier bed in the course of petroleum migration.

4.1.2 Classification and Character

The composition of the petroleum sporo-pollen assemblages may be divided into three types. They are mono-type, mixed-type, and special-type petroleum sporo-pollen assemblages.

The mono-type petroleum sporo-pollen assemblage only contains the spores and pollen of the reservoir bed. The character of this type is that the composition of the assemblage belongs to only one geological age or stratigraphic position. Examples of the mono-type are the petroleum sporo-pollen assemblages of the Jurassic reservoir bed of the Qigu oil field in the Junggar Basin; the Jurassic reservoir bed of the Qiktim oil field in the Turpan Basin; and the Oligocene reservoir bed of the Xinglongtai oil field in the Liaohe Basin.

The mixed-type petroleum sporo-pollen assemblages contain the spores and pollen belonging to two or several geological ages or stratigraphic positions. The character of this type is that the composition of assemblage includes the spores and pollen of the reservoir bed as well as the earlier or later reservoir bed. Examples of the mixed-type are the petroleum sporo-pollen assemblages of the Neogene reservoir bed of the Kelatu oil field in the Tarim Basin; the Neogene reservoir bed of the Kekeya oil field also in the Tarim Basin; and the Neogene reservoir bed of the Laojunmiao oil field in the Jiuquan Basin.

The special-type petroleum sporo-pollen assemblages contain the spores and pollen derived from source bed and carrier bed, but lack fossils derived from the reservoir bed. The character of this type is that the composition of assemblage contains only the spores and pollen younger than reservoir bed. Examples of the special-type are the petroleum sporo-pollen assemblages of the Carboniferous igneous rock reservoir bed of the Beisantai oil field in the Junggar Basin; the Silurian metamorphic rock reservoir bed of the Yaerxia oil field in the Jiuquan Basin; and the Ordovician dolomite reservoir bed of the Yakela oil field in the Tarim Basin.

The special-type petroleum sporo-pollen assemblage reflects specific petroleum pools, such as igneous rocks, metamorphic rocks, or carbonate rocks. Because these do not contain naturally occurring spores and pollen, and they have fissures or caves that may concentrate petroleum, spores, and pollen in petroleum of such reservoirs obviously were brought in during the course of petroleum migration, they are indicative for source bed and carrier bed floras.

4.2 Tarim Basin

4.2.1 North Tarim Upheaval

One hundred and thirty-six species of fossil spores and pollen referred to 66 genera were found in crude oil samples from Ordovician, Carboniferous, Triassic, Jurassic, Cretaceous, and Tertiary petroleum reservoirs in the petroliferous region of the North Tarim Upheaval (Table 3.1). The sporo-pollen assemblages in the petroleum may be used as reliable evidence for judging the ages of the petroleum source rocks.

4.2.1.1 Petroleum Sporo-pollen Assemblages

The petroleum sporo-pollen assemblage of the Ordovician reservoir bed of the Yakela oil field and the Lunnan oil field is composed of 85 species of Carboniferous to Jurassic spores and pollen. These are all later in age than the reservoir bed itself. They represent fossils carried from the petroleum source bed and carrier bed. As such, the fossils in the Ordovician carbonate rock-cave reservoir represent the special-type of petroleum sporo-pollen assemblage. Carboniferous index species such as *Punctatosporites minutus*, *Vesicaspora wilsonii*, and *Calamospora pedata* are typical of Carboniferous System rocks in this region, so they indicate a Carboniferous petroleum source rock. Triassic index species such as *Punctatisporites triassicus*, *P. ambiguus*, *Retusotriletes mesozoicus*, *Verrucosisporites contactus*, *Osmundacidites alpinus*, *Apiculatisporis parvispinosus*, *Apiculatisporis spiniger*, *Conbaculatisporites mesozoicus*, *Lycopodiacidites kuepperi*, *Lycopodiacidites rhaeticus*, *Tigrisporites halleinis*, *Zebrasporites kahleri*, *Limatulasporites dalongkouensis*, *L. parvus*, *Aratrisporites fischeri*, *A. coryliseminis*, *A. scabratus*, *A. granulatus*, *A. strigosus*, *A. paenulatus*, *Enzonalasporites tenuis*, *Lueckisporites triassicus*, *Chordasporites singulichorda*, *Alisporites parvus*, *Parcisporites rarus*, and *Minutosaccus parcus* are typical of Triassic System rocks, especially those of the Middle and Upper Triassic series in this region, thereby indicating a Triassic petroleum source rock. Jurassic index species such as *Dictyophyllidites harrisii*, *Undulatisporites pflugii*, *Marattisporites scabratus*, *Piceites expositus*, *Cedripites minor*, *Podocarpidites multicinus*, and *Callialasporites minus* are widely distributed in the Jurassic System in this region, thereby indicating a Jurassic petroleum source

rock. In addition, Carboniferous to Permian species such as *Granulatisporites adnatoides*, and *Cordaitina uralensis* are found in Permian System rocks in the region. Jurassic to Cretaceous species such as *Cyathidites australis*, *C. minor*, *Osmundacidites wellmanii*, *Cycadopites nitidus*, *C. typicus*, and *C. minimus* are widely distributed in the Lower Cretaceous series in the region. Therefore, it is possible that there are the Permian and Lower Cretaceous petroleum source rocks in this region.

The petroleum sporo-pollen assemblage of the Carboniferous reservoir bed of the Donghetang oil field contains only Carboniferous species. These include *Lycospora pusilla*, *Retispora florida*, and *Punctatosporites minutus*, so the fossils are a mono-type petroleum sporo-pollen assemblage, indicating a Carboniferous petroleum source rock.

The petroleum sporo-pollen assemblage of the Triassic reservoir bed of the Yakela and Lunnan oil fields is composed of 99 species of Carboniferous to Jurassic spores and pollen. This is a complicated mixed-type petroleum sporo-pollen assemblage. With the exception of the Triassic spores and pollen, the rest of the assemblage indicates source bed and carrier bed. The Carboniferous index species *Triquitrites desperatus* and *T. subrotundus* are found in the Carboniferous System in this region; their presence supports a Carboniferous petroleum source rock. The Jurassic index species such as *Deltoidospora perpusilla*, *Dictyophyllidites harrisii*, *Gleicheniidites rouseii*, *G. nilssonii*, *G. conflexus*, *Undulatisporites pflugii*, *Cibotiumspora paradoxa*, *Granulatisporites jurassicus*, *G. minor*, *Apiculatisporis variabilis*, *Lycopodiumsporites paniculatoides*, *L. subrotundum*, *Duplexisporites scanicus*, *Vitreisporites itunensis*, *Piceites expositus*, *P. pseudorotundiformis*, *Cedripites minor*, *Podocarpidites multicinus*, *Callialasporites minus*, *Bennettiteaepollenites lucifer*, and *Classopollis qiyangensis* are distributed in the Jurassic System, especially the Lower and Middle Jurassic series in this region; they may indicate a Lower to Middle Jurassic petroleum source rock. In addition, Permian to Triassic species such as *Lueckisporites virkkiae*, *L. tattooensis*,

Taeniaesporites pellucidus, Protohaploxypinus microcorpus, P. samoilovichii, and *Striatoabietites duivenii* are all found in the Permian System in this region; Jurassic to Cretaceous species such as *Cyathidites australis, C. minor, Osmundacidites wellmanii, Piceites latens, Podocarpidites multesimus, Cycadopites nitidus,* and *C. typicus* are widely distributed in the Lower Cretaceous series in this region. Therefore, it is possible that there are Permian and Lower Cretaceous petroleum source rocks in the region.

The petroleum sporo-pollen assemblage of the Jurassic reservoir bed of the Yakela and Lunnan oil fields contains 43 species of Triassic to Jurassic spores and pollen. This is a mixed-type petroleum sporo-pollen assemblage. With the exception of the Jurassic spores and pollen, the rest of the assemblages are useful to determine source bed and carrier bed. Triassic index species, such as *Punctatisporites triassicus, Lundbladispora nejburgii, Aratrisporites scabratus, A. strigosus, Alisporites australis,* and *A. parvus,* are widely distributed in the Triassic System of this region; they support a Triassic petroleum source rock.

The petroleum sporo-pollen assemblage of the Cretaceous reservoir bed of the Yakela oil field is composed of 33 species of Triassic to Cretaceous spores and pollen. This is a mixed-type petroleum sporo-pollen assemblage. With the exception of the Cretaceous spores and pollen, the rest of the assemblages are indicative for source bed and carrier bed. Triassic index species, such as *Retusotriletes mesozoicus, Osmundacidites alpinus, Lycopodiacidites rhaeticus, Aratrisporites scabratus, A. strigosus,* and *Colpectopollis pseudostriatus,* are widely distributed in the Triassic System of the region; they support a Triassic petroleum source rock. Jurassic index species, such as *Dictyophyllidites harrisii, Gleicheniidites rouseii, Undulatisporites pflugii, Cibotiumspora paradoxa, Granulatisporites jurassicus, Duplexisporites scanicus, Piceites pseudorotundiformis,* and *Cedripites minor,* are widely distributed in the Jurassic System of the region; they may indicate a Jurassic petroleum source rock.

The petroleum sporo-pollen assemblage of the Tertiary reservoir bed of the Yingmaili oil field is a simple mixed-type petroleum sporo-pollen assemblage. With the exception of Tertiary spores and pollen, the rest of the assemblages are Jurassic index species such as *Dictyophyllidites harrisii, Gleicheniidites conflexus, Cibotiumspora paradoxa, Piceites expositus,* and *Cedripites minor;* they may indicate a Jurassic petroleum source rock.

4.2.1.2 Analyses of Petroleum Source Rocks

The composition of the above six petroleum sporo-pollen assemblages indicates that there are Carboniferous, Triassic, and Jurassic petroleum source rocks, as well as possible Permian and Lower Cretaceous petroleum source rocks in the petroliferous region of the North Tarim Upheaval. The spores and pollen in the petroleum of this petroliferous region coincide with those in Carboniferous, Triassic, and Jurassic dark gray and black mudstones; they match in morphology, structure, size range, and especially burial color (Jiang and Yang 1999b). Therefore, these spores and pollen are derived from Carboniferous, Triassic, and Jurassic petroleum source rocks. In addition to the petroleum sporo-pollen assemblage, the correlations between palynomorphs in the crude oils and those in potential petroleum source rocks can be used to determine geological ages and strato-horizons of the actual petroleum source rocks.

It is possible to also quantify the relative contribution by age. The distribution of 53 species of Triassic spores and pollen extracted from crude oils in Triassic strata in the North Tarim Upheaval (Table 4.1) shows that the Triassic System made an important contribution to the source of petroleum. The distribution of 27 species of Jurassic spores and pollen extracted from crude oils in Jurassic strata in the North Tarim Upheaval (Table 4.2) shows that the Jurassic System contributed to the petroleum source. The results of correlations between spores and pollen in crude oils and those in potential source rocks (Tables 4.1 and 4.2) indicate that the black and dark gray mudstones of the Lower Triassic Ehuobulake Formation, the Middle Triassic Karamay Formation, the Upper Triassic Huangshanjie and Taliqike formations, the Lower

Table 4.1 Distribution of Triassic spores and pollen in crude oils from Triassic strata in the North Tarim Upheaval

Spores and pollen	Ehoubulake Formation T_1e	Karamay Formation T_2k	Huangshanjie Formation T_3h	Taliqike Formation T_3t
Spores				
Punctatisporites triassicus	+	+	+	+
P. ambiguous			+	+
Calamospora tener			+	+
Retusotriletes mesozoicus		+	+	+
R. arcticus	+	+		
Verrucosisporites contactus		+	+	+
V. remyanus		+	+	+
Osmundacidites alpinus			+	+
Lophotriletes corrugatus	+			
Apiculatisporis globosus		+	+	+
A. parvispinosus		+	+	+
A. spiniger	+	+	+	+
Conbaculatisporites mesozoicus		+	+	+
Lycopodiacidites kuepperi			+	+
L. rhaeticus			+	+
Tigrisporites halleinis		+	+	+
Zebrasporites kahleri			+	+
Camarozonosporites rudis			+	+
Multinodisporites junctus	+	+		
Limatulasporites parvus	+	+		
L. dalongkouensis	+			
Lundbladispora playfordi	+			
L. nejburgii	+	+		
L. subornata	+			
Aratrisporites fischeri		+	+	+
A. coryliseminis		+	+	+
A. scabratus	+	+	+	+
A. granulatus	+	+	+	+
A. strigosus		+	+	+
A. paenulatus		+	+	+
A. parvispinosus			+	+
A. paraspinosus			+	+
A. tenuispinosus		+	+	+

(continued)

Table 4.1 (continued)

Spores and pollen	Ehoubulake Formation T_1e	Karamay Formation T_2k	Huangshanjie Formation T_3h	Taliqike Formation T_3t
Pollen				
Enzonalasporites tenuis			+	+
Lueckisporites triassicus		+	+	+
Taeniaesporites rhaeticus			+	+
T. divisus	+	+		
Colpectopollis pseudostriatus		+	+	+
C. scitulus		+	+	+
Chordasporites singulichorda		+	+	+
C. orientalis	+			
C. impensus	+			
Alisporites australis		+	+	+
A. parvus		+	+	+
A. aequalis			+	+
A. fusiformis	+			
Platysaccus undulates	+			
Podocarpidites queenslandi		+	+	+
Parcisporites solutus			+	+
P. rarus	+	+		
Minutosaccus parcus		+	+	+
Cedripites parvisaccus	+	+		
Rugubivesiculites lepidus		+	+	

Jurassic Ahe and Yangxia formations, and the Middle Jurassic Kezilenuer and Qiakemake formations should be the main petroleum source rocks. The black shales of the Upper Carboniferous Bijingtawu Formation and the Carboniferous to Permian Muziduke Group also contributed as petroleum source rocks in the petroliferous region of the North Tarim Upheaval.

4.2.2 Kuqa Depression

Thirty-four species of fossil spores and pollen referred to 23 genera were found in crude oil samples from Jurassic, Cretaceous, and Tertiary petroleum reservoirs in the petroliferous region of the Kuqa Depression (Table 3.2). The petroleum sporo-pollen assemblages composed of these microfossils may be used as reliable evidence for judging the ages of the petroleum source rocks.

The petroleum sporo-pollen assemblage of the Jurassic reservoir bed of the Yiqikelike oil field is composed of 22 species of Jurassic spores and pollen. This is a mono-type petroleum sporo-pollen assemblage. Jurassic index species such as *Deltoidosporna perpusilla*, *D. gradata*, *Concavisporites toralis*, *Dictyophyllidites harrisii*, *Cibotiumspora paradoxa*, *Duplexisporites scanicus*, *Protopicea exilioides*, *Podocarpidites*

Table 4.2 Distribution of Jurassic spores and pollen in crude oils from Jurassic strata in the North Tarim Upheaval

Spores and pollen	Ahe Formation J_1a	Yangxia Formation J_1y	Kezilenuer Formation J_2k	Qiakemake Formation J_2q
Spores				
Deltoidospora perpusilla	+		+	+
Dictyophyllidites harrisii	+	+	+	+
Gleicheniidites rouseii			+	
G. nilssonii			+	+
G. conflexus			+	+
Undulatisporites pflugii			+	
Cibotiumspora paradoxa	+	+	+	+
Granulatisporites jurassicus		+	+	+
G. minor	+	+		
Apiculatisporis variabilis			+	
Lycopodiumsporites paniculatoides			+	+
L. subrotundum		+	+	+
Duplexisporites anagrammensis			+	+
D. scanicus			+	+
Marattisporites scabratus		+	+	+
Pollen				
Vitreisporites itunensis			+	+
Alisporites lowoodensis		+		
Piceites expositus			+	+
P. pseudorotundiformis	+	+	+	+
Cedripites minor		+	+	+
Podocarpidites multicinus	+	+	+	
P. rousei				+
P. unicus			+	+
Callialasporites dampieri			+	+
C. minus			+	+
Classopollis qiyangensis			+	+
Bennettiteaepollenites lucifer		+	+	+

multicinus, *Parvisaccites enigmatus*, *Abietineaepollenites dunrobinensis*, *Cedripites minor*, and *Cycadopites subgranulosus*, and the common Jurassic species such as *Cyathidites australis*, *C. minor*, *Osmundacidites wellmanii*, *Podocarpidites multesimus*, *Abietineaepollenites minimus*, *Cycadopites nitidus*, *C. typicus*, and *C. minimus* are distributed in the Jurassic System, especially the Lower and Middle Jurassic series, in the Kuqa Depression (Jiang and Yang 1983; Liu 2003); they may indicate a Lower to Middle Jurassic petroleum source rock.

The petroleum sporo-pollen assemblage of the Cretaceous reservoir bed of the Yiqikelike oil field is composed of 17 species of Jurassic and Cretaceous spores and pollen. This is a

Table 4.3 Distribution of Jurassic spores and pollen in crude oils from Jurassic strata in the Kuqa Depression

Spores and pollen	Ahe Formation J_1a	Yangxia Formation J_1y	Kezilenuer Formation J_2k	Qiakemake Formation J_2q
Spores				
Deltoidospora perpusilla		+	+	+
D. gradata		+	+	+
Cyathidites australis	+		+	+
C. minor	+	+	+	+
Concavisporites toralis	+			+
Dictyophyllidites harrisii	+	+		
Cibotiumspora paradoxa	+	+	+	+
Duplexisporites scanicus	+	+	+	
Pollen				
Protopicea exilioides		+	+	+
Abietineaepollenites dunrobinensis		+	+	+
A. minimus			+	+
Cedripites minor		+	+	+
Podocarpidites multicinus		+	+	+
P. multesimus	+		+	+
Parvisaccites enigmatus		+	+	
Classopollis annulatus		+	+	+
Cycadopites nitidus	+	+	+	+
C. typicus		+	+	+
C. minimus			+	+
C. subgranulosus		+		

mixed-type petroleum sporo-pollen assemblage. The Cretaceous spores and pollen *Hymenophyllumsporites deltoidus* and *Cedripites cretaceus* are derived from the reservoir bed. Jurassic spores and pollen such as *Deltoidospora perpusilla, Dictyophyllidites harrisii, Cibotiumspora paradoxa, Podocarpidites multicinus, Parvisaccites enigmatus, Cedripites minor,* and *Cycadopites subgranulosus* may indicate a Jurassic petroleum source rock.

The petroleum sporo-pollen assemblage of the Tertiary reservoir bed of the Kuqatawu structure at Kangcun and the Dongqiulitake structure at Jilishen is composed of 18 species of Jurassic and Tertiary spores and pollen. This is a mixed-type petroleum sporo-pollen assemblage. Tertiary spores and pollen such as *Polypodiaceaesporites haardti, Pinuspollenites labdacus, Piceaepollenites alatus, Ephedripites tertiarius, Quercoidites henrici, Q. microhenrici, Chenopodipollis multiporatus,* and *Artemisiaepollenites sellularis* are derived from the reservoir bed. Jurassic spores and pollen such as *Deltoidospora perpusilla, D. gradata, Dictyophyllidites harrisii, Cibotiumspora paradoxa,* and *Cedripites minor* may indicate a Jurassic petroleum source rock.

The composition of the above three petroleum sporo-pollen assemblages reflects the important contribution of the Jurassic System to petroleum sources in the Kuqa Depression. Jurassic spores and pollen in the crude oils coincide with those in the Jurassic dark-colored mudstones of the region in morphology, structure, and color (Jiang and Yang 1983). No reworked Jurassic spores or pollen resedimented into the Cretaceous and Tertiary sediments have been found in the region, so Jurassic spores and pollen in crude oils should be derived from the petroleum source rock. The results of correlations between spores and pollen in crude oils and those in potential source rocks (Table 4.3) indicate that the dark-colored mudstones of the Lower Jurassic Ahe and Yangxia formations and the Middle Jurassic Kezilenuer and Qiakemake formations should be the petroleum source rocks of the Kuqa Depression.

4.2.3 Southwest Tarim Depression

4.2.3.1 Kashi Sag

Forty-six species of fossil spores and pollen referred to 31 genera were found in crude oil samples from the Neogene petroleum reservoir in the Kelatu oil field in the Kashi Sag (Table 3.3). These microfossils in the petroleum sporo-pollen assemblage from the Neogene reservoir bed of the Kelatu oil field may be used as reliable evidence for judging the age of the petroleum source rocks in the Kashi Sag.

The petroleum sporo-pollen assemblage of the Neogene reservoir bed of the Kelatu oil field is composed of 46 species of Jurassic and Neogene spores and pollen. This is a mixed-type petroleum sporo-pollen assemblage. The Neogene pollen such as *Pinuspollenites labdacus, Piceaepollenites alatus, Ephedripites tertiarius, Caryapollenites simplex, Chenopodipollis multiplex,* and *Artemisiaepollenites sellularis* are derived from the reservoir bed. The Jurassic spores and pollen such as *Deltoidospora lineata, Gleicheniidites rouseii, Dictyophyllidites harrisii, Cibotiumspora paradoxa, Todisporites major, Granulatisporites minor, Leptolepidites major, Klukisporites variegatus, Murospora jurassica, M. minor, Paleoconiferus asaccatus, Vitreisporites jurassicus, V. shouldicei, V. jansonii, Pteruchipollenites thomasii, Alisporites lowoodensis, Protopinus scanicus, Protopicea exilioides, Piceites expositus, Cedripites minor, Podocarpidites florinii, P. langii, P. multicinus, Chasmatosporites major, C. elegans, C. minor, C. canadensis, Cycadopites subgranulosus,* and *Cerebropollenites carlylensis* may indicate a Jurassic petroleum source rock.

Jurassic spores and pollen in the crude oils coincide with those in the Jurassic dark-colored mudstones of the region in morphology, structure, size range, and color (Jiang and Yang 1996). No reworked Jurassic microfossils resedimented into Tertiary sediments have been found in the region, so the Jurassic spores and pollen of the assemblage should be derived from a petroleum source rock. The results of correlations between spores and pollen in crude oils and those in potential source rocks (Table 4.4) indicate that the dark-colored mudstones of the Lower Jurassic Kansu Formation and the Middle Jurassic Yangye and Taerga formations should be the petroleum source rocks of the Kashi Sag in the southwest Tarim Depression.

4.2.3.2 Yecheng Sag

Fifty-three species of fossil spores and pollen referred to 36 genera were found in crude oil samples from the Neogene petroleum reservoir of the Kekeya oil field in the Yecheng Sag (Table 3.4). The petroleum sporo-pollen assemblage composed of these microfossils may be used as reliable evidence for judging the age of the petroleum source rocks in the Yecheng Sag.

The petroleum sporo-pollen assemblage of the Neogene reservoir bed of the Kekeya oil field is composed of 53 species of Jurassic and Neogene spores and pollen. This is a mixed-type petroleum sporo-pollen assemblage. The Neogene pollen such as *Pinuspollenites labdacus, Piceaepollenites alatus, Ephedripites tertiarius, Quercoidites microhenrici, Chenopodipollis multiplex,* and *Artemisiaepollenites sellularis* are derived from the reservoir bed. The Jurassic spores and pollen such as *Deltoidospora perpusilla, D. gradata, Gleichenllidites rouseii, Dictyophyllidites harrisii, Dictyophyllum rugosum, Undulatisporites concavus, Cibotiumspora paradoxa, C. jurienensis, Leptolepidites major, L. verrucatus, Murospora minor, Paleoconiferus asaccatus, Pteruchipollenites thomasii, Protopicea exilioides, Piceites expositus, Cedripites*

Table 4.4 Distribution of Jurassic spores and pollen in crude oils from Jurassic strata in the Kashi Sag

Spores and pollen	Kansu Formation J_1k	Yangye Formation J_2y	Taerga Formation J_2t
Spores			
Deltoidospora lineata		+	+
Cyathidites australis	+	+	+
C. minor	+	+	+
Gleicheniidites rouseii		+	+
Dictyophyllidites harrisii	+	+	+
Cibotiumspora paradoxa	+	+	+
Todisporites major		+	+
Granulatisporites minor	+		
Leptolepidites major		+	+
Klukisporites variegatus	+	+	+
Murospora jurassica		+	+
M. minor		+	+
Pollen			
Paleoconiferus asaccatus	+	+	+
Vitreisporites jurassicus		+	+
V. shouldicei		+	+
V. jansonii		+	+
Pteruchipollenites thomasii	+	+	+
Alisporites lowoodensis	+	+	
Protopinus scanicus	+	+	
Protopicea exilioides		+	+
Piceites expositus		+	+
Cedripites minor	+	+	+
Podocarpidites florinii		+	+
P. langii		+	+
P. multicinus	+	+	+
Chasmatosporites major	+	+	
C. elegans	+	+	
C. minor	+	+	
Cycadopites subgranulosus	+	+	+
Cerebropollenites carlylensis		+	+
Quadraeculina limbata	+	+	+
Classopollis annulatus	+	+	+

minor, *Platysaccus lopsinensis*, *Podocarpidites major*, *P. wapellensis*, *P. multicinus*, *Parvisaccites enigmatus*, *Chasmatosporites major*, *C. elegans*, *C. minor*, *Cycadopites subgranulosus*, and *Bennettiteaepollenites lucifer* may indicate a Jurassic petroleum source.

Jurassic spores and pollen in the crude oils coincide with those in the Jurassic dark-colored mudstones of the region in morphology, structure, size range, and color (Jiang and Yang 1986). No reworked Jurassic spores and pollen resedimented into the Tertiary sediments have been found in the region, so the Jurassic spores and pollen of the assemblage should be derived from the petroleum source rock. The results of correlations between spores and pollen in the crude oils and those in the potential source rocks (Table 4.5) indicate that the black shales and dark gray mudstones of the Lower Jurassic Kansu Formation and the Middle Jurassic Yangye and Taerga formations should be the petroleum source rocks of the Yecheng Sag in the southwest Tarim Depression.

Based on the petroleum sporo-pollen assemblages and the correlations between spores and pollen in crude oils and those in potential source rocks, it can be concluded that the Carboniferous, Triassic, and Jurassic systems, particularly the Middle and Upper Triassic series and the Lower and Middle Jurassic series, should contain petroleum source rocks. The Permian and Cretaceous systems might contain petroleum source rocks in the Tarim Basin.

4.3 Junggar Basin

4.3.1 East Junggar Depression

4.3.1.1 Beisantai Petroliferous Region

The petroleum sporo-pollen assemblage of the Carboniferous igneous reservoir of the Beisantai oil field is composed of 96 species of Carboniferous to Jurassic spores and pollen (Table 3.6). This is a typical special-type petroleum sporo-pollen assemblage. It is highly impossible that crystalline igneous rocks would contain biological fossils. The spores and pollen in the crude oils must have been carried from the sedimentary source rocks to the reservoir in the course of petroleum migration. The spores and pollen of the assemblage are mostly distributed in the Permian and Triassic systems. Some are distributed in the Carboniferous and Jurassic systems. The Carboniferous spores are *Lophotriletes gibbosus*, *Punctatosporites minutus*, and *Wilsonia vesicata*. The Permian spores and pollen are *Calamospora densirugosus*, *Punctatisporites cathayensis*, *P. parasolidus*, *Lophotriletes humilus*, *Apiculatisporis decorus*, *A. xiaolongkouensis*, *Kraeuselisporites argutus*, *K. spinulosus*, *Torispora medius*, *Endosporites multirugulatus*, *Cordaitina rotata*, *C. duralimita*, *Florinites minutus*, *Vestigisporites dalongkouensis*, *Gardenasporites xinjiangensis*, *Lueckisporites virkkiae*, *L. prolixus*, *Striatopodocarpites tojmensis*, *S. crassus*, *Protohaploxypinus dvinensis*, *Limitisporites rectus*, *L. rhombicorpus*, *Labiisporites granulatus*, *Vittatina cryptosaccata*, *Vesicaspora xinjiangensis*, *Triangulisaccus primitivus*, *Entylissa spinosa*, and *Cycadopites granulatus*. The Triassic spores and pollen are *Punctatisporites triassicus*, *Verrucosisporites morulae*, *Lophotriletes delicatus*, *Discisporites psilatus*, *Multinodisporites junctus*, *Limatulasporites dalongkouensis*, *L. parvus*, *Lundbladispora subornata*, *Aratrisporites parvispinosus*, *A. coryliseminis*, *A. fischeri*, *A. strigosus*, *Florinites millotti*, *Chordasporites orientalis*, *C. impensus*, *Taeniaesporites kraeuseli*, *Vitreisporites signatus*, *Cedripites parvisaccus*, *Parcisporites rarus*, *Minutosaccus parcus*, *Podocarpidites queenslandi*, *P. transversus*, *Alisporites australis*, *A. parvus*, and *A. auritus*. The common Jurassic species are *Calamospora nathorstii* and *Cycadopites subgranulosus*.

The composition of the petroleum sporo-pollen assemblage reflects the important contribution of the Permian and Triassic systems to petroleum sources in the Beisantai Petroliferous Region. The species of spores and pollen in crude oils and those in dark-colored mudstones are identical in morphology, structure, and color, so it is believed that the spores and pollen in petroleum should come from these petroleum source rocks (Jiang and Yang 1994). The results of correlations between spores and pollen in crude oils and those in potential source rocks (Tables 4.6 and 4.7) indicate that the Upper Permian Lucaogou, Hongyanchi, Quanzijie, Wutonggou formations should be the main petroleum source rocks; the Lower Triassic Jiucaiyuan and Shaofanggou formations, the Middle Triassic Karamay Formation, and the Upper

Table 4.5 Distribution of Jurassic spores and pollen in crude oils from Jurassic strata in the Yecheng Sag

Spores and pollen	Kansu Formation J_1k	Yangye Formation J_2y	Taerga Formation J_2t
Spores			
Deltoidospora perpusilla	+	+	+
D. gradata	+	+	+
Cyathidites australis	+	+	+
C. minor	+	+	+
Gleicheniidites rouseii		+	+
Dictyophyllidites harrisii	+	+	+
Dictyophyllum rugosum	+	+	+
Undulatisporites concavus		+	+
Cibotiumspora paradoxa	+	+	+
C. jurienensis	+	+	
Leptolepidites major		+	+
L. verrucatus	+	+	+
Apiculatisporis ovalis		+	+
Murospora minor		+	+
Pollen			
Paleoconiferus asaccatus	+	+	+
Pteruchipollenites thomasii	+	+	+
Protopicea exilioides		+	+
Piceites expositus		+	+
Cedripites minor	+	+	+
Platysaccus lopsinensis		+	+
Podocarpidites major	+	+	
P. wapellensis		+	+
P. multicinus	+	+	+
Parvisaccites enigmatus		+	+
Chasmatosporites major	+	+	
C. elegans	+	+	
C. minor	+	+	+
Cycadopites subgranulosus	+	+	+
Bennettiteaepollenites lucifer		+	+
Quadraeculina limbata	+	+	+
Eucommiidites troedssonii	+	+	+
Classopollis classoides	+	+	+
C. annulatus	+	+	+

Triassic Huangshanjie and Haojiagou formations should be the secondary petroleum source rocks in the region. The Carboniferous and Jurassic systems might contain some potential source rocks. And the Upper Carboniferous series and the Lower and Middle Jurassic series possibly might contain oil or gas source rocks in the East Junggar Depression.

Table 4.6 Distribution of Permian spores and pollen in crude oils from Permian strata in the East Junggar Depression

Spores and pollen	Lucaogou Formation P_2l	Hongyanchi Formation P_2h	Quanzijie Formation P_2q	Wutonggou Formation P_2w
Spores				
Calamospora densirugosus				+
Lophotriletes humilus			+	+
Apiculatisporis decorus				+
A. xiaolongkouensis				+
Kraeuselisporites argutus				+
K. spinulosus			+	+
Endosporites multirugulatus				+
Pollen				
Cordaitina uralensis	+	+	+	+
C. rotata		+	+	+
C. duralimita				+
Florinites minutus				+
Vestigisporites dalongkouensis				+
Lueckisporites virkkiae	+			+
Gardenasporites xinjiangensis			+	+
Limitisporites rhombicorpus	+	+	+	+
Vesicaspora xinjiangensis				+
Alisporites australis	+	+		
A. parvus	+	+		
Protohaploxypinus samoilovichii	+	+		+
Striatopodocarpites tojmensis	+	+		
Vittatina cryptosaccata	+	+		
Triangulisaccus primitivus				+

4.3.1.2 Huoshaoshan Petroliferous Region

The petroleum sporo-pollen assemblage of the Permian reservoir bed of the Huonan oil field is composed of 60 species of Carboniferous to Jurassic spores and pollen (Table 3.7). This is a mixed-type petroleum sporo-pollen assemblage. The Carboniferous spores *Leiotriletes microtriangulus, Convolutispora grandicornis*, and *Punctatosporites minutus* were reported from the Carboniferous System of northern Xinjiang (Ouyang et al. 2003). The Permian spores and pollen such as *Calamospora densirugosus, Lophotriletes humilus, Apiculatisporis decorus, A. xiaolongkouensis, A. setaceformis, Kraeuselisporites argutus, Tuberculatosporites homotubercularis, Endosporites multirugulatus, Gardenasporites xinjiangensis, G. latisectus,*

Table 4.7 Distribution of Triassic spores and pollen in crude oils from Triassic strata in the East Junggar Depression

Spores and pollen	Jiucaiyuan Formation T_1j	Shaofanggou Formation T_1s	Karamay Formation T_2k	Huangshanjie Formation T_3h	Haojiagou Formation T_3ha
Spores					
Calamospora nathorstii			+		+
Punctatisporites triassicus	+		+	+	+
Verrucosisporites morulae		+	+		
Discisporites psilatus	+		+		+
Limatulasporites dalongkouensis	+	+		+	
L. parvus	+				
Lundbladispora subornata	+				
Aratrisporites fischeri			+		
A. parvispinosus		+			
A. coryliseminis	+				
A. strigosus			+		+
Pollen					
Florinites millotti	+				
Taeniaesporites kraeuseli			+		+
Minutosaccus parcus			+		
Podocarpidites queenslandi		+			+
P. transversus	+				
Alisporites australis	+		+		
A. parvus	+	+	+		+

Limitisporites rhombicorpus, Triangulisaccus primitivus, Cordaitina rotata, Hamiapollenites bullaeformis, H. obliquus, Vittatina subsaccata, and *V. cryptosaccata* were reported from the Permian System of Jimsar of Xinjiang and Wucaiwan of the eastern Junggar Basin (Institute of Geology, Chinese Academy of Geological Sciences, and Institute of Geology, Xinjiang Bureau of Geology and Mineral Resources 1986; Ouyang et al. 2003). The Triassic spores such as *Punctatisporites triassicus, Lophotriletes flavus,* and *Apiculatisporis spiniger* were also reported from the Triassic System of Jimsar of Xinjiang.

The results of correlations between spores and pollen in the crude oils and those in potential source rocks indicate that the Upper Permian Lucaogou, Hongyanchi, Quanzijie, and Wutonggou formations should be the main petroleum source rocks, and Carboniferous and Triassic source rocks should be important as well in the Huoshaoshan Petroliferous Region.

In addition, the megaspores *Triangulatisporites junggarensis* found in the crude oil of the Huonan oil field were reported from the Upper Permian Quanzijie and Wutonggou formations of Jimsar of Xinjiang (Institute of Geology, Chinese

4.3.2 South Junggar Depression

4.3.2.1 Qigu Petroliferous Region

The petroleum sporo-pollen assemblage of the Jurassic reservoir bed of the Qigu oil field is composed of 38 species of Jurassic spores and pollen (Table 3.8). This is a mono-type petroleum sporo-pollen assemblage. Most of the spores and pollen are Jurassic index species such as *Gleicheniidites rouseii*, *G. nilssonii*, *G. conflexus*, *Dictyophyllidites harrisii*, *Cibotiumspora paradoxa*, *C. jurienensis*, *Granulatisporites jurassicus*, *G. minor*, *Concavissimisporites delcourtii*, *C. southeyensis*, *C. subgranulosus*, *Verrucosisporites variabilis*, *Converrucosisporites minor*, *C. elegans*, *Apiculatisporis variabilis*, *Ceratosporites jurassicus*, *C. varispinosus*, *Paleoconiferus asaccatus*, *Pteruchipollenites thomasii*, *Pinuspollenites globosaccus*, *Protopicea vilujensis*, *P. exilioides*, *Piceites expositus*, *Cedripites minor*, and *Podocarpidites multicinus*. The other spores and pollen are common Jurassic species such as *Cyathidites australis*, *C. minor*, *Piceites latens*, *P. podocarpoides*, *Podocarpidites multesimus*, and *Pinuspollenites alatipollenites*.

The above spores and pollen were reported from the Lower Jurassic Badaowan and Sangonghe formations and the Middle Jurassic Xishanyao and Toutunhe formations in the southern Junggar Basin (Zhang 1990; Liu 1993). The results of correlations between spores and pollen in the crude oils and those in potential source rocks indicate that Lower and Middle Jurassic dark-colored mudstones should be good petroleum source rocks in the Qigu Petroliferous Region.

4.3.2.2 Dushanzi Petroliferous Region

The petroleum sporo-pollen assemblage of the Tertiary reservoir bed of the Dushanzi oil field is composed of 51 species of Jurassic and Tertiary spores and pollen (Table 3.9). This is a mixed-type petroleum sporo-pollen assemblage. The Tertiary pollen such as *Pinuspollenites labdacus*, *Piceaepollenites alatus*, *Tsugaepollenites viridifluminipites*, *Ephedripites tertiarius*, and *Artemisiaepollenites sellularis* are derived from the reservoir bed. The Jurassic spores and pollen *Leiotriletes parvus*, *Dictyophyllidites harrisii*, *Murospora jurassica*, *M. minor*, *M. scabratus*, *Aratrisporites xiangxiensis*, *Schizosporis microreticulatus*, *Paleoconiferus asaccatus*, *Pseudowalchia ovalis*, *P. landesii*, *Pteruchipollenites thomasii*, *Abietineaepollenites dunrobinensis*, *Protopicea exilioides*, *Piceites pseudorotundiformis*, *Cedripites minor*, *Podocarpidites multicinus*, *Parvisaccites enigmatus*, and *Callialasporites radius* may indicate a Jurassic petroleum source rock. In addition, the Paleogene pollen such as *Ephedripites eocenipites*, *Salixipollenites discoloripites*, *Juglanspollenites verus*, *Quercoidites henrici*, *Q. microhenrici*, *Q. asper*, and *Rhoipites pseudocingulum* are distributed in the Oligocene Anjihaihe Formation in the region.

In accordance with the composition of the petroleum sporo-pollen assemblage and the results of correlations between spores and pollen in the crude oils and those in potential source rocks (Yang and Jiang 1989), the Lower Jurassic Badaowan and Sangonghe formations, the Middle Jurassic Xishanyao and Toutunhe formations, and the Oligocene Anjihaihe Formation should be the petroleum source rocks in the Dushanzi Petroliferous Region.

Based on the petroleum sporo-pollen assemblages and the correlations between spores and pollen in the crude oils and those in potential source rocks, it can be concluded that the Carboniferous, Permian, Triassic, Jurassic, and

Paleogene systems are petroleum source rocks in the Junggar Basin.

4.4 Turpan Basin

4.4.1 Qiktim Petroliferous Region

The petroleum sporo-pollen assemblage of the Jurassic reservoir bed of the Qiktim oil field is composed of 68 species of Jurassic spores and pollen (Table 3.10). This is a typical mono-type petroleum sporo-pollen assemblage. Most of the spores and pollen are Jurassic index species such as *Leiotriletes medius*, *Deltoidospora gradata*, *D. lineata*, *D. magna*, *Dictyophyllidites harrisii*, *Cibotiumspora paradoxa*, *C. jurienensis*, *Gleicheniidites rouseii*, *G. conflexus*, *Undulatisporites pflugii*, *Undulatisporites concavus*, *Divisisporites undulatus*, *Tripartina variabilis*, *Granulatisporites jurassicus*, *G. minor*, *Todisporites minor*, *Verrucosisporites variabilis*, *Converrucosisporites minor*, *Leptolepidites major*, *Apiculatisporis variabilis*, *Protoconiferus minor*, *Paleoconiferus asaccatus*, *Pseudowalchia ovalis*, *P. landesii*, *Pteruchipollenites thomasii*, *Abietineaepollenites dunrobinensis*, *Pityosporites similis*, *Piceaepollenites omoriciformis*, *Protopices exilioides*, *Piceites expositus*, *Cedripites minor*, *Vitreisporites jurassicus*, *V. jansonii*, *Platysaccus lopsinensis*, *Podocarpidites major*, *P. multicinus*, *P. rousei*, *P. wapellaensis*, *Parvisaccites enigmatus*, *Inaperturopollenites dettmannii*, *Callialasporites radius*, and *Chasmatosporites elegans*. The other spores and pollen are common Jurassic species such as *Cyathidites australis*, *C. minor*, *Osmundacidites wellmanii*, *Apiculatisporis ovalis*, *Pseudopicea variabiliformis*, *Piceites latens*, *P. podocarpoides*, *Podocarpidites multesimus*, *Callialasporites dampieri*, *Classopollis classoides*, *C. annulatus*, *Cycadopites nitidus,* and *C. minimus*.

The above spores and pollen were reported from the Lower Jurassic Badaowan and Sangonghe formations and the Middle Jurassic Xishanyao and Toutunhe formations in the Turpan Basin (Sun 1989; Wang et al. 1998). The species of spores and pollen in the crude oils and in the rocks are identical in morphology, structure, size range, and color (Jiang and Yang 1989). In accordance with the composition of petroleum sporo-pollen assemblage and the results of correlations between spores and pollen in crude oils and those in potential source rocks (Table 4.8), the Lower Jurassic Badaowan and Sangonghe formations and the Middle Jurassic Xishanyao and Toutunhe formations should be the petroleum source rocks in the Turpan Basin. The Lower and Middle Jurassic dark-colored mudstones should be good petroleum source rocks, and the Jurassic sandstones should be reservoir rocks.

Table 4.8 Distribution of Jurassic spores and pollen in crude oils from Jurassic strata in the Turpan Depression

Spores and pollen	Badaowan Formation J_1b	Sangonghe Formation J_1s	Xishanyao Formation J_2x	Toutunhe Formation J_2t
Spores				
Leiotriletes medius			+	+
Deltoidospora gradata	+	+	+	+
D. magna	+	+	+	+
Cyathidites australis	+	+	+	+
C. minor	+	+	+	+
Gleicheniidites rouseii	+	+	+	+
G. conflexus	+	+	+	+

(continued)

4.4 Turpan Basin

Table 4.8 (continued)

Spores and pollen	Badaowan Formation J_1b	Sangonghe Formation J_1s	Xishanyao Formation J_2x	Toutunhe Formation J_2t
Dictyophyllidites harrisii		+	+	
Cibotiumspora paradoxa	+	+	+	+
C. jurienensis	+	+	+	+
Undulatisporites pflugii	+	+	+	+
Divisisporites undulatus		+	+	+
Granulatisporites jurassicus	+	+	+	+
G. minor		+	+	+
Osmundacidites wellmanii	+	+	+	+
Todisporites minor	+	+	+	+
Leptolepidites major	+	+	+	
Apiculatisporis variabilis	+	+	+	
A. ovalis	+	+		
Pollen				
Paleoconiferus asaccatus	+	+	+	+
Pteruchipollenites thomasii		+	+	
Abietineaepollenites dunrobinensis		+	+	
Protopicea exilioides		+	+	+
Piceites expositus	+	+	+	+
P. latens		+	+	+
P. podocarpoides	+	+	+	
Pseudopicea variabiliformis		+	+	+
Piceaepollenites omoriciformis			+	+
P. complanatiformis			+	+
Cedripites minor			+	+
Platysaccus lopsinensis			+	+
Podocarpidites multicinus		+	+	+
P. multesimus	+	+	+	+
P. rousei			+	+
P. wapellaensis				+
Parvisaccites enigmatus	+	+	+	+
Callialasporites dampieri			+	+
C. radius			+	+
Chasmatosporites elegans	+	+	+	
Cycadopites subgranulosus	+	+	+	
C. nitidus	+	+	+	+
C. minimus	+	+	+	+
Eucommiidites troedssonii			+	
Classopollis classoides		+	+	+
C. annulatus	+	+	+	+

4.4.2 Shengjinkou Petroliferous Region

The petroleum sporo-pollen assemblage of the Jurassic reservoir bed of the Shengjinkou oil field is composed of 23 species of Jurassic spores and pollen (Table 3.11). This is a mono-type petroleum sporo-pollen assemblage. Most of the spores and pollen are Jurassic index species such as *Leiotriletes medius, Deltoidospora lineata, Dictyophyllidites harrisii, Cibotiumspora paradoxa, Gleicheniidites rouseii, Granulatisporites jurassicus, Todisporites minor, Leptolepidites major, Apiculatisporis variabilis, Paleoconiferus asaccatus, Protopicea exilioides, Piceites expositus, Cedripites minor*, and *Podocarpidites multicinus*. The other spores and pollen are common Jurassic species such as *Cyathidites minor, Osmundacidites wellmanii, Piceites latens, Podocarpidites multesimus, Classopollis classoides*, and *C. annulatus*.

The above spores and pollen are widely distributed in the Lower and Middle Jurassic series in the region. The composition of this petroleum sporo-pollen assemblage is similar to that of the Qiktim oil field, both have the same character. Therefore, the Shengjinkou Petroliferous Region possesses the same petroleum sources as that of the Qiktim Petroliferous Region.

4.5 Qaidam Basin

4.5.1 North Border Block-fault Zone

The petroleum sporo-pollen assemblage of the Jurassic reservoir bed of the Lenghu oil field in the northwestern North Border Block-fault Zone is composed of 44 species of Jurassic spores and pollen (Table 3.12). This is a mono-type petroleum sporo-pollen assemblage. Most of the spores and pollen are Jurassic index species such as *Dictyophyllidites harrisii, Gleicheniidites rouseii, G. conflexus, Osmundacidites elegans, Leptolepidites major, Apiculatisporis variabilis, Acanthotriletes midwayensis, Lycopodiacidites rugulatus, Duplexisporites scanicus, Paleoconiferus asaccatus, Pityosporites similis, Protopinus scanicus, Protopicea vilujensis, P. minutereticulata, P. exilioides, Piceaepollenites omoriciformis, Piceites expositus, Cedripites minor, Protopodocarpus mollis, Podocarpidites multicinus, P. wapellaensis, P. langii, P. paulus, Parvisaccites enigmatus, Callialasporites radius, Cerebropollenites carlylensis, Chasmatosporites major*, and *C. elegans*. The other spores and pollen are common Jurassic species such as *Cyathidites australis, C. minor, Osmundacidites wellmanii, Piceites latens, Podocarpidites multesimus, Quadraeculina limbata, Classopollis annulatus, Cycadopites subgranulosus*, and *C. nitidus*.

The above spores and pollen are distributed in the Lower Jurassic Xiaomeigou Formation and the Middle Jurassic Dameigou Formation in the North Border Block-fault zone. The same species of spores and pollen found in the crude oils are found in the dark-colored mudstones; they are identical in morphology, structure, size range, and color (Jiang and Yang 1997a). According to the composition of the petroleum sporo-pollen assemblage and the results of correlations between spores and pollen in the crude oils and those in potential source rocks (Table 4.9), the Lower Jurassic Xiaomeigou Formation and the Middle Jurassic Dameigou Formation should be the petroleum source rocks of the region.

The petroleum sporo-pollen assemblage of the Jurassic reservoir bed of the Yuka oil field in the eastern North Border Block-fault Zone is composed of 20 species of Jurassic spores and pollen (Table 3.13). This is a mono-type petroleum sporo-pollen assemblage. Most of the spores and pollen are Jurassic index species such as *Dictyophyllidites harrisii, Gleicheniidites conflexus, Paleoconiferus asaccatus, Pityosporites similis, Protopicea exilioides, Piceites expositus, Cedripites minor, Podocarpidites multicinus, P. paulus, Callialasporites radius, Cerebropollenites carlylensis*, and *Chasmatosporites major*. The other spores and pollen are common Jurassic species such as *Cyathidites australis, C. minor, Classopollis annulatus, Cycadopites nitidus*, and *C. subgranulosus*.

The composition of this petroleum sporo-pollen assemblage is similar to that of the

Table 4.9 Distribution of Jurassic spores and pollen in crude oils from Jurassic strata in the North Border Block-fault Zone

Spores and pollen	Lenghu		Yuka	
	Xiaomeigou Formation J_1x	Dameigou Formation J_2d	Xiaomeigou Formation J_1x	Dameigou Formation J_2d
Spores				
Cyathidites australis		+	+	+
C. minor		+	+	+
Dictyophyllidites harrisii	+	+	+	+
Gleicheniidites conflexus			+	+
Osmundacidites wellmanii	+	+	+	+
O. elegans	+		+	+
Leptolepidites major		+		
Apiculatisporis variabilis		+	+	+
Acanthotriletes midwayensis	+		+	+
Lycopodiacidites rugulatus		+	+	+
Duplexisporites gyratus		+	+	+
D. amplectiformis		+	+	+
D. anagrammensis			+	
D. scanicus		+	+	+
Pollen				
Paleoconiferus asaccatus	+	+	+	
Protopicea exilioides	+	+	+	+
Piceites expositus	+		+	+
P. lateens	+	+		
Piceaepollenites omoriciformis		+		
Protopinus scanicus	+			
Cedripites minor		+		+
Protopodocarpus mollis	+			
Podocarpidites multesimus	+	+	+	+
P. paulus	+		+	+
Parvisaccites enigmatus		+	+	+
Quadraeculina limbata			+	+
Callialasporites radius			+	+
Classopollis annulatus		+	+	+
Chasmatosporites major			+	+
Cycadopites subgranulosus	+	+	+	+
C. nitidus	+	+	+	+
C. carpentieri			+	+

Lenghu oil field, both having the same character. Therefore, the Yuka oil field has the same petroleum source as the Lenghu oil field.

Based on the investigations of the petroleum sporo-pollen assemblages, it may be concluded that the Lower Jurassic Xiaomeigou and the Middle Jurassic Dameigou formations should be the petroleum source rocks in the North Border Block-fault Zone.

4.5.2 Mangnai Depression

The petroleum sporo-pollen assemblage of the Pliocene reservoir bed of the Youquanzi oil field in the middle part of the Mangnai Depression contains 59 species of Tertiary spores and pollen (Table 3.14). This is a mixed-type petroleum sporo-pollen assemblage. The Neogene spores and pollen, such as *Granulatisporites pteridiumoides, Echinosporis qaidamensis, Ephedripites neogenicus, Betulaepollenites lenghuensis, Artemisiaepollenites sellularis,* and *A. communi,* are derived from the reservoir bed. Paleogene spores and pollen such as *Lygodiumsporites pseudomaximus, Podocarpidites nageiaformis, P. paranageiaformis, Abietineaepollenites cembraeformis, Pinuspollenites banksianaeformis, Piceaepollenites tobolicus, Cedripites deodariformis, C. microsaccoides, Keteleeriaepollenites dubius, Ephedripites eocenipites, Salixipollenites elegans, Momipites coryloides, Quercoidites asper, Cupuliferoipollenites pusillus,* and *Chenopodipollis multiporatus* may indicate a Paleogene petroleum source rock.

Tertiary spores and pollen found in the crude oils of the Youquanzi oil field were reported from the Tertiary System in the Qaidam Basin (Research Institute of Exploration and Development, Qinghai Petroleum Administration, and Nanjing Institute of Geology and Palaeontology, Academia Sinica 1985). The results of correlations between spores and pollen in the crude oils and those in potential source rocks (Table 4.10) indicate that the Paleocene to Eocene Lulehe Formation and the Oligocene to Lower Miocene Ganchaigou Formation should be the main petroleum source rocks, and the Upper Miocene to Lower Pliocene Youshashan Formation should be an additional petroleum source rock in the Mangnai Depression (Jiang and Yang 1998).

The petroleum sporo-pollen assemblage of the Pliocene reservoir bed of the Xianshuiquan oil field in the northwestern part of the Mangnai Depression is composed of 23 species of Tertiary spores and pollen (Table 3.15). This is a mixed-type petroleum sporo-pollen assemblage. The Neogene spores and pollen such as *Granulatisporites pteridiumoides, Ephedripites neogenicus,* and *Artemisiaepollenites sellularis* are derived from the reservoir bed. The Paleogene pollen such as *Cedripites deodariformis, C. ovatus,* and *Momipites coryloides* may indicate a Paleogene petroleum source rock (Jiang and Yang 1998).

The composition of this petroleum sporo-pollen assemblage is similar to that of the Youquanzi oil field. Therefore, the Xianshuiquan oil field has the same petroleum source as the Youquanzi oil field.

4.6 West Jiuquan Basin

4.6.1 Laojunmiao Anticlinal Zone

The petroleum sporo-pollen assemblage of the Neogene reservoir bed of the Laojunmiao oil field in the middle part of the Laojunmiao Anticlinal Zone is composed of 25 species of Lower Cretaceous and Neogene spores and pollen (Table 3.16). This is a mixed-type petroleum sporo-pollen assemblage. The Neogene pollen such as *Pinus, Picea, Ephedra, Potamogeton, Graminidites, Cyperaceaepollis, Lilium, Ulmus, Chenopodium, Nymphaea, Tamarix,* and *Artemisia* are derived from the reservoir bed. The Lower Cretaceous spores and pollen such as *Cibotiumspora juncta, Osmundacidites wellmanii, Schizaeoisporites zizyphinus, S. gansuensis, Cingulatisporites ruginosus, Cycadopites minimus, Bennettiteaepollenites* sp., *Psophosphaera grandis, P. tenuis,* and *Classopollis annulatus* may indicate a Lower Cretaceous petroleum source rock.

The Lower Cretaceous spores and pollen were reported from the Lower Cretaceous Lower

4.6 West Jiuquan Basin

Table 4.10 Distribution of Paleogene spores and pollen in crude oils from the Paleogene System in the Mangnai Depression

Spores and pollen	Lulehe Formation $E_{1-2}l$	Ganchaigou Formation E_3g
Spores		
Lygodiumsporites pseudomaximus		+
Polypodiaceaesporites haardti	+	+
Polypodiisporites favus	+	+
Pollen		
Podocarpidites nageiaformis		+
P. paranageiaformis		+
P. verrucorpus		+
P. qigequanensis		+
Abietineaepollenites cembraeformis		+
A. microalatus major		+
Pinuspollenites banksianaeformis		+
P. labdacus maximus		+
P. labdacus minor		+
Piceaepollenites tobolicus	+	+
P. planoides	+	+
Cedripites deodariformis	+	+
C. ovatus	+	+
C. microsaccoides		+
C. pachydermus		+
Keteleeriaepollenites dubius		+
K. megasaccus		+
K. mangnaiensis	+	+
Tsugaepollenites igniculus major		+
T. igniculus minor		+
T. spinulosus		+
Abiespollenites lenghuensis		+
Ephedripites eocenipites	+	+
E. fusiformis	+	+
E. mangnaiensis	+	+
E. dafengshanensis	+	+
E. ganchaigouensis		+
Salixipollenites elegans		+
Momipites coryloides	+	+
Quercoidites henrici	+	+
Q. microhenrici	+	+
Q. asper	+	+
Cupuliferoipollenites pusillus	+	+
Meliaceoidites rhomboiporus	+	+

(continued)

Table 4.10 (continued)

Spores and pollen	Lulehe Formation $E_{1-2}l$	Ganchaigou Formation E_3g
Sparganiaceaepollenites sparganioides	+	+
Tubulifloridites macroechinatus		+
T. minispinulosus		+
Chenopodipollis multiporatus	+	+

Xinminbu Formation in the Jiuquan Basin and the Huahai Basin of Gansu (Hsü et al. 1974; Chiang and Young 1978). The species of spores and pollen in the crude oil samples and the black shales of the Lower Xinminbu Formation are identical in morphology, structure, size range, and color (Jiang and Yang 1980, 1981). Lower Cretaceous spores and pollen in the black shales were found in the Neogene reservoir. This fact reflects a result of petroleum migration and implies that the Lower Cretaceous black shales should be petroleum source rocks.

The petroleum sporo-pollen assemblage of the Neogene reservoir bed of the Yaerxia oil field in the northern part of the Laojunmiao Anticlinal Zone is composed of 11 species of Lower Cretaceous and Neogene spores and pollen. This is a mixed-type petroleum sporo-pollen assemblage. The Neogene pollen such as *Lilium, Chenopodium, Nymphaea, Tamarix, Artemisia*, and *Cyperaceaepollis* are derived from the reservoir bed. The presence of the Lower Cretaceous spores and pollen *Schizaeoisporites zizyphinus, Cycadopites minimus*, and *Psophosphaera tenuis* may indicate a Lower Cretaceous petroleum source rock.

The petroleum sporo-pollen assemblage of the Lower Cretaceous reservoir bed of the Yaerxia oil field is composed of five species of Lower Cretaceous spores and pollen. This is a mono-type petroleum sporo-pollen assemblage. The spores and pollen such as *Cibotiumspora juncta, Schizaeoisporites zizyphinus, Cycadopites minimus, Bennettiteaepollenites* sp., and *Psophosphaera tenuis* may indicate the petroleum source rock.

The Lower Cretaceous pollen *Cycadopites minimus* and *Psophosphaera tenuis* are also found in the crude oil samples taken from the Silurian metamorphic rock reservoir bed of the Yaerxia oil field. Because metamorphic rocks preserve only carbonized pollen fossils, the identifiable pollen fossils should be from Lower Cretaceous petroleum source rocks.

The different types of petroleum sporo-pollen assemblages in the Yaerxia oil field all contain the Lower Cretaceous spores and pollen found in the Lower Xinminbu formations. Therefore, the Lower Xinminbu Formation should be the petroleum source rock of this petroliferous region.

4.6.2 Baiyanghe Monoclinal Zone

The petroleum sporo-pollen assemblage of the Tertiary reservoir bed of the Baiyanghe oil field in the middle part of the Baiyanghe Monoclinal Zone is composed of 13 species of Lower Cretaceous and Tertiary spores and pollen. This is a mixed-type petroleum sporo-pollen assemblage. The Tertiary pollen such as *Pinus, Ephedra, Potamogeton, Graminidites, Chenopodium, Nitraria*, and *Artemisia* are derived from the reservoir bed. The Lower Cretaceous spores and pollen such as *Cibotiunspora juncta, Schizaeoisporites zizyphinus*, and *Cycadopites minimus* may indicate a Lower Cretaceous petroleum source rock.

The composition of this petroleum sporo-pollen assemblage is similar to that of the Laojunmiao oil field, both having the same character. It indicates that the Baiyanghe oil field has the same petroleum source as that of the Laojunmiao oil field.

According to the composition of petroleum sporo-pollen assemblages and the results of correlations between spores and pollen in the crude oils and those in potential source rocks, the black shales of the Lower Cretaceous Lower Xinminbu Formation should be the petroleum source rocks in the West Jiuquan Basin.

4.7 Liaohe Basin

The petroleum sporo-pollen assemblage of the Oligocene reservoir bed of the Xinglongtai oil field in the middle part of the Liaohe Basin is composed of 21 species of Paleogene spores and pollen (Table 3.17). This is a mono-type petroleum sporo-pollen assemblage. Most of the spores and pollen are Paleogene index species and common species such as *Pterisisporites undulatus*, *Plicifera decora*, *Pinuspollenites strobipites*, *Cedripites pachydermus*, *Keteleeriaepollenites dubius*, *Q. microhenrici*, *Q. asper*, *Chenopodipollis multiporatus*, and *C. microporatus*. The other pollen grains are new Paleogene species or new named combinations found in the coastal region of Bohai, such as *Cedripites diversus*, *Abietineaepollenites cembraeformis*, *A. microsibiricus*, and *Ephedripites cheganicus*.

Paleogene spores and pollen found in crude oils of the Xinglongtai oil field were reported from the Paleogene System in the coastal region of Bohai (Research Institute of Petroleum Exploration and Development, Ministry of Petroleum Chemistry Industry, and Nanjing Institute of Geology and Palaeontology, Chinese Academy of Sciences 1978a). In accordance with the composition of the petroleum sporo-pollen assemblage and the results of correlations between spores and pollen in the crude oils and those in potential source rocks (Yang and Jiang 1981), the Lower Oligocene Shahejie Formation and the Upper Oligocene Dongying Formation should be the petroleum source rocks in the Liaohe Basin.

4.8 Beibu Gulf Basin

The petroleum sporo-pollen assemblage of the Paleogene reservoir bed of the Weizhou oil field in the Beibu Gulf Depression is composed of 36 species of Paleogene spores and pollen (Table 3.18). This is a mono-type petroleum sporo-pollen assemblage. Most of the spores and pollen are Paleogene index species and common species such as *Leiotriletes adriensis*, *Polypodiaceaesportes ovatus*, *Podocarpidites andiniformis*, *Pinuspollenites strobipites*, *Cedripites eocenicus*, *C. cedroides*, *Salixipollenites discoloripites*, *Momipites coryloides*, *Quercoidites minor*, *Cupuliferoipollenites pusillus*, *Ulmipollenites granopollenites*, and *Liquidambarpollenites minutus*. The other spores and pollen are new Paleogene species found in the North Shelf of South China Sea, such as *Crassoretitriletes nanhaiensis*, *Ilexpollenites membranous*, *Tricolpites tenuicolpus*, *Retimultiporopollenites liushaensis*, *Trilobapollis leptus*, *T. ellipticus*, and *Verrutricolporites pachydermus* (Sun et al. 1980; Hou et al. 1981).

Paleogene spores and pollen found in the crude oils of the Weizhou oil field are widely distributed in the Paleogene Liushagang and Weizhou formations in the Beibu Gulf Bain. Comparisons between palynomorphs in crude oils and those in potential source rocks indicate that the Liushagang and Weizhou formations should contain favorable petroleum source rocks (Jiang and Yang 1999a). According to the composition of the petroleum sporo-pollen assemblage and the results of correlations between spores and pollen in the crude oils and those in potential source rocks (Table 4.11), the Eocene to Lower Oligocene Liushagang Formation and the middle to Upper Oligocene Weizhou Formation should be the petroleum source rocks of the Beibu Gulf Depression in the western North Shelf of South China Sea.

4.9 Zhujiang Mouth Basin

The petroleum sporo-pollen assemblage of the Upper Oligocene reservoir bed of the Zhuhai oil field in the Zhujiang Mouth Depression is composed of 30 species of Paleogene spores and pollen (Table 3.19). The composition of the assemblage includes the spores and pollen of the reservoir bed and pollen older than the reservoir

Table 4.11 Distribution of Paleogene spores and pollen in crude oils from the Paleogene System in the Beibu Gulf Depression

Spores and pollen	Liushagang Formation $E_2 - E_3^1 l$			Weizhou Formation
	E_2^2	E_2^3	E_3^1	$E_3^{2-3} w$
Spores				
Leiotriletes adriensis		+	+	+
Osmundacidites primarius		+	+	+
Crassoretitriletes nanhaiensis	+	+	+	+
Polypodiaceaesporites haardti	+	+	+	+
P. ovatus	+	+		
Polypodiisporites afavus		+	+	
Pollen				
Podocarpidites andiniformis		+	+	+
Abietineaepollenites microalatus minor			+	+
Pinuspollenites labdacus minor	+	+	+	
P. minutus	+	+	+	+
P. strobipites	+	+	+	+
Piceaepollenites alatus		+	+	+
P. planoides				+
Cedripites eocenicus	+	+		
C. cedroides		+	+	
Salixipollenites discoloripites	+	+		
Caryapollenites simplex	+	+	+	+
Juglanspollenites verus	+	+	+	+
Momipites coryloides	+	+		+
Quercoidites microhenrici	+	+	+	+
Q. minor	+	+	+	+
Cupuliferoipollenites pusillus	+	+	+	+
Ulmipollenites undulosus	+	+	+	+
U. granopollenites		+	+	+
Corylopsis princeps	+	+	+	
Liquidambarpollenites stigmosus	+	+	+	+
L. minutus			+	
Ilexpollenites membranous	+	+	+	+
Myrtaceidites parvus		+		
Tricolpites tenuicolpus		+	+	
Tricolporopollenites minutus			+	
Multiporopollenites punctatus			+	
Retimultiporopollenites liushaensis		+		
Trilobapollis leptus				+
T. ellipticus				+
Verrutricolporites pachydermus				+

Table 4.12 Distribution of Paleogene spores and pollen in crude oils from the Paleogene System in the Zhujiang Mouth Depression

Spores and pollen	Wenchang Formation E_2^2w	Enping Formation $E_2^3 + E_3^2e$	Zhuhai Formation E_3^3z
Spores			
Leiotriletes adriensis		+	+
Osmundacidites primarius		+	+
Polypodiaceaesporites haardti	+	+	+
P. gracilis	+	+	+
Polypodiisporites favus		+	+
P. afavus	+	+	+
Polypodiaceoisporites vitiosus		+	+
Pollen			
Abietineaepollenites microalatus minor		+	
Pinuspollenites labdacus minor	+	+	
P. minutus	+	+	
P. strobipites	+	+	
Piceaepollenites alatus		+	
Cedripites eocenicus	+	+	
C. cedroides		+	
Salixipollenites discoloripites	+		
S. hians		+	+
Caryapollenites simplex	+	+	+
Juglanspollenites verus	+	+	+
Alnipollenites verus	+	+	+
Momipites coryloides	+	+	
Quercoidites microhenrici	+	+	+
Cupuliferoipollenites pusillus	+	+	
Ulmipollenites undulosus	+	+	
Liquidambarpollenites stigmosus	+	+	+
Ilexpollenites margaritatus	+	+	+
I. membranous		+	+
Tricolpites tenuicolpus	+	+	
Tricolporopollenites minutus		+	
Operculumpollis operculatus	+	+	
Monocolpopollenites tranguillus	+		

bed, so this is a mixed-type petroleum sporo-pollen assemblage. All of the spores and pollen are Paleogene index species and common Paleogene species such as *Leiotriletes adriensis, Polypodiaceaesporites haardti, P. gracilis, Polypodiisporites favus, P. afavus, Alnipollenites verus,* and *Salixipollenites hians.* Some species of pollen such as *Pinuspollenites minutus, P. strobipites, Cedripites eocenicus, C. cedroides, Salixipollenites discoloripites, Momipites coryloides, Cupuliferoipollenites pusillus, Tricolpites tenuicolpus, Operculumpollis operculatus, Tricolporopollenites minutus,* and *Monocolpopollenites tranguillus* do not occur in the Upper Oligocene Zhuhai Formation (Lei 1985). They are therefore not derived from the reservoir bed. However, they were found in the Eocene series in the North Shelf of South China Sea (Hou et al. 1981), so they may indicate an Eocene petroleum source.

Study of palynomorphs in the petroleum indicates that the black and dark gray clay rocks of the Paleogene System should be the petroleum source rocks in the Zhujiang Mouth Basin (Jiang and Yang 2000). In accordance with the composition of the petroleum sporo-pollen assemblage and correlation between spores and pollen in the crude oils and those in potential source rocks (Table 4.12), the Middle Eocene Wenchang, the Upper Eocene to Middle Oligocene Enping, and the Upper Oligocene Zhuhai formations should be the petroleum source rocks of the Zhujiang Mouth Depression in the eastern North Shelf of South China Sea. This conclusion is supported by organic geochemical analyses (Jiang et al. 1994).

References

Chiang, T. C., & Young, H. C. (1978). Early Cretaceous palynological assemblage of the Huahai Basin, western Kansu. *Journal of Lanchow University (Natural Sciences), (2)*, 115–135 (in Chinese with English abstract).

Hou, Y. T., Li, Y. P., & Jin, Q. H. (1981). *Tertiary palaeontology of North Continental Shelf of South China Sea* (pp. 1–274). Guangzhou: Guangdong Science and Technology Press (in Chinese).

Hsü, J., Chiang, T. C., & Young, H. C. (1974). Sporo-pollen assemblage and geological age of the lower Xinminbu formation of Chiuchüan, Kansu. *Acta Botanica Sinica, 16*(4), 365–379. (in Chinese with English abstract).

Institute of Geology, Chinese Academy of Geological Sciences, Institute of Geology, Xinjiang Bureau of Geology and Mineral Resources. (1986). *Permian and Triassic strata and fossil assemblages in the Dalongkou area of Jimsar, Xinjiang* (pp. 1–262). Beijing: Geological Publishing House (in Chinese).

Jiang, D. X., & Yang, H. Q. (1980). Petroleum sporo-pollen assemblages and oil source rock of Yumen oil-bearing region in Gansu. *Acta Botanica Sinica, 22*(3), 280–285 (in Chinese with English abstract).

Jiang, D. X., Yang, H. Q. (1981). Spores and pollen in crude oils from West Jiuquan Basin. *Petroleum Geology, 5*(1), 1–9 (in Chinese).

Jiang, D. X., & Yang, H. Q. (1982). Spores/pollen and petroleum sources. In Division of Earth Science, Academia Sinica (Ed.), *Proceedings of petroleum earth science of Chinese Academy of Sciences* (pp. 157–162). Beijing: Science Press (in Chinese).

Jiang, D. X., Yang, H. Q. (1983). Petroleum sporo-pollen assemblages and oil source rock of Kuche Depression in Xinjiang. *Acta Botanica Sinica, 25*(2), 179–186 (in Chinese with English abstract).

Jiang, D. X., Yang, H. Q. (1986). Petroleum sporo-pollen assemblage and oil source rocks of Yecheng Seg in Xinjiang. *Acta Botanica Sinica, 28*(1), 111–116 (in Chinese with English abstract).

Jiang, D. X., & Yang, H. Q. (1989). Spores and pollen from crude oils of Turpan Basin, Xinjiang. *Acta Botanica Sinica, 31*(6), 477–483. (in Chinese with English abstract).

Jiang, D. X., & Yang H. Q. (1994). Spores and pollen in oil from igneous rock petroleum pool and petroleum origin of Junggar Basin. *Science in China (Series B), 24*(7), 774–778 (in Chinese) 37(12):1499–1505.

Jiang, D. X., & Yang, H. Q. (1996). Spores and pollen from crude oil of Kashi Seg, Xinjiang. *Acta Botanica Sinica, 38*(10), 809–813. (in Chinese with English abstract).

Jiang, D. X., & Yang, H. Q. (1997). Palynological evidence for Jurassic petroleum source of Qaidam Basin. *Acta Botanica Sinica, 39*(12), 1160–1164. (in Chinese with English abstract).

Jiang, D. X., & Yang, H. Q. (1998). Palynological evidence for Tertiary petroleum source of Qaidam Basin. *Acta Botanica Sinica, 40*(1), 77–82. (in Chinese with English abstract).

Jiang, D. X., & Yang, H. Q. (1999a). Spores and pollen from crude oils of Weizhou oil-field in Beibu Gulf of South China Sea. *Acta Botanica Sinica, 41*(1), 102–106. (in Chinese with English abstract).

Jiang, D. X., & Yang, H. Q. (1999b). Petroleum sporo-pollen assemblages and petroleum source rocks of North Tarim Upheaval in Xinjiang. *Acta Botanica Sinica, 41*(2), 213–218. (in Chinese with English abstract).

Jiang, D. X., & Yang, H. Q. (2000). Original environment of Eogene petroleum source rocks in the Pearl River Mouth Basin. *Acta Sedimentologica Sinica, 18*(3), 469–474. (in Chinese with English abstract).

Jiang, Z. X., Zeng, L., & Li, M. X. (1994). *Tertiary of petroliferous provinces of China (VIII) Petroliferous province in northern shelf of South China Sea* (pp. 1–145). Beijing: Petroleum Industry Press (in Chinese).

Lei, Z. Q. (1985). Tertiary sporo-pollen assemblage of Zhujiangkou (Pearl River Mouth) Basin and its stratigraphical significance. *Acta Botanica Sinica, 27*(1), 94–105. (in Chinese with English abstract).

Liu, Z. S. (1993). Jurassic sporo-pollen assemblages from the Beishan Coal-field, Qitai, Xinjiang. *Acta Micropalaeontologica Sinica, 10*(1), 13–36 (in Chinese with English abstract).

References

Liu, Z. S. (2003). Triassic and Jurassic sporo-pollen assemblages from the Kuqa Depression, Tarim Basin of Xinjiang, NW China. *Palaeontologia Sinica, whole Number 190, New Series A, No. 14* (pp. 1–244). Beijing: Science Press. (in Chinese with English summary).

Ouyang, S., Wang, Z., Zhan, J. Z., & Zhou, Y. X. (2003). *Palynology of the Carboniferous and Permian strata of northern Xinjiang, northwestern China* (pp. 1–700). Hefei: University of Science and Technology of China Press (in Chinese with English summary).

Research Institute of Exploration and Development, Qinghai Petroleum Administration, Nanjing Institute of Geology and Palaeontology, Academia Sinica. (1985). *A research on Tertiary palynology from the Qaidam Basin, Qinghai Province* (pp. 1–297). Beijing: Petroleum Industry Press (in Chinese with English abstract).

Research Institute of Petroleum Exploration and Development, Ministry of Petroleum Chemistry Industry, Nanjing Institute of Geology and Palaeontology, Chinese Academy of Sciences. (1978). *On the Paleogene Spores and Pollen from the Coastal Region of Bohai* (pp. 1–177). Beijing: Science Press (in Chinese with English abstract).

Sun, F. (1989). Early and Middle Jurassic sporo-pollen assemblages of Qiquanhu coalfield of Turpan, Xinjiang. *Acta Botanica Sinica, 31*(8), 638–646. (in Chinese with English abstract).

Sun, X. J., Kong, Z. C., & Li, M. X. (1980). Paleogene new pollen genera and species of South China Sea. *Acta Botanica Sinica, 22*(2), 191–197. (in Chinese with English abstract).

Wang, Y. D., Jiang, D. X., Yang, H. Q., & Sun, F. (1998). Middle Jurassic sporo-pollen assemblages from Turpan-Shanshan area, Xinjiang. *Acta Botanica Sinica, 40*(10), 969–976. (in Chinese with English abstract).

Yang, H. Q., & Jiang, D. X. (1981). Pollen and spores extracted from petroleum of Liaohe oil-field and their significance. *Acta Botanica Sinica, 23*(1), 52–57. (in Chinese with English abstract).

Yang, H. Q., & Jiang, D. X. (1989). Spores and pollen from crude oil of Dushanzi oil-field in Xinjiang. *Acta Botanica Sinica, 31*(12), 948–953. (in Chinese with English abstract).

Zhang, W. P. (1990). Jurassic sporo-pollen assemblages in Junggar Basin of Xinjiang. In Institute of Geology, Chinese Academy of Geological Sciences, and Research Institute of Petroleum Exploration and Development, Xinjiang Petroleum Administration (Ed.), *Permian to Tertiary strata and palynological assemblages in the North of Xinjiang* (pp. 57–121). Beijing: China Environmental Science Press (in Chinese).

5 Spore/Pollen Fossil Coloration and Petroleum Source Rock Quality

Abstract

The color of fossil spore/pollen exines is a useful measure of organic material maturity or carbonization. The colors relate directly to petroleum generation potential, thermal alteration index (TAI), and vitrinite reflectance (R_O). Orange and brown tissues are in the liquid petroleum and wet gas range, whereas dark brown ones are in the dry gas and condensate range. Having this information allows predictions for the petroleum source rock quality in each formation or group in all of the studied inland and coastal shelf basins.

Keywords

Spores and pollen fossil coloration · Organic material maturity · Hydrocarbon generation potential

5.1 Spores/Pollen Fossil Coloration and Maturity of Organic Material

Fresh exines of modern spores and pollen are pale yellowish to almost colorless. If exines are heated by deep burial of the enclosing sediment, the color intensifies from yellow to orange to brown, dark brown, and ultimately carbonized to black. This change of spores/pollen exine color in transmitted light reflects the thermal maturation of the dispersed organic matter (Staplin 1969).

Chemists named the dispersed insoluble organic matter "kerogen." The changing colors of kerogen including spores, pollen, algae, and amorphous organic matter are used as an index of organic material maturity. The changing colors are ranked (standards) 1–5, representing kerogen coloration from pale yellow to orange to brown, brownish black, and black, thereby indicating organic material maturity from immature to mature to overmature, and ultimately metamorphosed (Staplin 1969).

As shown in Table 5.1, pale yellow and yellow reflect immature organic matter, with burial temperatures ranging from 30 to 50 °C, thereby producing dry gas and heavy oil. Orange and brown reflect mature organic matter, with burial temperatures ranging from 100 to 150 °C, producing liquid oil and wet gas. Brownish black reflects overmature organic matter, with burial

Table 5.1 Kerogen coloration and maturity

Temperature (°C)	Color rank	Kerogen color	Maturity	Occurrence of hydrocarbons
30	1	Pale yellow	Immature	Dry gas
50	1^+	Yellow	Immature	Dry gas and heavy oil
100	2	Orange	Mature	Liquid oil and wet gas
150	3	Brown	Mature	Liquid oil and wet gas
175	4	Brown-black	Overmature	Condensate and dry gas
>200	5	Black	Metamorphic	Dry gas and barren

Annotation: After Staplin (1969), synthesized and simplified

temperature around 175 °C, producing condensate and dry gas. Black reflects metamorphosed organic matter, with temperatures in excess of 200 °C, producing only dry gas.

The initial application of kerogen coloration in Canada proved that this technique is useful for petroleum exploration. In the northwestern region of Alberta, spores from Devonian carbonate rocks are brown and reflect mature organic matter rank, and oil and wet gas occur in the region. In contrast, in British Columbia, spores from Devonian carbonate rocks are brownish black and black, reflecting overmature organic matter rank; in the metamorphosed rocks, only dry gas was found (Staplin 1969). The use of kerogen coloration correlated with organic material maturity for evaluation of hydrocarbon source potential has been further confirmed and developed (Batten 1980; Traverse 1988).

According to Traverse (1988) and Batten (1980, 1996), the relationship between fossil spores/pollen exine coloration and geothermal maturation ("coalification") is shown in Table 5.2. The observed changes in spores/pollen exine color in transmitted light are also called a carbonization–coalification route in reflected light. The thermal alteration index (TAI) as shown in the table is a scale based on both color of spores/pollen exines in transmitted light and vitrinite reflectance (R_O) measured by reflected light that is read through immersion oil. The relationship between coal rank and occurrence of hydrocarbons is used for prediction of hydrocarbon potential. Thus, spore/pollen exines showing pale yellow color have a TAI = 1, an R_O = 0.2 %, and the coal rank is peat; this indicates immature organic matter and as such has no potential for hydrocarbon. Thus, spore/pollen exines showing a light yellow to yellow color have a TAI = 1^+–2^-, an R_O = 0.3 %, and the coal rank is lignite; this indicates still immature organic matter and has potential for dry gas. Spore/pollen exines showing a dark yellow color have a TAI = 2, an R_O = 0.4 %, and the coal rank is subbituminous coal; this indicates marginally mature organic matter and has potential for wet gas. Spore/pollen exines showing orange color have a TAI = 2^+, an R_O = 0.5–0.7 %, and the coal rank is high-volatile C and B bituminous coal; this indicates low mature organic matter and has potential for liquid oil. Spore/pollen exines showing brown color have a TAI = 3^-–3^+, an R_O = 0.8–1.3 %, and the coal ranks are high-volatile A bituminous coal and medium-volatile bituminous coal; this indicates mature and high mature organic matter and has potential for liquid oil and wet gas. Spore/pollen exines showing dark brown color have a TAI = 4^-, an R_O = 2.0 %, and the coal rank is low-volatile bituminous coal; this indicates overmature organic matter and has potential for condensate and dry gas. Spore/pollen exines showing brown-black and black have a TAI = 4–5, an R_O ≥ 2.5 %, and the coal ranks are semi-anthracite and anthracite; this indicates metamorphosed organic matter and either has potential for dry gas or is barren (Table 5.2).

In the Tarim Basin, the fossil spores and pollen in crude oils as indicators of the Carboniferous, Triassic, and Jurassic petroleum source rocks are orange and brown (see Plates XXXVII-XL). Accordingly, the colors indicate TAI = 2^+–3^+,

5.1 Spores/Pollen Fossil Coloration and Maturity of Organic Material

Table 5.2 Fossil spore/pollen exine coloration with geothermal organic maturation

Spores/pollen exine color	Thermal alteration index (TAI)	Vitrinite reflectance (R_O) (%)	Coal rank	Organic material maturity	Hydrocarbon phase
Pale yellow	1	0.2	Peat	Immature	
Light yellow	1^+	0.3	Lignite	Immature	Dry gas
Yellow	2^-				
Dark yellow	2	0.4	Subbituminous coal	Submature	Wet gas
Orange	2^+	0.5 0.7	High-volatile C, B Bituminous coal	Low mature	Liquid oil
Light brown	3^-	0.8	High-volatile A Bituminous coal	Mature	Liquid oil
Brown	3	1.0		Mature	Liquid oil and wet gas
Brownish brown	3^+	1.3	Medium-volatile bituminous coal	High mature	Liquid oil and wet gas
Dark brown	4^-	2.0	Low-volatile bituminous coal	Overmature	Condensate and dry gas
Brown-black	4	2.5	Semi-anthracite	Metamorphic	Dry gas
Black	5		Anthracite	Metamorphic	Dry gas and barren

Annotation: After Traverse (1988) and Batten (1980, 1996), synthesized and simplified

$R_O = 0.5–1.3$ %, and the organic matter ranks from low mature to mature and high mature (see Table 5.2). The colors of the fossil spores and pollen indicate that the Carboniferous, Triassic, and Jurassic petroleum source rocks have potential for liquid petroleum and wet gas generation.

In the East Junggar Depression, the fossil spores and pollen in the crude oils that indicate Carboniferous, Permian, and Triassic petroleum source rocks are orange to brown (see Plates XLI-XLII). Accordingly, the colors indicate TAI = $2^+–3^+$, $R_O = 0.5–1.3$ %, and the organic matter ranks from low mature to mature and high mature, and liquid oil and wet gas generation can be expected (see Table 5.2). There is one exception to this; the Carboniferous to Permian spore *Laevigatosporites medius* is dark brown (see Plate XLI, Fig. 3). This color indicates TAI = 4^-, $R_O = 2.0$ %, overmature organic matter, and condensate and dry gas generation potential (see Table 5.2). Thus, the colors of the fossil spores and pollen indicate that Carboniferous, Permian, and Triassic petroleum source rocks possess potential for liquid petroleum generation and that other Carboniferous to Permian source rocks contain overmature organic matter which possess potential for condensate generation only.

In the South Junggar Depression, the fossil spores and pollen in the crude oils that indicate Jurassic petroleum source rocks are orange and light brown (see Plate XLIII). Accordingly, the colors indicate TAI = $2^+–3^-$, $R_O = 0.5–0.8$ %, organic matter from low mature to mature, and liquid oil generation potential (see Table 5.2). The colors of the fossil spores and pollen indicate that the Jurassic petroleum source rocks possess potential for liquid petroleum generation.

In the Turpan Depression, the fossil spores and pollen in crude oils that indicate Jurassic petroleum source rocks are orange (see Plate XLIV). Accordingly, the color indicates TAI = 2^+, $R_O = 0.5–0.7$ %, low mature organic matter, and liquid oil generation potential (see Table 5.2). The color of fossil spores and pollen indicates that the organic material maturity of the Jurassic petroleum source rocks is comparatively low, but the source rocks possess potential for liquid hydrocarbon generation.

In the North Border Block-fault Zone of the Qaidam Basin, the fossil spores and pollen in crude oils that indicate Jurassic petroleum source rocks are brown (see Plate XLV). Accordingly, the color indicates TAI = 3^--3^+, R_O = 0.8–1.3 %, organic matter mature to high mature, and liquid oil and wet gas generation potential (see Table 5.2). The color of fossil spores and pollen indicates that the organic material maturity of the Jurassic petroleum source rocks is comparatively high and within the main phase of liquid petroleum generation. In the Mangnai Depression of the Qaidam Basin, the fossil spores and pollen in crude oils that indicate Tertiary petroleum source rocks are light brown (see Plates XLVI-XLVII). Accordingly, the color indicates TAI = 3^-, R_O = 0.8 %, mature organic matter, and liquid oil generation potential (see Table 5.2). The color of fossil spores and pollen indicates that the Tertiary Lulehe, Ganchaigou, and lower Youshashan formations possess potential for liquid petroleum generation.

In the Beibu Gulf Depression of the North Shelf of South China Sea, the fossil spores and pollen in crude oils that indicate Paleogene petroleum source rocks are orange and light brown (see Plate XLVIII). Accordingly, the colors indicate TAI = 2^+-3^-, R_O = 0.5–0.9 %, organic matter low mature to mature, and liquid oil generation potential (see Table 5.2). The colors of fossil spores and pollen indicate that the Paleogene petroleum source rocks, such as mudstones and shales of the Liushagang and Weizhou formations, possess potential for liquid petroleum generation in the North Shelf of South China Sea.

5.2 Organic Material Type and Maturity with Hydrocarbon Generation Potential

The type and maturity of organic matter in sedimentary rock may be used for evaluation of hydrocarbon source potential. The organic matter in sedimentary rock is described differently, depending on the field of study. Palynologists use transmitted light microscopy, coal petrologists use reflected light microscopy, and petroleum chemists use chemicals to break down the component parts. As explained above, the dispersed insoluble organic matter in sedimentary rocks is called kerogen, a term from chemists.

Kerogen may be divided into the sapropelic type and the humic type. The former is the product of fatty, lipid matter, and the latter is probably the product of carbohydrates in peat.

Palynologists classify kerogens using transmitted light microscopy. The sapropelic type of kerogen to palynologists can be recognized as algal, amorphous, and herbaceous components. The humic type of kerogen is composed of the woody and coaly components.

Coal petrologists classify kerogens by reflected light microscopy. The sapropelic type of kerogen to coal petrologists is the exinites of pollen and spores. The humic type of kerogen is composed of vitrinite and inertinite.

Petroleum geochemists classify kerogens according to elemental analysis. The sapropelic type is the hydrogen-rich types I and II kerogen. The humic type is the hydrogen-poor type III kerogen. Type I kerogen has high H/C (1.7–1.4) and possesses great potential for oil and gas generation. Type III kerogen has low H/C (1.0–0.3) and possesses low potential for oil and gas generation (Hunt 1979).

In addition, the quality of a petroleum source rock is directly related to the quantity of organic matter in the rocks. Mudstone with organic carbon 0.5–1.0 % and a chloroform bitumen A content of 500–1000 ppm is referred to as medium oil-generating rock. Mudstone with organic carbon 1.0–2.0 % and chloroform bitumen A content of 1000–2000 ppm is referred to comparatively good oil-generating rock. Mudstone with organic carbon >2.0 % and chloroform bitumen A content of >2000 ppm is referred to as good oil-generating rock (Fan and Zhou 1990).

In the Tarim Basin, Triassic to Jurassic greyish black mudstones, which contain organic carbon 1.14–3.43 % and chloroform bitumen A content of 863–2649 ppm, should be referred to the rank of comparatively good and good oil-generating rocks (Fan and Zhou 1990).

In the Junggar Basin, the organic material type of the Upper Permian Lucaogou and Hongyanchi formations is the sapropelic kind belonging to type I and type II kerogen. Upper Permian greyish black and black mudstones contain organic carbon 1.46–14.73 % and chloroform bitumen A content of 1276–5569 ppm; the Upper Permian dark-colored oil shales contain organic carbon 12.76–22.09 % and chloroform bitumen A content of 3770–6300 ppm. They are referred to the good oil-generating rock (Hu et al. 1991). It is clear that they possess potential for oil and gas generation.

Based on sedimentology, organic geochemistry, and petroleum potential, it is concluded that the Lower and Middle Jurassic organic-rich mudstones in the Tarim, Junggar, and Turpan basins possess potential for liquid hydrocarbon generation (Hendrix et al. 1995). This conclusion was supported by the analysis of the Jurassic System in North China (Zhong et al. 2003).

In the Beibu Gulf Basin of the South China Sea, the dark-colored mudstones and shales of the Paleogene Liushagang Formation belong to the sapropelic type of kerogen and contain organic carbon 0.86–2.09 % and chloroform bitumen A content of 750–3345 ppm (Zeng and Guo 1981). The Paleogene dark-colored mudstones and shales are classified as good oil-generating rocks, thereby possessing great potential for hydrocarbon generation.

References

Batten, D. J. (1980). Use of transmitted light microscopy of sedimentary organic matter for evaluation of hydrocarbon source potential. In *Proceedings of 4th International Palynological Conference* (Vol. 2, pp. 589–594).

Batten, D. J. (1996). Palynofacies and petroleum potential. In J. Jansonius & D. C. McGregor (Eds.), *Palynology: Principles and applications* (Vol. 3, pp. 1065–1084). Dallas, Texas: American Association of Stratigraphic Palynologists Foundation.

Fan, S. F., & Zhou, Z. Y. (1990). *Paleogeotherm and petroleum in Tarim Basin* (pp. 1–77). Beijing: Science Press. (in Chinese).

Hendrix, M. S., Brassell, S. C., Carroll, A. R., & Graham, S. A. (1995). Sedimentology, organic geochemistry, and petroleum potential of Jurassic coal measures, Tarim, Junggar, and Turpan basins, Northwest China. *American Association of Petroleum Geologists Bulletin, 79*, 929–959.

Hu, B. L., Jiang, D. X., Yang, H. Q., Fu, H., & Sun, F. (1991). *Petroleum source rocks in East Xinjiang.* Lanzhou: Gansu Science and Technology Press. 1–140 (in Chinese).

Hunt, J. M. (1979). *Petroleum geochemistry and geology* (pp. 1–617). San Francisco: W. H. Freeman and Company.

Staplin, F. L. (1969). Sedimentary organic matter, organic metamorphism, and oil and gas occurrence. *Canadian Petroleum Geology, Bulletin, 17*(1), 47–66.

Traverse, A. (1988). *Paleopalynology* (pp. 1–600). Boston: Unwin Hyman.

Zeng, D. Q., & Guo, B. (1981). *Tertiary of North continental shelf of South China Sea* (pp. 1–258). Guangzhou: Guangdong Science and Technology Press. (in Chinese).

Zhong, X. C., Zhao, C. B., Yang, S. Z., & Shen, H. (2003). *Jurassic system in the North of China (II) Palaeoenvironment and oil-gas source* (pp. 1–201). Beijing: Petroleum Industry Press.

Palynological Evidence for Organic Petroleum Origin Theory

6

Abstract

The organic petroleum origin theory considers that petroleum is derived from organic materials produced by biological remains in sediments through thermal transformation. Dispersed insoluble organic matter in sedimentary rock is often called "kerogen," which is considered as the forerunner of petroleum. Because of the similarity of certain molecular formulas, it is inferred that some kerogen is transformed from sporopollenin. The inorganic petroleum origin theory considers that petroleum originates as a simple hydrocarbon that was dispersed from universal celestial bodies, or it is composed of carbon and hydrogen from magma derived from the depths of the Earth. The discovery of fossil spores and pollen in crude oil from the Beisantai igneous rock petroleum pool in the Junggar Basin provided reliable evidence for the organic petroleum origin theory. It is impossible that igneous rocks would contain uncarbonized biological fossils. The fossil spores and pollen in the crude oils must have been carried by oil, gas, and water from sedimentary source rocks to the igneous reservoir during petroleum migration. The Beisantai crude oil is rich in chemical biomarkers, including beta-carotene, gamma-carotene, cholestane, ergostane, sitostane, hopanes, and terpanes. Such biomarkers have no sources other than the breakdown of organisms, which confirms a biogenic hydrocarbon source.

Keywords

Organic petroleum origin · Inorganic petroleum origin · Sporopollenin · Fossil spores and pollen · Biomarkers

6.1 Sporopollenin and Petroleum Origin

Petroleum origin is an important subject that is closely related to the origin of life. The inorganic petroleum origin theory of Gold (1992) considers that petroleum is composed of carbon and hydrogen from magma evolved in the depths of the Earth producing inorganic patterns or else it originates as simple hydrocarbons from celestial bodies. Although a small amount of hydrocarbon in the Earth's crust might be of inorganic origin, the amount is not enough to form any commercial accumulations. Based on the studies of various hydrocarbon accumulations in crystalline igneous rocks and metamorphic rocks, it has been proved that all of the hydrocarbon accumulations originated from the surrounding sedimentary rocks, and no economic petroleum accumulations support an inorganic origin (Hedberg 1964).

The organic petroleum origin theory considers that petroleum is derived from organic materials produced by biological remains in sediments subjected to thermal transformations (Hunt 1979). Since Biot (1835) proved that petroleum fractions exhibit the characteristic rotary polarization of organic tissues, most scientists think that petroleum originates from organic materials. Historically, White and Stadnichenko (1923) discovered the sporangia of *Protosalvinia* and megaspores of Lycopodiales in Devonian black shales from Kentucky, USA, and suggested that these plants should be the petroleum-producing "mother plants." Treibs (1934) identified porphyrin derivatives of chlorophyll and hemochrome in petroleum and therefore implicated plants as the source of petroleum. Sanders (1937) extracted fossil spores and algae from crude oils and stated that the spores and algae had provided waxy, fatty, and oily secretions for petroleum generation, leaving a residue of decay-resistant exines. Therefore, the organic petroleum origin theory is supported by various lines of scientific evidence.

The elemental composition of petroleum is carbon, hydrogen, sulfur, nitrogen, and oxygen—on the average, carbon 85 %, hydrogen 13 %, sulfur 1 %, nitrogen 0.5 %, and oxygen 0.5 % (Hunt 1979). The elemental composition of lipids in general is carbon 76 %, hydrogen 12 %, and oxygen 12 % (Hunt 1979). Thus, once lipids lose some oxygen, lipids are easily transformed into petroleum. Spores of *Lycopodium* contain 50 % lipids. The fat content of plants in general is mainly found in seeds, spores, and pollen.

Sapropelic organic matter also lacks oxygen. It is the spores, pollen, and algae that become sedimented into the anoxic muck of lacustrine or marine environments. Spores and pollen are regarded as possible progenitors of petroleum, along with the lipid storage components of algae such as dinoflagellates and diatoms, and bacterial lipid membranes (Philip 1985). Sapropel is transformed into type II kerogen in sediments during diagenesis, which generates petroleum hydrocarbons upon catagenesis and metagenesis. Heating experiments on *Pinus* pollen grains showed that hydrocarbon generation from living pollen grains takes place in the temperature range of about 135 and 230 °C (Ujiié et al. 2003). Hydrogen-rich type I and type II kerogens are sapropelic organic materials. Type I kerogen contains alkanes, cycloalkanes, and aromatic hydrocarbons. Type II kerogen contains mainly cycloalkanes and aromatic hydrocarbons. The kerogen of the Green River Shale of Colorado, USA, contains cycloalkanes, aromatic hydrocarbons, alkanes, steranes, and terpanes (Gallegos 1973, 1975). In petroleum, the most prevalent molecular compound is cycloalkane (C_nH_{2n}), followed by alkane (C_nH_{2n+2}), alkene (C_nH_{2n-2}), and aromatic hydrocarbon (C_nH_{2n-6}). These molecular entities generally exist in the products of kerogen degradation. Thus, kerogen is considered as the forerunner of petroleum.

The chemical studies of various modern and fossil spore and pollen walls of pteridosperms, gymnosperms, angiosperms, fungi, and algae show a majority to be composed of sporopollenin (Brooks 1971). Sporopollenin is the very resistant and refractory organic substance that composes the exine and perine of spores and pollen, and apparently also the walls of dinoflagellates and acritarchs. This substance gives the sporomorph its extreme durability during geologic time, being readily destroyed only by oxidation or prolonged high temperature. It is a high molecular weight polymer of C-H-O, perhaps a carotenoid-like substance (Traverse 1988). The molecular formula of sporopollenin is similar to that of kerogen. For example, the *Lycopodium clavatum* spore wall is $C_{90}H_{144}O_{27}$; the *Selaginella kraussiana* spore wall is $C_{90}H_{124}O_{18}$; fossil sporopollenin from 250 million years ago is $C_{90}H_{129}O_{19}S_7N$; the kerogen of the Green River Shale of Colorado is $C_{90}H_{134}O_{225}NS$; and the kerogen of the Permian petroleum source rocks in Texas is $C_{67}H_{80}O_6N$ (Brooks 1971; Hunt 1979; Durand 1980; Ujiié et al. 2003). Because of the similarity in the molecular formula, it is inferred that kerogen is transformed from sporopollenin. Sporopollenin maturation can be used as a proxy for oil generation, because the chemical compositional changes are similar to those occurring during the maturation of amorphous organic matter, which is the dominant component of oil-prone kerogen. Sporopollenin composition and optical properties show a step function at the oil-window and allow direct observation of the generation of liquid hydrocarbons (Yule et al. 2000).

The most important factor affecting petroleum origin is the thermal history of the petroleum source rock. Organic materials in sediments undergoing the sequential process of thermal alteration known as diagenesis, catagenesis, and then metamorphism can evolve various organic products by bacterial transformation, increase of burial depth, and resulting increase in geotemperature. In the diagenetic stage, when geotemperatures rise to 50–60 °C, organic materials including lipids, proteins, carbohydrates, and woody materials become kerogen: first by way of bacterial transformations and chemical changes of reduction and then by catalysis and polymerization. In the catagenetic stage, when geotemperatures rise to 60–200 °C, many of the kerogen components are transformed into petroleum by way of thermal degradation. Eventually, in the metamorphic stage, when geotemperatures rise above 200 °C, petroleum and other organic residues are transformed into methane and graphite by way of thermal alteration (Hunt 1979).

Therefore, the thermal history of petroleum source rocks indicates that petroleum originates from organic materials of nature.

6.2 Fossil Spores and Pollen in Crude Oils from Sedimentary Rock Petroleum Reservoirs

Fossil spores and pollen that are extracted from crude oils in sedimentary rock petroleum reservoirs are the exine remains of the spores and pollen which provided waxy, fatty, and oily secretions that contributed to petroleum generation (Sanders 1937). The discovery of fossil spores and pollen in crude oils supports the organic petroleum origin theory. These fossil spores and pollen can bear witness to organic petroleum origins, because they have been proved to contribute to petroleum generation.

They also are incorporated intact into petroleum. Fossil spores, algae, and fungi were found in crude oils from the Cretaceous and Tertiary oil fields of Mexico and the Tertiary oil field of Romania (Sanders 1937). Fossil spores, pollen, and algae were found in crude oils from the Cambrian oil field of East Siberia, the Devonian, Carboniferous, and Permian oil fields of Volga-Ural, the Cretaceous oil field of Emba, and the Tertiary oil field of North Gorgaso and Kamchatka of Russia (Timofeev and Karimov 1953). Jurassic fossil spores and pollen were found in crude oils from the Permian, Triassic, and Jurassic petroleum reservoirs of the Moonie oil field in Queensland of Australia (de Jersey 1965).

In China, abundant fossil spores and pollen were found in crude oils from the Paleozoic, Mesozoic, and Tertiary petroleum reservoirs of the inland petroliferous basins and the Tertiary petroleum reservoirs of the coastal shelf petroliferous basins (Jiang 1988, 1991; Yang and Jiang 1981; Jiang and Yang 1999). As mentioned in Chap. 3, a large number of fossil spores and pollen were found in crude oils from 24 oil fields of the eight petroliferous basins of China (Tables 3.1–3.19). For example, 136 species of fossil spores and pollen referred to 66 genera were found in crude oils from the Ordovician, Carboniferous, Triassic, Jurassic, Cretaceous, and Tertiary petroleum reservoirs of the North Tarim Upheaval in the Tarim Basin (see Table 3.1). For another example, 60 species of fossil spores and pollen referred to 36 genera were found in crude oils from the Permian petroleum reservoir of the Huonan oil field in the East Junggar Depression of the Junggar Basin (see Table 3.7).

In addition, Triassic and Jurassic spores and pollen such as *Punctatisporites triassicus, P. textatus, Calamospora impexa, Marattisporites scabratus, Cyathidites minor, Gleicheniidites senonicus, Cycadopits nitidus, C. typicus, C. minimus, Classopollis classoides*, and *C. annulatus* were found in crude oils from the Jurassic petroleum reservoir of the Zhangqing oil field in the Ordos (Eerduosi) Basin (Jiang 1988). Paleogene spores and pollen such as *Plicifera decora, Polypodiaceaesporites haardti, Dacrydiumpollenites rarus, Cedripites eocenicus, C. deodariformis, Pinuspollenites labdacus, P. minutus, P. bankianaeformis, Momipites coryloides*, and *Quercoidites asper* were found in crude oils from the Paleogene Buxin Formation petroleum reservoir of the Sanshui Basin in Guangdong of China (Jiang 1988).

As the above data show, abundant fossil spores and pollen were extracted from crude oils. These have contributed to petroleum generation, so they are part of the organic petroleum origin.

6.3 Fossil Spores and Pollen in Crude Oils from Igneous Rock Petroleum Pools

Hydrocarbon accumulation in crystalline igneous rocks has been considered by Gold (1992) as support for an inorganic petroleum origin. No geochemical analyses or test borings have supported that hypothesis. Instead, the research of Jiang and Yang (1994), Jiang (1996) has shown that the petroleum in crystalline igneous rock petroleum pools has reached these reservoirs from nearby sedimentary rocks by way of petroleum migration. The Beisantai igneous rock petroleum pool in the East Junggar Depression of the Junggar Basin of China is a typical example that petroleum in crystalline igneous rock pools is from the surrounding sedimentary source rocks.

Ninety-six species of fossil spores and pollen referred to 52 genera were found in crude oil samples taken from the Carboniferous igneous rock petroleum reservoir of the Beisantai oil field in the Junggar Basin. The discovery of the abundant fossil spores and pollen in crude oil from the igneous rock petroleum reservoir of the Beisantai oil field is a perfect example of the organic petroleum origin theory. The petroleum originates from organic materials of the surrounding sedimentary rocks (Jiang and Yang 1994; Jiang 1996). Crystalline igneous rocks do not yield biological fossils. Instead, the spores and pollen in the crude oil had to have been carried by oil, gas, and water from the surrounding sedimentary petroleum source rocks to the igneous rock petroleum reservoir during petroleum migration.

The Beisantai oil field is located in the northern uplift of the Bogda Piedmont Depression. This is an uplifted area developed within a continental depression having a long geological history. Such uplift is the target of petroleum migration. The field is surrounded by the Bogda Piedmont Depression to the south, the Jimsar Depression to the east, and the Fukang Depression to the west. Therefore, the uplift potentially has Carboniferous, Permian, Triassic, and Jurassic petroleum sources. Furthermore, because of strong compression caused by the uplift of Bogda Mountain, faults and fissures are well developed in the region. These faults and fissures acted as available passages for the migration of petroleum that originated in the depressions and moved toward the reservoirs in the uplift.

There is chemical evidence to support the organic origin of the petroleum in the crystalline igneous rocks. Various chemical biomarkers, such as beta-carotene, gamma-carotene, cholestane, ergostane, sitostane, hopanes, and terpanes were isolated in the crude oil samples taken from the Beisantai igneous rock petroleum reservoir

(Hu et al. 1991). Carotanes are produced during the transformation of carotene occurring in the pigment of plants and animals. Furthermore, carotene is the main source of C_{40} terpene, and beta-carotene ($C_{40}H_{56}$) is a type of hydrocarbon. The steranes and terpanes are also derived from organic materials. For example, cholestane is derived from cholesterol in animal bodies, ergostane is derived from ergosterol in fungi and diatoms, and sitostane is derived from sitosterol in plants. These chemical biomarkers show that petroleum of the Beisantai igneous rock petroleum reservoir originates from organic materials produced by the remains of plants and animals in ancient sediments.

The data of both palynological studies and organic geochemical studies coincide. Therefore, the organic petroleum origin theory is strongly confirmed.

References

Biot, M. (1835). Memoir sur la polarisation circulaire et sur ses applications à la chimie organique. *Memoirs Academic of Royal Science Institution France, 13*, 39.

Brooks, J. (1971). Some chemical and geochemical studies on sporopollenin. In J. Brooks, P. G. Grant, M. Muir, P. van Gijzel, & G. Shaw (Eds.), *Sporopollenin* (pp. 351–407). London: Academic Press.

DeJersey, N. J. (1965). Plant microfossils in some Queensland crude oil samples. *Geological Survey of Queensland Publication, 329*, 1–9.

Durand, B. (1980). *Kerogen: Insoluble organic matter from sedimentary rocks* (pp. 1–519). Paris: Technip.

Gallegos, E. J. (1973). Identification of phenylcycloparaffin alkanes and other monoaromatics in Green River shale by gas chromatography-mass spectrometry. *Analytical Chemistry, 45*(8), 1399–1403.

Gallegos, E. J. (1975). Terpane-sterane release from kerogen by pyrolysis gas chromatography-mass spectrometry. *Analytical Chemistry, 47*(9), 1524–1528.

Gold, T. (1992). The deep, hot biosphere. *Proceedings of the National Academy of Sciences, 89*(13), 6045–6049.

Hedberg, H. D. (1964). Geological aspects of origin of petroleum. *American Association of Petroleum Geologists Bulletin, 48*(11), 1755–1803.

Hu, B. L., Jiang, D. X., Yang, H. Q., Fu, H., & Sun, F. (1991). *Petroleum source rocks in East Xinjiang* (pp. 1–140). Lanzhou: Gansu Science and Technology Press (in Chinese).

Hunt, J. M. (1979). *Petroleum geochemistry and geology* (pp. 1–617). San Francisco: W. H. Freeman and Company.

Jiang, D. X. (1988). Spores and pollen in oils as indicators of lacustrine source rocks. In A. J. Fleet, K. Kelts & M. R. Talbot (Eds.), *Lacustrine petroleum source rocks* (pp. 159–169). London: Blackwell Scientific Publications.

Jiang, D. X. (1991). Fossil spores and pollen in petroleum and their significance. *Chinese, Journal of Botany, 3*(1), 62–76.

Jiang, D. X. (1996). Fossil pollen and spores in crude oil from an igneous reservoir. In J. Jansonius & D. C. McGregor (Eds.), *Palynology: Principles and applications* (Vol. 3, pp. 1123–1128). Dallas, Texas: American Association of Stratigraphic Palynologists Foundation.

Jiang, D. X, Yang, H. Q. (1994). Spores and pollen in oil from igneous rock petroleum pool and petroleum origin of Junggar Basin. *Science in China (Series B), 24*(7), 774–778 (in Chinese) *37*(12), 1499–1505.

Jiang, D. X., & Yang, H. Q. (1999). Spores and pollen from crude oils of Weizhou oil-field in Beibu Gulf of South China Sea. *Acta Botanica Sinica, 41*(1), 102–106 (in Chinese with English abstract).

Philip, R. P. (1985). *Fossil fuel biomarkers: Application and spectra* (pp. 1–294). Amsterdam: Elsevier.

Sanders, J. M. (1937). The microscopical examination of crude petroleum. *Journal of Institute of Petroleum Technology, 23*(167), 525–573.

Timofeev, B. V., & Karimov, A. K. (1953). Spores and pollen in mineral oil. *Dokl. Akad. Nauk. SSSR, 92*(1), 151–152 (in Russian).

Traverse, A. (1988). *Paleopalynology* (pp. 1–600). Boston: Unwin Hyman.

Treibs, A. (1934). Chlorophyll and hemin derivate in bitumens, rocks, oil, waxes and asphalts. *Annals Chemistry, 510*, 42–62.

Ujiié, Y., Arata, Y., & Sugawara, M. (2003). Heating experiments on Pinus pollen grains and its relation to petroleum genesis. *Geochemical Journal, 37*, 367–376.

White, D., & Stadnichenko, T. (1923). Some mother plants of petroleum in the Devonian black shales. *Economic Geology, 18*, 238–252.

Yang, H. Q., & Jiang, D. X. (1981). Pollen and spores extracted from petroleum of Liaohe oil-field and their significance. *Acta Botanica Sinica, 23*(1), 52–57 (in Chinese with English abstract).

Yule, B. L., Roberts, S., & Marshall, J. E. A. (2000). The thermal evolution of sporopollenin. *Organic Geochemistry, 31*, 859–870.

Environment for the Formation of Petroleum Source Rocks

Abstract

Pollen and spores in crude oils and source rocks provide other types of useful geological information. The evolution, flourishing, and demise of plants and plant communities are closely related to climatic changes. The original plants that shed the spores and pollen found in crude oils from petroliferous basins of China mainly belong to mosses, ferns, cycads, conifers, and angiospermous dicotyledons. Most of the plants are classified as humidogene thermophytes. Geographical factors are also important for vegetational development. The geographical distribution of the plant and algal remains in crude oils can help interpret paleogeographical landscapes consisting of mountains, rivers, lakes, marshes, deltas, littoral lagoons, and oceans. Thirdly, the ecological characteristics of the original plants indicate that the petroleum source rocks in the inland petroliferous basins were probably formed in lacustrine and swamp/marsh sedimentary environments under warm and humid climatic conditions.

Keywords

Botanical relationships · Paleoclimate · Paleogeography · Paleoecology · Petroleum source rocks

7.1 Botanical Relationship of Dispersed Spores and Pollen

The spores of sporophytes and pollen of spermatophytes, which have relatively stable genetic traits, are the reproductive cells of these plants. Generally, related plants usually produce spores or pollen with recognizable morphological structure, whereas different plant taxa produce morphologically distinct spores or pollen grains. The spores and pollen fossils are almost always dispersed; the exception is the rare in situ spores or pollen preserved in the sporangia or pollen sacs. It is always difficult to determine the botanical relationships between dispersed spores or pollen and their original plants. By means of the discovery of more in situ spores or pollen, and the morphological study of modern spores and pollen, more and more relationships between dispersed spores or pollen and their original plants are being established. Based on the current achievements on the fossil in situ spores or pollen and the modern sporo-pollen morphology, the botanical affinity of the genera for dispersed sporo-pollen fossils ranging from the Carboniferous to the Paleogene can be generally determined. According to a large body of research (Couper 1958; Potonié 1962; Krutzsch 1971; Clement-Westerhof 1974; Litwin 1985; Traverse 1988; Balme 1995; Wang 1999; Beijing Institute of Botany, Academia Sinica 1976; Song et al. 1986; Wang et al. 1995; Ouyang et al. 2003), the botanical affinity of dispersed sporo-pollen fossils can be determined. The affinity of dispersed spore/pollen genera discussed in this chapter is summarized in Table 7.1.

Among the Carboniferous spores and pollen found in crude oils from the Tarim Basin, the genus *Calamospora* is horsetail type of plant. *Lycospora* is a lycopod. *Retispora* is assigned to the Selaginellales fern. *Punctatosporites* is always assigned to Marattiales ferns. *Vesicaspora* is assigned to the Pteridospermopsida of Gymnospermae.

Among Carboniferous spores and pollen found in crude oils from the Junggar Basin, *Calamospora, Laevigatosporites,* and *Latosporites* are assigned to horsetail-type plants. *Endosporites, Wilsonia,* and *Crassispora* are lycopods. *Convolutispora, Verrucosisporites, Torispora,* and *Punctatosporites* are members of the Marattiales ferns. *Leiotriletes, Punctatisporites, Acanthotriletes, Lophotriletes, Granulatisporites,* and *Cyclogranisporites* belong to Filicales ferns.

Among the Permian spores and pollen found in crude oils from the Junggar Basin, *Leiotriletes* and *Limatulasporites* are suggestive of Bryophytes. *Calamospora, Laevigatosporites,* and *Latosporites* are from horsetail-type plants. *Densosporites, Zonalosporites, Crassispora, Kraeuselisporites, Kraeuselisporites,* and *Lundbladispora* are assigned to Lycopsida. *Verrucosisporites* and *Toroispora* are Marattiales ferns. *Leiotriletes, Punctatisporites, Acanthotriletes, Lophotriletes, Apiculatisporis, Granulatisporites, Cyclogranisporites, Converrucosisporites,* and *Triquitrites* are Filicales ferns. *Vesicaspora, Protohaploxypinus, Striatoabietites, Striatopodocarpites, Hamiapollenites,* and *Vittatina* are Pteridospermopsida. *Cordaitina* and *Florinites* are Cordaitopsida. *Cycadopites* and *Entylissa* are Cycadales or Ginkgoales plants. *Limitisporites, Vestigisporites, Chordasporites, Lueckisporites, Gardenasporites, Pityosporites,* and *Klausipollenites* are Coniferales plants.

Among the Triassic spores and pollen found in crude oils from the Tarim and Junggar basins, *Limatulasporites* is suggestive of Bryophytes. *Calamospora* is horsetail-type plant. *Aratrisporites, Camarozonosporites, Lycopodiacidites,* and *Lundbladispora* are Lycopsida. *Verrucosisporites* and *Toroispora* are Marattiales ferns. *Punctatisporites, Apiculatisporis, Lophotriletes, Retusotriletes, Duplexisporites, Conbaculatisporites, Zebrasporites,* and *Tigrisporites* are Filicales ferns. *Protohaploxypinus* and *Striatoabietites* are Pteridospermopsida. *Podocarpidites* and *Platysaccus* are representatives of Podocarpaceae. *Cedripites* is a Pinaceae plant. *Chordasporites, Lueckisporites, Alisporites,* and *Taeniaesporites* are Coniferales plants.

Among the Jurassic spores and pollen found in crude oils from the Tarim, Junggar, Turpan, and Qaidam basins, *Lycopodiumsporites* and *Lycopodiacidites* are Lycopodiaceae plants. *Densoisporites* is a Selaginellaceae plant. *Marattisporites* is a Marattiales fern. *Osmundacidites, Todisporites,* and *Biretisporites* are Osmundaceae ferns. *Klukisporites* and *Concavissimisporites* are Lygodiaceae or Schizaeaceae ferns. *Gleicheniidites* and *Concavisporites* are Gleicheniaceae ferns. *Cyathidites* and *Deltoidospora* are Cyatheaceae ferns. *Cibotiumspora, Divisisporites,* and *Tripartina* are Dicksoniaceae ferns. *Dictyophyllidites* might be Dipteridaceae or Cheiropleuriaceae ferns. *Leiotriletes, Acanthotriletes, Apiculatisporis, Granulatisporites, Converrucosisporites, Duplexisporites, Leptolepidites,* and *Undulatisporites* are assigned to Filicales ferns. *Caytonipollenites* and *Vitreisporites* are Caytoniales of Pteridospermopsida. *Pteruchipollenites* is a Pteridospermopsida plant. *Cycadopites* is a Cycadales or Ginkgoales plant. *Bennettiteaepollenites* is Bennettitales plant.

7.1 Botanical Relationship of Dispersed Spores and Pollen

Table 7.1 Botanical relationships of selected Carboniferous–Tertiary dispersed spore/pollen genera

Bryophyta	Cordaitopsida
Musci	Cordaitales
Sphagnaceae	*Cordaitina* Samoilovich, 1953
Stereisporites Pflug 1953	*Florinites* Schopf, Wilson at Bentall, 1944
Incertae sedis	Cycadopsida
Leiotriletes Naumova, 1937 emend. Potonié et Kremp, 1954	Cycadales/Ginkgoales
Limatulasporites Helby et Foster, 1979	*Cycadopites* (Wodehouse) Wilson et Webster, 1946
Pteridophyta	*Entylissa* (Naumova) Ishchenko, 1952
Sphenopsida	*Monosulcites* Cookson ex Couper, 1953
Equisetales	Bennettiales
Calamariaceae	*Bennettiteaepollenites* Thiergart, 1949
Calamospora Schopf, Wilson et Bentall, 1944	Incertae sedis
Incertae sedis	*Chasmatosporites* Nilsson, 1958
Laevigatosporites Ibrahim, 1933	*Eucommiidites* Erdtman ex Potonié, 1958
Latosporites Potonié et Kremp, 1954	Coniferopsida
Lycopsida	Coniferales
Lepidodendrales	Podocarpaceae
Lepidodendraceae	*Podocarpidites* (Cookson) Potonié, 1958
Lycospora (S., W. et B., 1944) Potonié et Kremp, 1954	*Platysaccus* (Naumova) Potonié et Kremp, 1954
Pleuromeiales	*Parvisaccites* Couper, 1958
Pleuromeiaceae	Pinaceae
Aratrisporites (Leschik) Playford et Dettmann, 1965	*Pinuspollenites* Raatz 1938 ex Potonié, 1958
Lycopodiales	*Abiespollenites* Thiergart 1937 ex Raatz, 1938
Lycopodiaceae	*Piceaepollenites* Potonié, 1931
Lycopodiumsporites Thiergart, 1938	*Cedripites* Wodehouse, 1933
Camarozonosporites Pant ex Potonié, 1956	*Tsugaepollenites* Potonié et Venitz, 1934
Selaginellales	*Abietineaepollenites* potonié 1951
Selaginellaceae	*Keteleeriaepollenites* Nagy, 1969
Densoisporites Weyland et Krieger, 1953	Taxodiaceae
Lundbladispora Balme, 1963	*Taxodiaceaepollenites* Kremp, 1949
Incertae sedis	Araucariaceae
Lycopodiacidites Couper, 1953 emend. Potonié, 1956	*Araucariacites* Cookson, 1949
Densosporites (Berry) Potonié et Kremp, 1954	*Callialasporites* Sukh Dev, 1961
Endosporites Wilson et Coe, 1940	Cheirolepidiaceae
Crassispora Bharadwaj, 1957	*Classopollis* Pflug, 1953
Kraeuselisporites Leschik, 1955	Incertae sedis
Retispora Staplin, 1960	*Limitisporites* Leschik, 1956
Wilsonia Kosanke, 1950	*Vestigisporites* (Balme et Hennelly) Tiwari et Singh, 1984

(continued)

Table 7.1 (continued)

Filicopsida	*Chordasporites* Klaus, 1960
Marattiales	*Lueckisporites* (Potonié et Klaus) Potonié, 1958
Marattiaceae	*Gardenasporites* Klaus, 1963
Marattisporites Couper, 1958	*Taeniaesporites* (Leschik) Jansonius, 1962
Incertae sedis	*Lunatisporites* (Leschik) Scheuring, 1970
Verrucosisporites (Ibrahim) Potonié et Kremp, 1954	*Klausipollenites* Jansonius, 1962
Convolutispora Hoffmeister, staplin et Malloy, 1955	*Pityosporites* (Seward) Manum, 1960
Punctatosporites Ibrahim, 1933	*Alisporites* (Daugherty) Jansonius, 1971
Torispora Balme, 1952	*Protoconiferus* Bolchovitina, 1956
Filicales	*Paleoconiferus* Bolchovitina, 1956
Osmundaceae	*Protopodocarpus* Bolchovitina, 1956
Osmundacidites Couper, 1953	*Protopinus* Bolchovitina, 1956
Todisporites Couper, 1958	*Protopicea* Bolchovitina, 1956
Baculatisporites Thomson et Pflug, 1953	*Pseudopicea* Bolchovitina, 1956
Biretisporites Delcourt et Sprumont, 1955	*Piceites* Bolchovitina, 1956
Cyclogranisporites Potonié et Kremp, 1954	*Quadraeculina* Maljavkina ex Potonié, 1960
Lygodiaceae	*Cerebropollenites* Nilsson, 1958
Schizaeaceae	*Psophosphaera* (Naumova) Potonié, 1958
Lygodiumsporites (Pot., Thoms. et Thierg.) Potonié, 1956	Gnetopsida
Cicatricosisporites Potonié et Gelletich, 1933	Gnetales
Concavissimisporites Delcourt et Sprumont, 1955	Ephedraceae
Pilosisporites Delcourt et Sprumont, 1955	*Ephedripites* Bolchovitina ex Potonié, 1958
Klukisporites Couper, 1958	Angiospermae
Schizaeoisporites Potonié, 1951	Dicotyledoneae
Crassoretitriletes Germeraad, Hopping et Muller, 1968	Magnoliales
Leiotriletes (Naumova) Potonié et Kremp, 1654	Magnoliaceae
Gleicheniaceae	*Magnolipollis* Krutzsch, 1970
Gleicheniidites (Ross) Delcourt et Sprumont, 1955	Salicales
Plicifera Bolchovitina, 1966	Salicaceae
Cyatheaceae	*Salixipollenites* Srivastava, 1966
Cyathidites Couper, 1953	Juglandales
Deltoidospora (Miner) Potonié, 1956	Juglandaceae
Dicksoniaceae	*Caryapollenites* Raatz, 1937
Cibotiumspora Chang, 1965	*Juglanspollenites* Raatz, 1937
Divisisporites (Thomson) Potonié, 1956	Fagales
Tripartina (Maljavkina) Potonié, 1960	Betulaceae
Pteridaceae	*Alnipollenites* Potonié 1932 ex Potonié, 1960
Pterisisporites Sung et Zheng, 1976	*Betulaceoipollenites* Potonié, 1951
Leptolepidites Couper, 1953	*Momipites* Wodehouse, 1933

(continued)

Table 7.1 (continued)

Polypodiaceoisporites (Potonié) Potonié, 1956	Urticales
Dipteridaceae	Ulmaceae
Cheiropleuriaceae	*Ulmipollenites* Wolff, 1934
Dictyophyllidites Couper, 1958	Hamamelidales
Polypodiaceae	Hamamelidaceae
Polypodiaceaesporites Thiergart, 1938	*Liquidambarpollenites* Raatz, 1937
Polypodiisporites Ross, 1949	Myrtales
Echinosporis Krutzsch, 1967	Myrtaceae
Incertae sedis	*Myrtaceidites* (Cookson et Pike) Potonié, 1960
Leiotriletes (Naumova) Potonié et Kremp, 1954	Rutales
Punctatisporites (Ibrahim) Potonié et Kremp, 1954	Anacardiaceae
Retusotriletes Naumova, 1953	*Rhoipites* Wodehouse, 1933
Triquitrites (Wilson et Coe) Potonié et Kremp, 1954	Meliaceae
Acanthotriletes (Naumova) otonié et Kremp, 1954	*Meliaceoidites* Wang, 1980
Lophotriletes (Naumova) Potonié et Kremp, 1954	Celastrales
Apiculatisporis Potonié et Kremp, 1956	Aquifoliaceae
Granulatisporites (Ibrahim) Potonié et Kremp, 1954	*Ilexpollenites* Thiergart, 1937
Converrucosisporites Potonié et Kremp, 1954	Caryophyllales
Concavisporites (Pflug) Delcourt et Sprumont, 1955	Chenopodiaceae
Undulatisporites Pflug, 1953	*Chenopodipollis* Krutzsch, 1966
Cingulatisporites (Thomson) Potonié, 1956	Asterales
Duplexisporites (Deak) Playford et Dettmann, 1965	Compositae
Zebrasporites Klaus, 1960	*Compositoipollenites* Potonié, 1951
Tigrisporites Klaus, 1960	*Tubulifloridites* Cookson 1947 ex Potonié, 1960
Conbaculatisporites Klaus, 1960	*Artemisiaepollenites* Nagy, 1969
Gymnospermae	Monocotyledoneae
Pteridospermopsida	Liliales
Caytoniales	Liliaceae
Caytonipollenites Couper, 1958	*Liliacidites* Couper, 1953
Vitreisporites (Leschik) Jansonius, 1962	Graminales
Incertae sedis	Gramineae
Alisporites (Daugherty) Jansonius, 1971	*Graminidites* Cookson, 1947
Pteruchipollenites Couper, 1958	Cyperales
Vesicaspora Schemel, 1951	Cyperaceae
Protohaploxypinus (Samoilovich) Hart, 1964	*Cyperaceaepollis* Krutzsch, 1970
Striatoabietites (Sedova) Hart, 1964	
Striatopodocarpites (Zoricheva et Sedova) Hart, 1964	
Vittatina Luber, 1940	
Hamiapollenites (Wilson) Zhan, 2003	

Chasmatosporites and *Eucommiidites* are assigned to Cycadopsida. *Podocarpidites, Platysaccus,* and *Parvisaccites* are assigned to Podocarpaceae. *Pinuspollenites, Abietineaepollenites, Piceaepollenites, Piceites,* and *Cedripites* are Pinaceae plants. *Araucariacites* and *Callialasporites* are Araucariaceae plants. *Classopollis* is Cheirolepidiaceae plant. *Alisporites, Pitysporites, Protoconiferus, Paleoconiferus, Protopodocarpus, Protopinus, Protopicea,* and *Pseudopicea* are Coniferales plants. *Quadraeculina* and *Cerebropollenites* might be from Coniferales plants.

Among the Early Cretaceous spores and pollen found in crude oils from the Jiuquan Basin, *Osmundacidites* is Osmundaceae fern. *Schizaeoisporites* is a Lygodiaceae or Schizaeaceae fern. *Cibotiumspora* is Dicksoniaceae fern. *Cingulatisporites* is Filicales fern. *Cycadopites* is Cycadales or Ginkgoales plant. *Bennettiteaepollenites* is Bennettitales plant. *Psophosphaera* might be taxonomically related to the gymnospermae *Podozamites* (Coniferopsida).

Among the Paleogene spores and pollen found in crude oils in the Junggar, Qaidam, Liaohe, Beibu Gulf, and Zhujiang Mouth Basins, *Osmundacidites* is Osmundaceae fern. *Leiotriletes, Lygodiumsporites,* and *Crassoretitriletes* are ascribed to the Lygodiaceae ferns. *Pterisisporites* and *Polypodiaceaesporites* are Pteridaceae ferns. *Polypodiaceaesporites, Polypodiisporites,* and *Echinosporis* are Polypodiaceae ferns. *Podocarpidites* is Podocarpaceae plant. *Pinuspollenites, Abietineaepollenites, Piceaepollenites, Cedripites, Abiespollenites, Keteleeriaepollenites,* and *Tsugaepollenites* are Pinaceae plants. *Taxodiaceaepollenites* is a Taxodiaceae plant. *Ephedripites* is ascribed to Ephedraceae of Gnetales. *Salixipollenites* is assigned to Salicaceae of angiosperms. *Caryapollenites* and *Juglanspollenites* are Juglandaceae of angiosperms. *Alnipollenites, Betulaceoipollenites,* and *Momipites* are Betulaceae of angiosperms. *Quercoidites* and *Cupuliferoipollenites* are Fagaceae of angiosperms. *Ulmipollenites* is an Ulmaceae of angiosperm. *Liquidambarpollenites* is Hamamelidaceae of angiosperm. *Myrtaceidites* is Myrtaceae of angiosperm. *Rhoipites* is Anacardiaceae of angiosperm. *Meliaceoidites* is Meliaceae of angiosperm. *Ilexpollenites* is an Aquifoliaceae of angiosperm. *Chenopodipollis* is Chenopodiaceae of angiosperm. *Compositoipollenites, Tubulifloridites,* and *Artemisiaepollenites* are assigned to Compositae of angiosperm. *Liliacidites* is Liliaceae of Monocotyledoneae. *Graminidites* is Gramineae of Monocotyledoneae, and *Cyperaceaepollis* is Cyperaceae of Monocotyledoneae.

7.2 Paleoecology and Paleoclimate

Climate is the most significant and active environmental factor that plays an important role in the interactions between plants and environment. The growth and extirpation of the plant community is closely related to climatic conditions such as temperature and humidity. The Quaternary sporo-pollen diagram in the Qaidam Basin indicates that the plants had obvious responses to climate variation in cold, warm, dry, and wet conditions. The succession of the plant community is a reflection of glacial and interglacial climate change there (Jiang and Robbins 2000). The vegetation is very sensitive to the climate change in the Qinghai–Tibet Plateau (Yao et al. 1997; Jiang and Yang 2001).

The ecological types of original plants of fossil spores and pollen can provide information about the paleoclimate of each period (Table 7.2). As mentioned above, Carboniferous spores and pollen found in crude oils from the Tarim and Junggar basins were mainly ferns. Among them, Lycopsida, Marattiales, and Filicales are dominant; Sphenopsida rank second. Representatives of Pteridospermopsida are also present. Of the ferns, Lycopsida were important coal-forming plants; the Paleozoic arboreal Lycopsida were hydrophytic plants that flourished in the Carboniferous. Living Marattiales ferns are large terrestrial plants native chiefly to tropical and subtropical regions. Extant Filicales ferns are ecologically thermophytic plants mainly distributed in the tropics, subtropics, and warm temperate regions. Drought-tolerant Calamitales

7.2 Paleoecology and Paleoclimate

Table 7.2 Paleoclimatic characteristic reflected by the plant ecological types

Ecotype	Vegetation	Sporo-pollen	Temperature and humidity
Thermophytes	Lycopodiales, Marattiales, Filicales, Cycadales, Coniferales, Podocarpaceae, Araucariaceae, Myrtaceae of Myrtales, Anacardiaceae of Rutales, Meliaceae	*Lycopodiumsporites, Camarozonosporites, Marattisporites, Osmundacidites, Todisporites, Lygodiumsporites, Gleicheniidites, Cyathidites, Cibotiumspora, Pterisisporites, Polypodiisporites, Dictyophyllidites, Cycadopites, Podocarpidites, Araucariacites, Myrtaceidites, Rhoipites, Meliaceoidites*	Hot tropical to subtropical
Thermophytes Mild type	Filicales, Cordaitales, Ginkgoales, Coniferales, Salicaceae, Juglandaceae, Betulaceae, Fagaceae, Ulmaceae, Aquifoliaceae, Hamamelidaceae	*Osmundacidites, Schizaeoisporites, Polypodiaceaesporites, Cordaitina, Entylissa, Keteleeriaepollenites, Tsugaepollenites, Taxodiaceaepollenites, Salixipollenites, Caryapollenites, Juglanspollenites, Alnipollenites, Momipites, Quercoidites, Cupuliferoipollenites, Ulmipollenites, Ilexpollenites, Liquidambarpollenites*	Warm Subtropical to warm temperate
Hardy plants	*Cedrus, Abies, Picea*	*Cedripites, Abiespollenites, Piceaepollenites*	Cold temperate
Hydrophytes	Lycopodiales, Marattiales, Filicales, mosses	*Lycopodiumsporites, Lycopodiacidites, Marattisporites, Osmundacidites, Todisporites, Gleicheniidites, Plicifera, Cyathidites, Cibotiumspora, Dictyophyllidites, Stereisporites*	Humid
Mesophytes	Cordaitopsida, Cycadopsida, Coniferopsida,	*Cordaitina, Florinites, Cycadopites, Monosulcites, Bennettiteaepollenites, Podocarpidites, Platysaccus, Pinuspollenites, Piceaepollenites, Tsugaepollenites, Cedripites, Taxodiaceaepollenites*	Semi-humid to semi-arid
Xerophytes	Cheirolepidiaceae, Ephedraceae, Chenopodiaceae, Compositae	*Classopollis, Ephedripites, Chenopodipollis, Artemisiaepollenites, Tubulifloridites*	Arid

plants of Sphenopsida are thought to have mainly inhabited fluvial sandy soils, mountain stream clefts, and outer edges of swamps. Paleobotanists have speculated that Paleozoic Pteridospermopsida mainly inhabited relatively arid highlands. Generally, the Carboniferous paleoclimate of the Tarim and Junggar basins was a warm and humid climate with localized aridity. Permian palynoflora in the Junggar Basin is characterized by diverse spores and pollen of mosses, ferns (e.g., Sphenopsida, Lycopsida, Marattiales, Filicales), and gymnosperms (e.g., Pteridospermopsida, Cordaitopsida, Cycadopsida, and Coniferopsida). In this Permian palynoflora, gymnosperms such as Pteridospermopsida, Cordaitopsida, and Coniferopsida gradually overcame the ferns in dominance. Pteridospermopsida are generally considered to have been indicative of a relatively arid climate. In contrast, the coal-forming Cordaitopsida plants are indicators of a warm and humid climate. The extant Coniferopsida plants are mainly evergreen trees or shrubs that tolerate moderate humidity and are native chiefly to tropical, subtropical, and temperate regions.

Triassic spores and pollen found in crude oils from the Tarim and Junggar basins are related to mosses, Sphenopsida, Lycopsida, Marattiales, Filicales, Pteridospermopsida, and Coniferopsida. In the Triassic palynoflora, gymnosperms are the dominant group; ferns rank second in diversity, while mosses are rare. Extant gymnosperms, mainly represented by conifers, are mostly arboreal plants. Podocarpaceae plants of the Coniferopsida are evergreen plants, which are native today chiefly in the tropics, subtropics, and south temperate zones. Pinaceae plants are mainly arboreal plants although there are some shrubby varieties; Pinaceae is mainly distributed in the temperate and subtropical regions of the Northern Hemisphere. Both of these families are adaptable to moderate humidity. Pteridospermopsida plants mainly inhabit relatively dry highlands. As humidogene thermophytes, Lycopsida, Marattiales, and Filicales are mainly distributed in the tropical, subtropical, and warm temperate regions. Sphenopsida horsetail plants such as Calamitaceae mainly inhabit tropical and subtropical swamps; extant horsetails are able to adapt to relatively dry climate. Mosses are typical hydrophytes that always grow in damp environments. The floral compositions in Xinjiang Autonomous Region during the Triassic period indicate that a warm and humid to semi-humid climate prevailed.

Jurassic spores and pollen found in crude oils from the Tarim, Junggar, Turpan, and Qaidam basins are quite abundant and diverse. Their original plants mainly belong to the Lycopsida, Marattiales, Filicales, and the gymnospermous Pteridospermopsida, Cycadopsida, and Coniferopsida. According to the diversity and richness of spores and pollen of these basins, the Jurassic palynofloras of the Xinjiang Autonomous Region and the Qinghai Province are dominated by gymnosperms, while ferns rank second. Gymnosperms are represented by Caytoniales of the Pteridospermopsida, Cycadales/Ginkgoales and Bennettitales of the Cycadopsida, and several families of Coniferales of Coniferopsida such as Podocarpaceae, Pinaceae, Araucariaceae, and Cheirolepidiaceae. The main fern taxa include Lycopodiaceae (Lycopodiales, Lycopsida), Selaginellaceae (Selaginellales), Marattiaceae (Marattiales), and several members of the Filicales, including Osmundaceae, Lygodiaceae, Schizaeaceae, Gleicheniaceae, Cyatheaceae, Dicksoniaceae, Dipteridaceae, and Cheiropleuriaceae. The order Cycadales, which flourished in the Mesozoic, is mainly distributed in the tropical and subtropical regions today and is represented by several relicts. Ginkgoales has one single living species (*Ginkgo biloba*) that is only found naturally in China and Japan. Evergreen arboreal and shrub Podocarpaceae of Coniferales mainly grow in the tropical or subtropical regions of Asia, Australia, and South America. In China, Podocarpaceae is only found to the south of the Yangtze River and the Taiwan Province. Pinaceae plants are mostly evergreen arboreal trees and distributed in the temperate and subtropical mountainous areas of the Northern Hemisphere. Araucariaceae is mainly found in tropical or subtropical regions of the Southern Hemisphere. The seasonal xerophyte Cheirolepidiaceae is able to adapt arid climate and inhabits highland slopes. The extant Lycopsida is mostly herbaceous, though there are a few woody climbing representatives. Lycopsida plants are widely distributed in the tropical, subtropical, and temperate regions. Modern Lycopodiaceae mainly inhabit humic acid soils in the understory of coniferous forests and coniferous broad-leaved mixed forests. Selaginellaceae ferns are distributed in the subtropical and temperate regions and grow under forests or in wetlands having slowly moving water. The hydrophytic Marattiales has a wide distribution in tropical and subtropical regions. In China, Marattiales is mainly found in the jungles of the Yunnan, Fujian, Taiwan, and Hainan provinces. Osmundaceae of the Filicales are medium-sized ferns, which commonly inhabit stream banks, wet valleys, and lowland swamps in tropical, subtropical, and temperate regions. Lygodiaceae are mainly found in tropical and subtropical regions. As a representative of the Lygodiaceae, the genus *Lygodium* is a climbing plant that inhabits mixed forests, woodlands, or valley streams. Schizaeaceae are small-sized terrestrial ferns with erect growth habits and are mainly found in the Southern Hemisphere and equatorial zones. The genus *Schizaea* of Schizaeaceae commonly grows in tropical acid soils. The family Gleicheniaceae is characterized by having a protostele and grows in shady moist places such as in ravines or under forest canopies. *Cyathea* ferns of the Cyatheaceae and Dicksoniaceae are tree ferns which grow in tropical and subtropical humid areas. Dipteridaceae, a hydrophyte, flourished in the Mesozoic, is now found in the tropics and subtropics, such as South Asia and the Yunnan, Guizhou, Guangxi, and Taiwan provinces of China. Cheiropleuriaceae are hydrophytic understory plants that are mainly distributed in the tropical regions, such as Indonesia, Philippines, and the Guangdong, Guangxi, Taiwan, and Hainan provinces of China (Beijing Institute of Botany, Academia Sinica 1976; Wang et al. 1995; Jiang and Wang 2002b). Judging from the Jurassic floral composition in the hydrocarbon-bearing regions, a subtropical to warm temperate humid climate was dominant in the Early–Middle Jurassic; then, the climate shifted to drier conditions by the end of the Middle Jurassic (i.e., the Callovian epoch). Finally, an arid climate prevailed in the Late Jurassic (Jiang et al. 2009).

Early Cretaceous spores and pollen found in crude oils of the Jiuquan Basin are relatively low in diversity. Their original plants mainly belong to the Filicales of the ferns (represented by Osmundaceae, Lygodiaceae, Schizaeaceae, and Dicksoniaceae) and the gymnospermous Cycadopsida and Coniferopsida. Extant Osmundaceae ferns grow in swamps, wet valleys, or in the understory of acid soils in tropical, subtropical, and temperate regions. Extant Lygodiaceae are mainly found in the tropics and subtropics and inhabit mixed coniferous broad-leaved forests or grow under valley bushes. Living Schizaeaceae are xerophytes which grow in arid, sandy soils in tropical regions. Extant Dicksoniaceae are tree ferns that mainly inhabit tropical damp environments and acid soils. Modern Ginkgoales and Cycadopsida have a tropical to subtropical distribution. Podocarpaceae are thermophilous plants that are distributed in the tropical and subtropical regions, whereas Pinaceae have a temperate and subtropical distribution pattern. Judging from ecological characteristics of this palynoflora, a subtropical hot and humid climate prevailed in the Jiuquan Basin during the Early Cretaceous.

In the Junggar, Qaidam, Liaohe, Beibu Gulf, and Zhujiang Mouth basins, Paleogene spores and pollen found in crude oils are quite diverse and rich. They are mainly shed by ferns, gymnosperms, and angiosperms. Among them, ferns are Osmundaceae, Lygodiaceae, Pteridaceae, and Polypodiaceae; gymnosperms are represented by Podocarpaceae, Pinaceae, and Taxodiaceae of the Coniferales and Ephedraceae of the Gnetales (Gnetopsida). Angiosperms in these basins are composed of both representatives of Dicotyledoneae (i.e., Salicaceae of the Salicales, Juglandaceae of the Juglandales, Betulaceae and Fagaceae of the Fagales, Ulmaceae of the Urticales, Hamamelidaceae of the Hamamelidales, Myrtaceae of the Myrtales, Anacardiaceae and Meliaceae of the Rutales, Aquifoliaceae of the Celastrales, Chenopodiaceae of the Caryophyllales, and Compositae of the Asterales), and Monocotyledoneae (i.e., Liliaceae of the Liliales, Gramineae of the Graminales, and Cyperaceae of the Cyperales). As mentioned above, the extant fern families, Osmundaceae and Lygodiaceae, are warm hydrophytic plants. The large terrestrial Pteridaceae having heights of 1–2 m mainly inhabit tropical and subtropical mountainous sunny slopes, dense valley forests, and the understory of forests in general. Representatives of Polypodiaceae are mostly epiphytes that are mainly low lying. With a tropical distribution, extant Polypodiaceae grow by means of climbing trees and rocks, or grow in the understory of forests or in damp valleys. Among gymnosperms, extant Podocarpaceae are mainly distributed in the tropical and subtropical regions. Taxodiaceae are chiefly native to subtropics, the genera *Tsuga* and *Keteleeria* of Pinaceae are also philotherms, while Ephedraceae are xerophytic shrubs. Dicotyledoneae are arboreal, shrubs, or herbaceous. Anacardiaceae are mainly distributed in the tropics. Meliaceae have a tropical and subtropical distribution. Myrtaceae are mostly arboreal or shrubs and are mainly distributed in the tropical America and Australia. Fagaceae are always found in subtropical and temperate regions. Aquifoliaceae are evergreen shrubs or small-sized arboreal trees that are mainly distributed in Central and South America. Juglandaceae are tall arboreal trees that are mainly distributed in the north temperate zone although some representatives can be found in tropics. Salicaceae, Betulaceae, and Ulmaceae are mainly distributed in the north temperate zone, while several representatives are found in South America and South Africa. Hamamelidaceae are mainly found along the Atlantic coast of North America, the Himalayas, and Southeast Asia; a few representatives of this family grow in Madagascar and South Africa. Also among the Dicotyledoneae, Chenopodiaceae and Compositae (Asteraceae) are mostly herbaceous plants, although a few are shrubs or small trees. The Chenopodiaceae and Asteraceae exhibit high drought tolerance and cosmopolitan distribution. According to the ecological types of all of the above plants, the Paleogene climate in the Junggar and Qaidam basins was subtropical, semi-humid to semi-arid. In the Liaohe Basin, the Paleogene climate was warm and humid. And the Paleogene climate in the Beibu Gulf and Zhujiang Mouth basins was tropical or subtropical and humid.

7.3 Paleoecology and Paleogeography

Besides climatic factors, geographical conditions are also important factors that affect survival and reproduction of vegetation. The relationships between plant ecology and geographical factors mainly manifest themselves in the typical geographical distribution of the plants. A particular plant always belongs to a particular horizontal zone or vertical zone, and therefore, it can reflect the physical geographical landscape of that horizontal or vertical zone. The floral composition is not only relevant to latitude, but also to longitude. For example, tropical and subtropical rain forests are mainly distributed near the equator. Ignoring deserts, vegetation zones are distributed across the Earth from the equator to the poles as a gradual transition from broad-leaved evergreen forests, to coniferous forest, to forest steppe, and to tundra. At mid-latitudes, the broad-leaved forest zone is distributed on both sides of the continent near the oceans, while the inner continent is occupied by the forest steppe zone and steppe zone. At high latitudes, the oceanic coniferous forest zone occupies both sides of the continent near the ocean, while the plants of the inner continent belong to the continental coniferous forest zone (i.e., taiga). This is the general pattern of the horizontal zonality of the vegetation.

Vegetation distribution is also greatly affected by landforms. In mountainous regions, vertical zonality of vegetation is very typical. The altitude, exposure, and slope gradient affect vertical vegetation zonality. In Transcaucasia, for example, the vegetation at an elevation of 3200–4200 m is alpine desert; 2500–3200 m is the alpine meadow; 1900–2500 m is mountain woodlands, elfin wood forest, and subalpine meadow; 1200–1900 m is a spruce, fir, and beech forest; 600–1200 m is a forest zone dominated by oaks; and 200–600 m is a grassland and xerophytic woodland. The regularity of such vegetation zonation today helps us to understand the paleogeographical environment from the view of plant ecology.

Many of the plant families are still present, which helps interpret aspects of the paleoenvironment. Reconstruction of the paleoenvironment from extinct plant families relies on various clues that vary from sedimentological analysis to associations of macrofossils.

According to the ecological characteristics of late Early Cretaceous palynoflora in the Mazong Mountain region of the Gansu Province, Wang and Tang (2000) proposed that the area was divided into various types of geomorphic units, such as high mountains, lowlands, lakes, and swamp, all of which constituted a paleogeographical model for an intermountain basin. Ecological characteristics of the Late Triassic palynoflora in the Tongchuan area of the Shaanxi Province reflected an area geomorphically characterized by high terrains and steep elevation differences, wide drainage areas, and complicated topography, representing a paleogeographical model of rivers, lakes, delta, and swamp (Jiang et al. 2006). Judging from the ecological characteristics of the Early Cretaceous palynoflora in Wuqia and Baicheng regions of the Xinjiang Autonomous Region, the plant community was composed of understory and canopy layers with a distinct vertical zonality. This indicates that there were complex paleogeographical landscapes from the mountains to swamps and marshes in the studied area (Jiang et al. 2007, 2008). All these examples demonstrate that palynoflora can be used to understand the ancient local geographical environment and therefore contribute to paleogeographical reconstructions.

The plants that shed the Carboniferous spores and pollen found in crude oils from the Tarim and Junggar basins mainly belonged to Lycopsida, Sphenopsida, Filicales, and Marattiales of Filicopsida. Lycopsida were important coal-forming plants that were commonly tall woody trees. Sphenopsida today mainly grow in marshes. Marattiales ferns were terrestrial macrophytes having heights of 1–2 m or more and therefore grew in the understory layer of thick forests. Herbaceous and tree ferns grow along river banks or as the understory layer of forests. The plant communities indicate that Carboniferous landscapes were composed of lakes, rivers, and marshes in the studied area.

Among the original plants of the Permian spores and pollen found in crude oils from the eastern Xinjiang Autonomous Region, gymnosperms dominated and ferns ranked second. In this flora, Pteridospermopsida, Cordaitopsida, primitive conifers, Lycopsida, and Filicopsida flourished. Some of these plants are still present, which helps interpret aspects of the Permian paleoenvironment. Pteridospermopsida grew mainly in the highlands, while coal-forming Cordaitopsida formed dense forests. Arborescent and shrubby conifers today can occupy various altitudinal belts. Cycadophytes mainly grew under the tall arborescent trees. Tree ferns grew along the banks of lakes and rivers, while herbaceous ferns grew in the understory of forests. The vegetation mainly consists of plants representing highland, mid-altitude plains, and lowland swamps. The vertical vegetation zonation reflects the paleogeographical landscapes of mountains, valleys, rivers, plains, lakes, and swamps.

The original plants of the Triassic spores and pollen found in crude oils from the Junggar and Tarim basins were mainly composed of Lycopsida, Sphenopsida, and Filicopsida (Marattiales, Filicales), Pteridospermopsida, and Coniferopsida, as well as mosses. Woody Filicopsida declined at the end of the Paleozoic, whereas in the Triassic, herbaceous and climbing woody Filicopsida grew in the warm and humid forest understory or bushes. Terrestrial Marattiales mainly inhabited hot and humid dense forests or rain forests. Tree ferns of Filicales grew along the banks of rivers or lakes, and herbaceous ferns grew in the understory of forests or in swamps. Sphenopsida (Equisetales) mainly inhabited marshes. Mosses grew under the herbaceous layer. Seed ferns and conifers that were adapted to the cool and dry climate occupied the mountainous regions, while Podocarpaceae that were adapted to the hot and humid climate inhabited lower altitude mountains or plains. The vertical zonality of the Triassic flora reflects that there were various paleogeographical landscapes in the studied area, represented by mountains, rivers, lakes, fluviolacustrine deltas, alluvial plains, and swamps.

Jurassic spores and pollen found in crude oils of the Tarim, Junggar, Turpan, and Qaidam basins are rich and diverse. Their original plants mainly composed of Lycopsida, Filicopsida (Marattiales, Filicales), and gymnospermous Pteridospermopsida, Cycadopsida, and Coniferopsida. Pteridophytes were mainly represented by Lycopsida (Lycopodiales, Selaginellales) and numerous Filicopsida, such as Marattiales (e.g.,

Marattiaceae) and Filicales (e.g., Osmundaceae, Lygodiaceae, Schizaeaceae, Gleicheniaceae, Cyatheaceae, Dicksoniaceae, and Dipteridaceae). Gymnosperms were represented by Pteridospermopsida (Caytoniales), Cycadopsida (Cycadales, Bennettitales), and several families of Coniferopsida (Podocarpaceae, Pinaceae, Araucariaceae, and Cheirolepidiaceae). Currently, both the Tarim and Junggar basins are surrounded by high mountains. The snow line of these mountains reaches an altitude of 3500 m. In the Mount Bogda, the extant coniferous forest is dominated by spruce, fir, and cedar that mainly grow at an altitude of 2000–2800 m. In the Qilian Mountains, the spruce forest occupies an altitude of 2000–3500 m. Importantly, the Jurassic vegetation in the four basins can be divided vertically into six layers. Vegetation of the first layer at the higher altitudes was composed of pteridosperms and hardy conifers, such as cedar, fir, and spruce. The drought-tolerant Cheirolepidiaceae constituted the dominant vegetation of the second layer, especially in drainages along sunny slopes. The Podocarpaceae, which favor a hot and humid environment, together with Araucariaceae, constituted the vegetation of the third layer at lower altitudes. Cycads and tree ferns formed the vegetation of the fourth layer in valleys and plains. Coniferous shrubs constituted the vegetation of the fifth floor in lowlands or surrounding mountain streams. Lycopodiales and herbaceous Filicales constituted the sixth layer of herbs growing beneath a shrubby layer. The biodiversity and distinct vertical zonality of these Jurassic palynofloras reflected great variety in the paleogeographical landscapes in these four basins, ranging from mountains to marshes. Specifically, mountains, deep-cutting valleys, alluvial plains formed by complex river networks, swamps, wetlands, lakes, rivers, deltas, and lacustrine deltas formed the Jurassic landscapes of the studied areas.

Early Cretaceous spores and pollen found in crude oils from the Jiuquan Basin belong to the Filicales of Filicopsida (e.g., Osmundaceae, Lygodiaceae, Schizaeaceae, and Dicksoniaceae) and the gymnospermous Cycadopsida and Coniferopsida. In the Early Cretaceous, Dicksoniaceae mainly inhabited humid tropical areas. Cycadopsida grew in low mountains, valleys, or riverbanks. Schizaeaceae ferns mainly grew in the forest by climbing on trees. Osmundaceae ferns always grew in swamps and wetlands. Ecological characteristics of this palynoflora suggested that rivers, lakes, and swamps were the major paleogeographical landscapes of the Jiuquan Basin during the Early Cretaceous.

The original plants of the Paleogene spores and pollen found in crude oils from the Junggar, Qaidam, Liaohe, Beibu Gulf, and Zhujiang Mouth basins are mainly ferns, gymnosperms, and angiosperms. Among them, Filicales is represented by Osmundaceae, Lygodiaceae, Pteridaceae, and Polypodiaceae; gymnosperms include Podocarpaceae, Pinaceae, and Taxodiaceae of Coniferales and Ephedraceae of Gnetales. Angiosperms are represented by a wide variety of modern families including Salicaceae, Juglandaceae, Betulaceae, Fagaceae, Ulmaceae, Hamamelidaceae, Myrtaceae, Anacardiaceae, Meliaceae, Aquifoliaceae, Chenopodiaceae, Compositae, Liliaceae, Gramineae, and Cyperaceae. In the studied areas, the Paleogene vegetation can vertically be divided into five layers, namely the cold-tolerant taiga forest, the coniferous and broad-leaved mixed forest, the evergreen broad-leaved forest, the shrub layer, and the herb layer. In inland basins, coniferous forests flourished, while broad-leaved forests were sparse. In contrast, in coastal shelf basins, broad-leaved forests flourished, while coniferous forests were sparse. In the Tarim and Junggar basins, the hardy coniferous cedar, fir, and spruce forests indicate that the mountains surrounding these two basins had relatively high altitudes of approximately 2000–3000 m. Mountain snowmelt was the major source for the rivers in these basins. The scarcity of cedar and spruce pollen in the Liaohe, Beibu Gulf, and Zhujiang Mouth basins suggests that low-altitude regions surrounded these basins. Meanwhile, abundant tropical evergreen broad-leaved plant pollen (e.g., Anacardiaceae, Meliaceae, and Myrtaceae) demonstrates a tropical rain forest landscape then. Today, in the mountainous regions surrounding the Liaohe Plain, spruce forests mainly inhabit an altitude of 1600–1800 m, while deciduous forests inhabit the zone below 1600 m. In the Paleogene, the mixed forest composed of conifers (e.g., Podocarpaceae, Taxodiaceae, *Keteleeria*, and *Tsuga*) and broad-leaved trees (e.g., Fagaceae, Betulaceae, and Ulmaceae) occupied low mountains and hilly terrains. The evergreen broad-leaved forests consisted of Anacardiaceae, Meliaceae, Juglandaceae, Fagaceae, Betulaceae, Ulmaceae, and Salicaceae, all of which occupied the open plains. In the Beibu Gulf Basin and other coastal basins, broad-leaved forests particularly flourished where rainfall was sufficient. Herbaceous Filicales grew in the understory of forest trees or bushes. The extant genus *Lygodium* grows by climbing on trees in the forest, a habit interpreted for *Lygodiumsporites*. Polypodiaceae plants commonly grow as epiphytes on edge trees or on rocks, or in damp valleys. Osmundaceae today often grow in clusters in lowland swamps. In the tree understory, the shrubby layer consisted of holly, myrtle, ephedra, and other shrubs that were distributed in the open areas. Lake edges were optimal habitat for *Sparganium* and other aquatic plants. The optimal habitat for drought-tolerant *Ephedra* was the relatively arid areas within the Qaidam Basin. Under the shrubby layer, herbaceous Chenopodiaceae, Compositae (Asteraceae), Gramineae (Poaceae), and Cyperaceae constituted grasslands growing in soils having different humidities. The ecological environments of the above-mentioned plant community reflected various paleogeographical landscapes, such as mountains, rivers, lakes, swamps, and fluvial deltas.

7.4 Sedimentary Environment and Petroleum Source

The plant community and natural environment together constitute a closely related and interactive unity, as well as a synthesis of matter circulation and energy transformation. Warm and humid lacustrine environments can create a vibrant plant community, within which dead organisms eventually are recycled or become oil and gas sources. Lacustrine and swamp environments are not only favorable places for the growth of hydrophytic vegetation, but also favorable zones for the enrichment and preservation of terrigenous organic matter. Therefore, lacustrine and swamp sedimentary environments play a critical role in the formation of oil and gas.

Judging from investigations of sporo-pollen fossils in crude oil samples from 17 Chinese continental oil fields, lacustrine and swamp environments under a humid climate contributed to the formation of lacustrine oil source rocks (Jiang 1988). In the Tarim Basin, sporo-pollen and algae in the crude oils support this thesis that the same conditions were conducive to the formation of oil–gas source rocks (Jiang and Yang 1997). Similarly, in the Tongchuan region of Shaanxi Province, Late Triassic palynoflora shows the same hot and humid lacustrine environment contributed to the formation of petroleum; there, a temperate swamp environment was also conducive to the formation of coal and natural gas (Jiang et al. 2006). Additionally, the palynofloras show a swamp environment in the Dongsheng region of the northeastern Ordos Basin in Middle Jurassic coal-bearing strata which formed both coal and coal gas (Jiang and Wang 2002a).

The petroleum sporo-pollen assemblages and the correlations between spores and pollen found in crude oils and in potential source rocks suggest that there are three sets of petroleum source rocks in the Tarim Basin, namely the Carboniferous–Permian, Triassic, and Jurassic. The plants that bore the spores and pollen in Carboniferous–Permian rocks were mainly Lycopsida, Filicopsida, Pteridospermopsida, and Cordaitopsida. Those in the Triassic were mostly Lycopsida, Sphenopsida, Marattiales, and Filicales of Filicopsida, Coniferales of Coniferopsida, and Cycadales of Cycadopsida. As arborescent plants, the Paleozoic Lycopsida and Cordaitopsida were important coal-forming plants that indicated warm and humid swamp environments. Podocarpaceae and Araucariaceae of Coniferales, and Cycadales are all philotherm arborescent trees having a tolerance for moderate humidity. Pinaceae of Coniferales, which tolerate moderate humidity, are mostly evergreen trees that mainly inhabit temperate and subtropical mountainous regions. Therefore, the three age sets of petroleum source rocks in the Tarim Basin were all formed under warm and humid climate.

Besides spores and pollen, some algal fossils have also been described in crude oils of northern and southwestern Tarim Basin. As representative of Dinophyceae (Pyrrophyta), the genera *Rhombodella* and *Dinogymnium* are only recorded in marine deposits, as expected. *Hungarodiscus* of the phylum Chlorophyta has also been reported in marine deposits. As the resting cysts of Dinophyceae, dinoflagellates with spinate surface ornamentations are indicators of marine environments, whereas unornamented ones are useful indicators for lacustrine environments (Traverse 1988). As expected, ornamented fossil dinoflagellates were very scarce in crude oils of the Tarim Basin, supporting the idea of a small-scale seawater transgression into the basin.

Generally, environments such as inland lakes and lowland swamps are suitable for the growth of warm hydrophytic plants, while coastal lagoon environments favor the growth of aquatic plants and algae. The sporo-pollen and algae composition suggests that the petroleum source rocks of the Tarim Basin were mainly formed in the warm and humid inland lakes and lowland swamps, but a coastal lagoon environment with brackish water or semi-saline water is also indicated.

The petroleum sporo-pollen assemblages and the correlations between spores and pollen found in crude oils and in potential source rocks indicate that there are five sets of petroleum source rocks in the Junggar Basin, namely the Carboniferous, Permian, Triassic, Jurassic, and Paleogene. According to the ecological characteristics of the plants that bore the spores and pollen, the environment that formed the Mesozoic petroleum source rocks in the Junggar Basin is similar to that of the Tarim Basin, namely a warm and humid lacustrine and marsh environment. Permian spores and pollen from eastern Junggar Basin were very diverse and included mosses, ferns, and gymnospermous Pteridospermopsida, Cordaitopsida, and Coniferopsida. Judging from this ecological association, the Permian petroleum source rocks were formed in warm and semi-humid or semi-arid lacustrine and marsh environments. The plants that bore the Paleogene spores and pollen collected from southern Junggar Basin were quite diverse and included Lygodiaceae, Pinaceae, Taxodiaceae, Ephedraceae, Salicaceae, Juglandaceae, Fagaceae, Anacardiaceae, Chenopodiaceae, and Compositae (Asteraceae). The ecological characteristics of the plants that bore pollen and spores in the Paleogene petroleum source rocks indicate that the source rocks were formed in warm and semi-humid or semi-arid lacustrine environments.

The spores and pollen found in crude oil of the Qiketai and Shengjinkou oil fields of the Turpan Basin are all common elements of the Jurassic found elsewhere. Judging from the correlations of sporo-pollen assemblages and petroleum sources, both the Lower Jurassic and Middle Jurassic are all petroleum source rock series in this inland basin. The spores and pollen are very abundant and diverse and represent Marattiales and Filicales of Filicopsida, Pteridospermopsida, Cycadopsida, as well as Coniferopsida such as the Coniferales Podocarpaceae, Pinaceae, and Cheirolepidiaceae. The ecological characteristics of this palynoflora indicate that Lower and Middle Jurassic petroleum source rocks were all formed in warm and humid limnetic and swamp environments.

The spores and pollen found in crude oils of the Lenghu and Yuka oil fields in northern Qaidam Basin are all common Jurassic plants. Correlations of the sporo-pollen assemblages with petroleum sources suggest that both the Lower Jurassic and Middle Jurassic are petroleum source rock series in this inland basin. The original plants of these spores and pollen include Lycopodiales of Lycopsida, Filicales of Filicopsida (e.g., Cyatheaceae, Osmundaceae, and Gleicheniaceae), Cycadopsida, as well as representatives of Coniferopsida such as the Coniferales Podocarpaceae, Pinaceae, and Cheirolepidiaceae. The ecological characteristics of this palynoflora reflect that the Lower and Middle Jurassic petroleum source rocks were all formed in temperate and humid lacustrine and marsh environments.

Diverse Paleogene spores and pollen were found in the crude oils of the Youquanzi and Xianshuiquan oil fields, which lie in the Mangnai Depression Belt of the Qaidam Basin. The plants that produced the microfossils are Lygodiaceae and Polypodiaceae of Filicales, Podocarpaceae and Pinaceae of Coniferales, and Ephedraceae of Gnetales, as well as several angiospermous families such as Salicaceae, Betulaceae, Fagaceae, Hamamelidaceae, Meliaceae, Chenopodiaceae, and Compositae. The ecological characteristics of this palynoflora indicate that the Paleogene petroleum source rocks were formed in a warm and semi-humid or semi-arid lacustrine sedimentary environment.

Sporo-pollen fossils in crude oils of the Xinglongtai oil field of the Liaohe Basin are always in Paleogene strata in the coastal areas of the Bohai Sea. Based on the correlations of oil and sources, the Oligocene Shahejie and Dongying formations should be the oil source rock series in this basin. The plants that produced these spores and pollen were mainly Osmundaceae and Pteridaceae of Filicales, Pinaceae of Coniferales, Ephedraceae of Gnetales, and angiosperms, such as Fagaceae, Chenopodiaceae, Compositae, Sparganiaceae, Liliaceae, Gramineae, and Cyperaceae. Besides sporo-pollen, some algal fossils, e.g., *Campenia irregularis*, *Dictyotidium reticulatum*, *Palaeostomocystis minor*, and *Bohaidina laevigata*, were also found in crude oils from the Xinglongtai oil field. The genera *Campenia*, *Dictyotidium*, and Paleostomocystis were mostly found in freshwater, rarely in saline water. These sporo-pollen and algal fossils indicate that the Oligocene petroleum source rocks were formed in a warm and humid, fresh or brackish lacustrine environment, as well as a brackish paralic to neritic environment.

In the Beibu Gulf Basin, the petroleum sporo-pollen assemblages and the correlations between spores and pollen found in crude oils and in potential source rocks indicate that the Eocene to Oligocene Liushagang and Weizhou formations should be the oil source rocks. The plants that bore the palynoflora are mainly Osmundaceae, Lygodiaceae, and Polypodiaceae of Filicales, Podocarpaceae and Pinaceae of Coniferales, and several angiosperms, such as Salicaceae, Juglandaceae, Betulaceae, Fagaceae, Ulmaceae, Hamamelidaceae, Myrtaceae, and Aquifoliaceae. Additionally, algal fossils, e.g., *Leiosphaeridia hyaline* and *Granodiscus granulatus*, were also found in crude oils of the Weizhou oil field in the Beibu Gulf Basin. The dark mudstone of the Liushagang Formation is as thick as 500 m, and its genesis was that of a deepwater lacustrine deposit (Zeng and Guo 1981). The above plant taxa are mostly hydrophytes, indicating a hot and humid paleoclimate with sufficient rainfall for these plants. The ecological characteristics of this palynoflora implied that during the depositional period of the Liushagang and Weizhou formations, broad-leaved trees, such as willow, walnut, birch, and elm, flourished along the lake shore and river-banks; other broad-leaved trees, such as walnut, pecan, oak, chestnut, and sweet gum, grew in the lowlands; a coniferous and broad-leaved forest composed of coniferous podocarps, pine, spruce, and broad-leaved pecan occupied hills, accompanied by bushes such as hazelnut, chestnut, elm, myrtle, and holly; the understory was occupied by osmundaceous, polypodiaceous, and lygodiaceous ferns. The biodiversity and ecological landscape indicate that the oil source rocks were formed in a tropical hot or warm and humid lacustrine sedimentary environment. The two fossil species, *Leiosphaeridia hyaline* and *Granodiscus granulatus*, have been recognized to be systematically related to the Chlorophyta which are mostly found in freshwater and indicative of lacustrine environments. Considering the fact that many *Granodiscus granulatus* fossils were also recorded in marine strata, the oil source rocks near the offshore of the Beibu Gulf were thus inferred to have been formed in a brackish paralic to neritic sedimentary environment.

In the Zhujiang Mouth Basin, the petroleum sporo-pollen assemblages and the correlations between spores and pollen in crude oils and in potential source rocks suggest that the dark mudstones of the Eocene to Oligocene Wenchang, Enping, and Zhuhai formations should be the oil source rocks. The plants that bore these spores and pollen are mainly Osmundaceae and

Polypodiaceae of Filicales, Pinaceae of Coniferales, as well as the angiospermous Salicaceae, Juglandaceae, Betulaceae, Fagaceae, Ulmaceae, Hamamelidaceae, and Aquifoliaceae. The ecological characteristics of the palynoflora are similar to that of Beibu Gulf Basin and therefore indicates that the Eocene to Oligocene petroleum source rocks were formed in a tropical warm and humid lacustrine sedimentary environment.

The petroleum source rocks of the above eight petroliferous basins were mostly formed in subtropical or warm temperate, humid or semi-humid to semi-arid, and lacustrine and swamp sedimentary environments. Some were formed in a paralic to neritic or lagoon sedimentary environment.

As explained in detail in Chap. 5, the dispersed insoluble organic matter in sedimentary rocks can be divided into the sapropelic type (Type I, II) and the humic type (Type III) kerogens. The sapropelic type is often formed in anoxic deepwater lakes or the oceans, while the humic type is often formed in the oxic swamp or neritic environment. Most oil and natural gas is formed by sapropel kerogen or mixed sapropel-humic kerogen. Humic kerogen mainly forms natural gas.

Sporo-pollen fossils, which are structured microfossils, are sapropelic-type kerogen. The outer walls, named exines, may play a minor role in petroleum formation. But the fossils play a large role in understanding the climatic and ecological environments in which oil and gas form. The spores and pollen found in the crude oils in China's basins indicate that a warm and humid, deep lacustrine environment is the best sedimentary environment for the formation of terrestrial oil–gas source rocks, while warm and semi-humid or semi-arid, lacustrine, swamp, and costal lagoon environments are also implicated in the formation of terrestrial oil–gas source rocks. As a generalization, a warm and humid lacustrine environment contributed to the formation of oil, while a warm and humid swamp environment favored the formation of gas.

References

Balme, B. E. (1995). Fossil in situ spores and pollen grains: an annotated catalogue. *Review of Palaeobotany and Palynology, 87*, 81–323.

Beijing Institute of Botany, Academia Sinica. (1976). *Sporae Pteridophytorum Sinicorum* (pp. 1–451). Beijing: Science Press (in Chinese).

Clement-Westerhof, J. A. (1974). In situ pollen from gymnospermous cones from the upper permian of the Italian alps, a preliminary account. *Review of Palaeobotany and Palynology, 17*, 63–73.

Couper, R. A. (1958). British mesozoic microspores and pollen grains. A systematic and stratigraphic study. *Palaeontographica, 103*, 75–179.

Jiang, D. X. (1988). Spores and pollen in oils as indicators of lacustrine source rocks. In A. J. Fleet, K. Kelts, & M. R. Talbot (Eds.), *Lacustrine petroleum source rocks* (pp. 159–169). London: Blackwell Scientific Publications.

Jiang, D. X., & Robbins, E. I. (2000). Quaternary palynofloras and paleoclimate of the Qaidam Basin, Qinghai province, northwestern China. *Palynology, 24*, 95–112.

Jiang, D. X., & Wang, Y. D. (2002a). Middle jurassic sporo-pollen assemblage form the Yanan formation of Dongsheng, Nei Monggol, China. *Acta Botanica Sinica, 44*(2), 230–238.

Jiang, D. X., & Wang, Y. D. (2002b). Middle jurassic palynoflora and its environmental significance of Dongsheng, Inner Mongolia. *Acta Sedimentologica Sinica, 20*(1), 47–54. (in Chinese with English abstract).

Jiang, D. X., Wang, Y. D., & Wei, J. (2006). Palynoflora and its environmental significance of the late Triassic in Tongchuan, Shaanxi Province. *Journal of Palaeogeography, 8*(1), 23–33. (in Chinese with English abstract).

Jiang, D. X., Wang, Y. D., & Wei, J. (2007). Palynofloras and their environmental significance of the early cretaceous in Wuqia, Xinjiang autonomous region. *Journal of Palaeogeography, 9*(2), 185–196. (in Chinese with English abstract).

Jiang, D. X., Wang, Y. D., & Wei, J. (2008). Palynoflora and its environmental significance of the early cretaceous in Baicheng, Xinjiang autonomous region. *Journal of Palaeogeography, 10*(1), 77–86. (in Chinese with English abstract).

Jiang, D. X., Wang, Y. D., & Wei, J. (2009). Palynoflora and its environmental significance of the middle jurassic in Wuqia, Xinjiang autonomous region. *Journal of Palaeogeography, 11*(2), 205–214. (in Chinese with English abstract).

Jiang, D. X., & Yang, H. Q. (1997). Microfossils from crude oils of Tarim basin and sedimentary environment of petroleum source rocks. *Acta Sedimentologica Sinica, 15*(3), 84–90. (in Chinese with English abstract).

References

Jiang, D. X., & Yang, H. Q. (2001). Palynological evidence for climatic changes in Dabuxun Lake of Qinghai Province during the Past 500,000 Years. *Acta Sedimentologica Sinica, 19*(1), 101–106. (in Chinese with English abstract).

Krutzsch, W. (1971). Atlas der mittel-und jungtertiaren dispersen Sporen und Pollen. Sowie der Mikroplanktonformen des nordlichen. Mitteleuropas. Lief. VI, Coniferenpollen. Volkseigener Betrieb Deutscher Verlag der Wissenschaften. Berlin 1–274.

Litwin, R. J. (1985). Fertile organs and in situ spores of ferns from the late triassic chinle formation of Arizona and New Mexico, with discussion of the associated dispersed spores. *Review of Palaeobotany and Palynology, 44*, 101–146.

Ouyang, S., Wang, Z., Zhan, J. Z., Zhou, Y. X. (2003). Palynology of the Carboniferous and Permian strata of northern Xinjiang, northwestern China. Hefei: University of Science and Technology of China Press. 1–700 (in Chinese with English summary).

Potonié, R. (1962). Synopsis der Sporae in situ. *Beih Geol Jahrb, 52*, 1–204.

Song, Z. C., Li, M. Y., Zhong, L. (1986). Cretaceous and early tertiary sporo-pollen assemblages from the Sanshui Basin, Guangdong Province. Palaeontologia Sinica, Whole Number 171, New Series A, No. 10. Beijing: Science Press. 1–170 (in Chinese with English summary).

Traverse, A. (1988). *Paleopalynology* (pp. 1–600). Boston: Unwin Hyman.

Wang, Y. D. (1999). Fertile organs and in situ spores of marattia asiatica (Kawasaki) Harris (Marattiales) from the lower jurassic Hsiangchi formation in Hubei, China. *Review of Palaeobotany and Palynology, 107*, 125–144.

Wang, F. H., Qian, N. F., Zhang, Y. L., Yang, H. Q. (1995). Pollen flora of China (2nd edition). Beijing: Science Press. 1–461 (in Chinese with English abstract).

Wang, X. Z., & Tang, F. (2000). Late mesozoic spores and pollen from Mazongshan area, Gansu. In Z. H. Zhu (Ed.), *Collected papers of palynology in petroliferous basins of China* (pp. 105–110). Beijing: Petroleum Industry Press. (in Chinese).

Yao, T. D., Thompson, L. G., Shi, Y. F., Qin, D. H., Jiao, K. Q., Yang, Z. H., et al. (1997). Climate variation since the last interglaciation recorded in the Guliya ice core. Science in China (Series D), *27*, 447–452 (in Chinese) *40*, 662–668 (in English).

Zeng, D. Q., Guo, B. (1981). Tertiary of north continental shelf of South China Sea. Guangzhou: Guangdong Science and Technology Press. 1–258 (in Chinese).

Mechanisms of Petroleum Migration

Abstract

The mechanisms of petroleum migration are important for understanding the distribution of petroleum in the inland and coastal shelf basins. Spores and pollen found in crude oils are similar to those in petroleum source rocks in terms of species and color. This indicates that the spores and pollen in crude oils still retain their original exine morphology after migration. So, the spores and pollen in crude oils can bear witness to petroleum migration. Microfissures formed by abnormal high pressure and undercompaction during the process of diagenesis are common, and such microfissures are presumably available for initial migration and expulsion of petroleum and microfossils such as spores and pollen. Connective pores, interstratified openings, joints, and fissures in the carrier and petroleum reservoir beds are all available passages for secondary migration. Fault surfaces and unconformity interfaces provide larger passageways for secondary migration. The phase state of petroleum migration depends on the openings through which petroleum migrates. Because microfissures are wide enough for the passage of spores and pollen, the passageways are equally unimpeded for the passage of oil droplets. It follows that the migration of petroleum in the liquid phase is fully possible during the course of primary migration. Liquid phase migration is also common during the course of secondary migration, because the passageways are much wider than microfissures. It is well known that the migration of petroleum in gaseous phase has never met any resistance, because gas can penetrate through any pores. Spores and pollen in crude oils can also provide information about the direction and route of petroleum migration. The main direction of petroleum migration could be either vertical or lateral migration for different reservoir types and in different tectonic regimes.

Keywords

Primary migration · Secondary migration · Passages · Phase states · Directions · Spores and pollen

8.1 Introduction

Most geologists consider that petroleum has a three-stage sequence composed of origin, migration, and accumulation into pools. Many theories have been proposed to explain these. Levorsen (1956) succinctly wrote "We do not know just how oil and gas originated, nor how they have moved and accumulated into pools. These problems, if solved, would aid greatly in the main job of the petroleum geologist—the search for new pools." Petroleum migration is obviously a highly complicated process. Migration is central to the processes that link generation–migration–accumulation of petroleum. Because the original oil and gas occur in source rocks in the dispersal state, oil and gas must have undergone migration to accumulate and form oil and gas pools. The mechanisms of migration are important for understanding the regularity of distribution of petroleum. Moreover, it is significant for exploration and development of petroleum resources.

Oil and gas seepages at the surface bear macroscopic witness to petroleum migration, reflecting the course through which oil and gas migrated. Spores, pollen, and algae in crude oils bear microscopic witness to petroleum migration, reflecting the tracks through which oil and gas migrated from source rocks to petroleum pools. The palynological evidence for petroleum migration in the Tarim Basin showed that fossil spores and pollen in crude oils could provide reliable information about passageways, phase states, directions, routes, and distances of petroleum migration (Jiang et al. 2002).

8.2 Primary Migration

Primary migration of petroleum is defined here as the migration of the generated petroleum from the petroleum source bed outward.

8.2.1 Passageways for Primary Migration

It is unlikely that the original petroleum exited through petroleum source rock pore networks, because the pore diameters of very fine-grained source rocks are generally less than 0.01 μm. The passage of oil droplets of petroleum is a key point in understanding the mechanisms of petroleum migration.

Spores and pollen in crude oils are generally larger than 15 μm in diameter, so the pores of petroleum source rocks are too small for their passage. The presence of spores and pollen derived from petroleum source rocks in crude oils supports the ideas that the passage of petroleum primary migration could be via microfissures in the source rocks, not via pore space. Microfissures formed by abnormal high pressures and undercompaction during the process of diagenesis are common, and such microfissures are presumably available for initial migration and expulsion of petroleum (Tissot and Welte 1978; Hua and Lin 1989; Li 2004). The width of microfissures ranges from 30–100 μm. Hua and Lin (1989) reported that microfissures filled with bitumen were observed under microscope in the Lower Cretaceous mudstones of the West Jiuquan Basin of China. This is a reliable evidence for microfissures serving as passageways for petroleum primary migration.

Jurassic spores and pollen, such as *Deltoidospora perpusilla*, *D. gradata*, *Gleicheniidites rouseii*, *Dictyophyllidites harrisii*, *Dictyophyllum rugosum*, *Undulatisporites concavus*, *Cibotiumspora paradoxa*, *C. jurienensis*, *Leptolepidites major*, *L. verrucatus*, *Murospora minor*, *Paleoconiferus asaccatus*, *Pterchipollenites thomasii*, *Protopicea exilioides*, *Piceites expositus*, *Cedripites minor*, *Platysaccus lopsinensis*, *Podocarpidites major*, *P. wapellensis*, *P. multicinus*, *Parvisaccites enigmatus*, *Chasmatosporites major*, *C. elegans*, *C. minor*, *Cycadopites subgranulosus*, and *Bennettiteaepollenites lucifer*, have been found in crude oils from the Neogene petroleum reservoir of the Kekeya oil field in the Tarim Basin. In general, spores are 18–70 μm in equatorial diameter, disaccate pollen grains are 30–80 μm in overall breadth, and monosulcate pollen grains are 15–42 μm in breadth. They could obviously not exit through petroleum source rock pore spaces, so they must have been expelled from the petroleum source rock bed by passageway of microfissures in the source rocks. Furthermore, Lower Cretaceous spores and pollen including *Cibotiumspora juncta, Osmundacidites wellmanii, Schizaeoisporites zizyphinus, S. gansuensis, Cingulatisporites ruginosus, Cycadopites minimus, Psophosphaera grandis, P. tenuis,* and *Classopolls annulatus* are found in crude oils (see Table 3.16) from the Neogene petroleum reservoir of the Laojunmiao oil field in the West Jiuquan Basin. These spores and pollen derived from a Lower Cretaceous petroleum source bed must also have been expelled through petroleum source rock microfissures. The discovery of spores and pollen derived from a petroleum source bed in crude oils explains that microfissures in source rocks should be the important passageways for primary migration of petroleum.

One shocking find was a species of the fossil megaspore *Triangulatisporites junggarensis* in crude oil in the Upper Permian petroleum reservoir of the Huonan oil field in the Junggar Basin. This megaspore has diameters between 428 and 475 μm. It could not have exited through microfissures. Its presence implies that there are structural fissures in source rock beds formed by tectonic stress. Therefore, structural fissures might be additional passageways for primary migration of petroleum.

8.2.2 Phase States of Primary Migration

Petroleum migrates in a variety of phase states. This includes water-soluble, oil-soluble, gas-soluble, oil, gas, and diffusion phases. Of these, oil phase migration is the most important, having the highest efficiency during primary migration (Li 2004).

Most petroleum originates from indigenous organic materials through thermal transformation, where petroleum source rocks are buried from 1500 to 4500 m deep, and geotemperatures reach 60–150 °C (Hunt 1979). It is logical to suggest that hydrocarbon generation must cause and raise pressure in the rocks (Law and Spencer 1998). Such abnormal high pressure is an important cause for formation of microfissures. Therefore, microfissures are to be expected during petroleum generation.

The phase state of petroleum migration essentially depends on the passageways for petroleum migration. Because microfissures are wide enough for the passage of spores and pollen, the passageways must also be unimpeded for the passage of oil droplets. Thus, oil phase migration is fully possible during the course of primary migration. Migration of hydrocarbon in the gas phase can be expected to not meet resistance, because gas can penetrate through any size pores. Therefore, petroleum migrates in the original phase state, either oil phase or gas phase, during primary migration. As mentioned in Chap. 5, the colors of fossil spores and pollen indicate that the petroleum source rocks in the Tarim, Junggar, Turpan, Qaidam, and Beibu Gulf basins of China all possess potential for liquid petroleum generation. Associate with liquid petroleum generation, microfissures caused by abnormal high pressure are common in the petroleum source rocks. Consequently, oil phase primary migration is fully possible in these basins.

Some fossil spores and pollen including *Cyathidites minor, Dictyophyllidites harrisii, Undulatisporites pflugii, Punctatisporites triassicus, Aratrisporites strigosus, A. scabratus, Cycadopites nitidus, C. subgranulosus, Cedripites parvisaccus,* and *Alisporites parvus* have been found not only in crude oils, but also in oil-associated natural gas and oil field water from the Yakela oil field in the Tarim Basin (Fu and Liu 1992; Jiang et al. 2002). It is generally considered that oil phase migration is usually accompanied by gas phase migration, so the mixed phases of oil, gas, and water might be the main phase state during primary migration in the Tarim Basin.

8.3 Secondary Migration

Secondary migration of petroleum is defined here by the migration of the petroleum from the carrier bed to the trap or from the reservoir bed to the trap. Lacking a trap in the pathway, secondary migration will continue until loss of the petroleum at the surface.

8.3.1 Passageways for Secondary Migration

The passageways for secondary migration include connective pore spaces, bedding voids, joints, fissures, faults, and unconformities in the carrier and petroleum reservoir beds. Of these, faults and unconformities are particularly important passageways for the secondary migration of petroleum.

In the South Junggar Depression of the Junggar Basin, the Dushanzi oil field lies in an anticlinal trap oil pool with a fault zone at the anticlinal axis. There are oil and gas seepages and mud volcanoes along the fault zone. These reflect faults that are available passageways for petroleum migration. The presence of large passageways is supported by the presence of many Jurassic spores and pollen including *Cyathidites minor, Dictyophyllidites harrisii, Murospora jurassica, Marattisporites scabratus, Aratrisporites xiangxiensis, Cycadopites nitidus, C. typicus, Callialasporites dampieri, C. radius, Paleoconiferus asaccatus, Podocarpidites multicinus, Piceites pseudorotundiformis,* and *Eucommiidites troedssonii* in crude oil samples collected from the Oligocene to Miocene Shawan Formation petroleum reservoir at an anticlinal axis. In contrast, only few spores and pollen have been found in crude oil samples collected from the petroleum reservoir at the anticlinal limb (Yang and Jiang 1989). These Jurassic spores and pollen, along with the original petroleum, left the petroleum source bed and accumulated into the Tertiary petroleum reservoir bed by way of faults during the course of petroleum migration. These data indicate that faults are efficient passageways for upward vertical migration of petroleum during secondary migration.

In the North Tarim Upheaval of the Tarim Basin, Carboniferous, Triassic, and Jurassic spores and pollen have been found in crude oil samples collected from the Ordovician petroleum reservoir bed of the Yakela oil field. Carboniferous spores and pollen including *Calamospora pedata, Punctatosporites minutus,* and *Vesicaspora wilsonii* are derived from a Carboniferous petroleum source bed. Triassic spores and pollen, such as *Punctatisporites triassicus, Retusotriletes mesozoicus, Apiculatisporis parvispinosus, A. spiniger, Lycopodiacidites kuepperi, Zebrasporites kahleri, Tigrisporites halleinis, Limatulasporites dalongkouensis, L. parvus, Aratrisporites fischeri, A. scabratus, A. strigosus, Lueckisporites triassicus,* and *Chordasporites singulichorda,* are derived from a Triassic petroleum source bed. Jurassic spores and pollen including *Dictyophyllidites harrisii, Undulatisporites pflugii, Cycadopites subgranulosus,* and *Cedripites minor* are derived from a Jurassic petroleum source bed. This complexity reflects the presence of six stratigraphic unconformities in this region. The above spores and pollen, along with the original petroleum, left the petroleum source beds and accumulated in the Ordovician oil pool by way of these unconformities during the course of petroleum migration. It also indicates that unconformity interface passageways are very important for lateral migration of petroleum during secondary migration.

8.3.2 Phase States and Flow Types of Secondary Migration

The phase states of petroleum secondary migration generally inherit the phase states of primary migration. After the original petroleum enters the carrier bed or the petroleum reservoir bed in an oil phase or gas phase migration, secondary migration involves moving from there to the trap. When oil and gas meet a trap, they will accumulate into oil pools and gas pools during the secondary migration of petroleum. In the course of secondary migration, oil and gas remain in their original phase state. It is clear that oil phase migration is the most efficient process for the formation of oil pools and gas phase migration is the most efficient process for the formation of gas pools. However, oil-soluble and gas-soluble phase migrations are also efficient processes for the formation of oil and gas pools during secondary migration.

The flow types of petroleum during secondary migration include buoyancy flow, seepage flow, and diffusion flow. Buoyancy flow is the most important flow type in secondary migration, because buoyancy is the simplest process. Seepage flow including oil–water, gas–water, and oil–gas–water phases occurs frequently during secondary migration. In dense rocks, diffusion flow is probably most important.

8.3.3 Directions, Routes, and Distances of Secondary Migration

The direction and route of petroleum secondary migration are mainly controlled by the sedimentary facies and structural elements. In a sedimentary facies belt, petroleum migration is always from argillaceous rocks to arenaceous rocks, namely from low porosity and permeability to high porosity and permeability.

Spores and pollen in crude oils can also provide information about direction, route, and distance of petroleum migration. For example, Lower to Middle Jurassic spores and pollen including *Cyathidites australis, C. minor, Gleicheniidites rouseii, Dictyophyllidites harrisii, Cibotumspora paradoxa, Klukisporites variegatus, Murospora jurassica, M. minor, Chasmatosporites elegans, Cerebropollenites carlylensis,* and *Cedripites minor* have been found in crude oil samples collected from the Neogene petroleum reservoir of the Kelatu oil field in the Kashi Sag of the Tarim Basin. These Jurassic spores and pollen derived from the petroleum source bed were carried by oil, gas, and water from the Jurassic mudstones into the Neogene sandstones by way of faults in the course of secondary migration. Their presence indicates that the direction of migration was upward vertical migration, the route of migration was from the Lower to Middle Jurassic Series to the Neogene System, and the distance of migration was less than 5 km.

In the North Tarim Upheaval, as mentioned above, spores and pollen derived from Carboniferous, Triassic, and Jurassic petroleum source beds have been found in crude oil samples collected from the Ordovician petroleum reservoir bed of the Yakela oil field in the Tarim Basin. These spores and pollen were carried by oil, gas, and water from the Carboniferous shales, Triassic mudstones, and Jurassic mudstones to the Ordovician dolomite-cave reservoir by way of unconformity interface passageways during the course of secondary migration. They indicate that the main direction of migration was lateral migration, the routes of migration were from the Carboniferous, Triassic, and Jurassic petroleum source beds respectively to the Ordovician petroleum reservoir, and the distances were in the range of 5–100 km.

Other age mixtures are diagnostic in the petroleum reservoirs of the North Tarim Upheaval. Carboniferous spores and pollen have been found in crude oils from the Carboniferous reservoir of the Donghetang oil field and the Triassic reservoir of the Yakela oil field. Triassic and Jurassic spores and pollen have been found in crude oils from the Triassic, Jurassic, and Cretaceous reservoirs of the Yakela oil field, and Jurassic spores and pollen have been found in crude oils in the Tertiary reservoir of the Yingmaili oil field. These multi-age spores and pollen were carried by oil, gas, and water from the Carboniferous shales and Triassic and Jurassic mudstones to Carboniferous, Triassic, Jurassic, Cretaceous, and Tertiary sandstones during the course of secondary migration. The directions of migration to these fields may be either upward/downward vertical migration or lateral migration. The routes of migration are very complicated and have not been fully analyzed to explain the routes from each petroleum source bed to each petroleum reservoir, nor have the distances of migration, which are dependent on the distances between traps and petroleum source beds, been assessed.

Secondary migration of petroleum is the process that accumulates dispersed oil droplets and concentrates them into petroleum accumulations.

However, if there are no traps, secondary migration is the process that results in the loss of petroleum accumulation. Oil and gas seepages at the ground surface are visual evidence of loss of petroleum accumulation caused by secondary migration. For example, in the Kuqa Depression of the Tarim Basin, Jurassic spores and pollen including *Deltoidospora perpusilla, D. gradata, Dictyophyllidites harrisii, Cibotiumspora paradoxa,* and *Cedripites minor* have been found in crude oil samples collected from the Miocene oil seepages at Kangcun and Jilishen. They indicate that at least part of the petroleum expelled from the Jurassic petroleum source bed in the Kuqa Depression was lost for lack of a trap during the course of secondary migration.

8.4 Period of Petroleum Deposit Formation

The availability of a trap is a necessary condition for petroleum deposit formation. Petroleum can only accumulate in traps after traps form. Therefore, petroleum deposit formation would be later than trap formation.

Having a petroleum source is the essential element for petroleum deposit formation. The age of a petroleum source rock and the main epoch of hydrocarbon expulsion are important for assessing the period of petroleum deposit formation. As mentioned in Chap. 4, in accordance with petroleum sporo-pollen assemblages, petroleum source rocks in the Liaohe Basin of the East China Sea and the Beibu Gulf Basin of the South China Sea all belong to the Paleogene System. The Himalayan orogenic movement promoted petroleum generation and hydrocarbon expulsion and resulted in structural trap formation. Thus, the earliest period of petroleum deposit formation in these basins should be the Neogene Period. In the Southwest Tarim Depression of the Tarim Basin, petroleum source rocks belong to the Lower to Middle Jurassic Series, the reservoir rocks belong to the Neogene System, and the Himalayan orogeny resulted in structural trap formation. It is suggested then that the earliest period of petroleum deposit formation in this depression should be the Neogene Period.

The Tarim and Junggar basins are large-sized superimposed composite basins, which host Carboniferous, Permian, Triassic, and Jurassic petroleum sources, and Carboniferous, Permian, Triassic, Jurassic, and Tertiary petroleum reservoirs. Because of multiple-period petroleum sources and multiple-period petroleum reservoirs, such basins had multiple times of petroleum deposit formation.

References

Fu, J. M., & Liu, D. H. (1992). *Migration, accumulation and trap condition of natural gas* (pp. 1–232). Beijing: Science Press (in Chinese).

Hua, B. Q., & Lin, X. X. (1989). Discussion on some problems of oil migration in Jiuxi Basin. *Acta Sedimentologica Sinica, 7*(1), 39–47 (in Chinese with English abstract).

Hunt, J. M. (1979). *Petroleum geochemistry and geology* (pp. 1–617). San Francisco: W. H. Freeman and Company.

Jiang, D. X., Wang, Y. D., & Wei, J. (2002). Palynological evidence for petroleum migration in Tarim Basin. *Acta Sedimentologica Sinica, 20*(3), 524–529 (in Chinese with English abstract).

Law, B. E., & Spencer, C. W. (1998). Abnormal pressure in hydrocarbon environments. In B. E. Law, G. F. Ulmishek & V. E. Slavin (Eds.), *Abnormal pressures in hydrocarbon environments*: AAPG Memoir 70, pp. 1–11.

Levorsen, A. L. (1956). *Geology of petroleum* (pp. 1–701). San Francisco: W. H. Freeman.

Li, M. C. (2004). *Petroleum and natural gas migration* (3rd ed., pp. 1–350). Beijing: Petroleum Industry Press (in Chinese).

Tissot, B. P., & Welte, D. H. (1978). *Petroleum formation and occurrence* (pp. 1–314). New York, Berlin: Springer.

Yang, H. Q., & Jiang, D. X. (1989). Spores and pollen from crude oil of Dushanzi oil-field in Xinjiang. *Acta Botanica Sinica, 31*(12), 948–953 (in Chinese with English abstract).

Geochronic and Geographic Distribution of Nonmarine Petroleum Source Rocks

Abstract

The study of the petroleum sporo-pollen assemblages and the correlations between spores and pollen in crude oils and those in potential source rocks is a powerful tool to determine the petroleum source rock series in the petroliferous basins of China. Certain generalizations can be made: Upper Paleozoic, Mesozoic, and Tertiary rocks in the petroliferous basins of China expelled petroleum. Tropical or subtropical lacustrine deposition was the favorable climatic and ecological setting for the formation of petroleum source rocks in the continental basins. In the inland petroliferous basins, such as the Turpan and Qaidam basins, Jurassic petroleum source rocks formed in lacustrine and swampy environments are important and abundant. In the large superimposed basins of the inland petroliferous basins, such as the Tarim and Junggar basins, the petroliferous regions possess Carboniferous, Permian, Triassic, Jurassic, and Paleogene petroleum source rocks formed in a limnetic sedimentary environment. The coastal shelf petroliferous basins, including the Liaohe, Beibu Gulf, Zhujiang Mouth, and Sanshui basins, contain Paleogene source rocks formed mainly in lacustrine settings. Dark-colored argillaceous rocks and oil shales, considered to be excellent petroleum source rocks, are common in these coastal shelf basins. Thus, the study of sporo-pollen in crude oils has added significant information to enhancing petroleum exploration in petroliferous basins.

Keywords

Petroliferous basins · Petroleum source rocks · Upper paleozoic · Mesozoic · Paleogene

9.1 Inland Petroliferous Basins

In the inland petroliferous basins of China, including the Tarim, Junggar, Turpan, Qaidam, and Jiuquan basins, the crude oils usually contain abundant fossil spores and pollen. It is believed that fossil spores and pollen may be used as indicators for correlation of oil and source to determine petroleum source rocks. In accordance with the petroleum sporo-pollen assemblages and the correlations between spores and pollen in crude oils and those in potential source rocks, the petroleum source rocks of the above petroliferous basins have been determined. The results of the case studies indicate that the Upper Paleozoic and Mesozoic Erathems and the Tertiary System of the inland petroliferous basin commonly contain petroleum source rocks. Lacustrine deposition under tropical or subtropical climates was the favorable condition for the formation of petroleum source rocks in these continental basins. In the Tarim Basin, the Carboniferous, Triassic, and Jurassic Systems contain petroleum source rocks, and the Permian and Cretaceous Systems might contain petroleum source rocks. In the Junggar Basin, the Carboniferous, Permian, Triassic, Jurassic, and Paleogene Systems contain petroleum source rocks. In the Turpan Basin, the Lower and Middle Jurassic dark-colored mudstones are petroleum source rocks. In the Qaidam Basin, the Lower and Middle Jurassic Series and the Paleogene System contain petroleum source rocks. And in the Jiuquan Basin, Lower Cretaceous black shales are petroleum source rocks.

In addition, in the Ordos Basin of North China, Triassic spores including *Punctatisporites triassicus, P. textatus,* and *Calamospora impexa* and Jurassic spores/pollen including *Cyathidites minor, Gleicheniidites senonicus, Marattisporites scabratus, Cycadopites nitidus, C. typicus, C. minimus, Classopollis classoides,* and *C. annulatus* have been found in crude oils from the Middle Jurassic reservoir of the Zhangqing oil field (Jiang 1988, 1991). The results of correlations between petroleum and source rocks indicate that the dark-colored shales and mudstones of the Upper Triassic Yanchang Formation and the Lower to Middle Jurassic Fuxian and Yanan formations are petroleum source rocks.

Mesozoic continental or nonmarine basins of different sizes, numbering in the hundreds, are common in East Asia, especially China. This is because (1) China was formed as a mosaic of continental blocks interspersed with Paleozoic geosynclinal foldbelts; (2) Indosinian, Yanshanian, and Himalayan tectonism affected most parts of China; and (3) As a consequence of these tectonic episodes, fault systems having different patterns and different extension occur in China, enabling the production of numerous fault troughs and downwarps favorable for sedimentation. Of these, the most important basins are Tarim, Junggar, Qaidam, Ordos, Sichuan, and Songliao basins (Huang and Chen 1987). In the Tertiary, the landscape of China was covered by continental or nonmarine basins. Large- and medium-sized basins of hereditary nature occur in the northwest, such as the Tarim, Junggar, Turpan, and Qaidam basins. The active Himalayan orogeny gives the stamp of the tectonic framework in the western part of China. The northward push of the Indian Plate was not only the cause of the folding and uplifting of the Himalayas and the Tibetan Plateau, but also the generative stress for the creation of the Kunlun, Tianshan, and the Qilianshan mountains.

The changes of relative motion, including changes in thermal regimes and tectonic forces are the active processes that produced petroliferous basin in China. Petroleum-generating organic matter, petroleum reservoir formation, and the formation of source rocks, reservoir rocks, and seal rocks are the products of such forces (Zhu 1965, 1986). The Mesozoic and Cenozoic petroliferous basins were formed in later geological historical stages and under new tectonic regimes. Most petroleum in the world formed in the post-Triassic, and more than half of the world's petroleum was produced from post-Early Jurassic sediments. Translated from the Chinese, this is termed "Post-Early Jurassic prosperous petroleum." The tectonic regime of this time was formation of fault troughs and downwarps, as well as combinations or transformations into fault troughs and downwarps.

Troughs and downwarps are favorable structures for formation of petroleum source rocks, reservoir rocks, and seal rocks. And the combination or transformation of fault troughs and downwarps are therefore favorable structures for the generation, migration, and accumulation of petroleum.

Many petroliferous basins, such as the Tarim and Junggar basins, are superimposed composite basins having Mesozoic, Cenozoic, and Paleozoic rocks. Such sedimentary and tectonic superimposition is unusually favorable for petroleum deposit formation, because the depressions of hereditary nature are favorable for generation of petroleum, and the uplifts of hereditary nature are favorable for migration and accumulation of petroleum. Thus, superimposition of sedimentation and tectonism is the cause of multiple-period petroleum sources in many basins. Such superimposition played an important role in the distribution and redistribution of Paleozoic petroleum. Research on the superimposition of the Mesozoic and Cenozoic Systems on the Paleozoic System is a useful model to open up new spheres for petroleum exploration (Zhu 1986).

9.2 Coastal Shelf Petroliferous Basins

In the coastal shelf petroliferous basins of China, including the Liaohe, the Beibu Gulf, and the Zhujiang Mouth basins, crude oils contain abundant Paleogene spores and pollen. In accordance with the petroleum sporo-pollen assemblages and the correlations between spores and pollen in crude oils and those in potential source rocks, the petroleum source rocks of these basins are the Paleogene dark-colored argillaceous rocks and oil shales formed mainly as lacustrine deposits. The Oligocene Shahejie and Dongying formations in the Liaohe Basin; the Eocene to lower Oligocene Liushagang Formation and the middle to upper Oligocene Weizhou Formation in the Beibu Gulf Basin; and the middle Eocene Wenchang Formation, the upper Eocene to middle Oligocene Enping Formation, and the upper Oligocene Zhuhai Formation in the Zhujiang Mouth Basin all contain excellent petroleum source rocks.

In addition, in the Sanshui Basin of the Guangdong Province of South China, Paleogene spores and pollen such as *Sphagnumsporites antiquasporites, Plicifera decora, Polypodiaceaesporites haardti, Dacrydiumpollenites rarus, Pinuspollenites bankianaeformis, P. minutus, Cedripites eocenicus, C. deodariformis, Momipites coryloides, Quercoidites minutus,* and *Q. asper* have been found in crude oil samples from the Paleogene reservoir (Jiang 1988). According to the petroleum sporo-pollen assemblage and the correlations of petroleum and source rocks, the black mudstones and oil shales of the Paleogene Buxin Formation should be the main petroleum source rocks in the basin.

Huang (1960) not only drew the attention of geoscientists to the great importance of Indosinian tectonism in Chinese geology and Tethyan geology, but his insights can be used for petroleum prospecting today. Indosinian tectonism was characterized by folding of different types. The first category was termed geosynclinal folding; the second, basement folding; the third, overprinting or rejuvenated folding; and the fourth, blanket folding (Huang and Chen 1987). Following Indosinian tectonism, the paleogeography of the Jurassic Period became quite different from that of the Triassic. In the Triassic, following the disappearance of the Tethys Ocean, Gondwanaland and Eurasia united together with the Americas, forming a true Pangaea. However, beginning in the Late Triassic, the Indus–Yarlung rifting zone opened widely in the Jurassic, and a Mesozoic Ocean or Mesotethys came into existence. The maximum opening of the Mesotethys appears to have been in the Cretaceous, whereas its closing stage was Late Cretaceous to Early Tertiary. Near the end of the Cretaceous and in the Early Tertiary, as a result of rifting followed by seafloor spreading, the oceanic and suboceanic provinces came into existence. In the

Tethys proper, owing to the closing of the Mesotethys, only a narrow shallow water embayment still lingered along the Indus–Yarlung Zone. In the Early Tertiary, Taiwan and the East China Sea were in direct connection, forming the Neotethys, parts of which continue to develop up to the present time. In the Oligocene to Miocene, rifting took place in the South China Sea. On Hainan Island and the Leizhou Peninsula, Miocene to Pliocene littoral deposits and shallow-sea deposits bear Foraminifera and marine Ostracoda. Cenozoic sedimentary basins of the East China and the South China Seas are connected by a shallow sea, so in addition to nonmarine petroleum source rocks, marine petroleum source rocks may also be forming and developing. Similar to the Gulf of Mexico Basin, petroleum exploration in these basins undoubtedly will be successful.

References

Huang, T. K. (1960). The main characteristics of the structure of China: Preliminary conclusions. *Scientia Sinica, 9*(4), 492–544.

Huang, J. Q., & Chen, B. W. (1987). *The evolution of the Tethys in China and adjacent regions* (pp. 1–109). Beijing: Geological Publishing House. (in Chinese and English).

Jiang, D. X. (1988). Spores and pollen in oils as indicators of lacustrine source rocks. In A. J. Fleet, K. Kelts, & M. R. Talbot (Eds.), *Lacustrine petroleum source rocks* (pp. 159–169). London: Blackwell Scientific Publications.

Jiang, D. X. (1991). Fossil spores and pollen in petroleum and their significance. *Chinese, Journal of Botany, 3*(1), 62–76.

Zhu, X. (1965). *Tectonics of China* (pp. 1–183). Beijing: Science Press. (in Chinese).

Zhu, X. (1986). *Tectonics of the petroliferous basins in China* (pp. 1–132). Beijing: Petroleum Industry Press. (in Chinese).

Concluding Remarks

Abstract

In *Petrolipalynology*, we have used palynological data to determine petroleum source rocks in many of the nonmarine petroliferous basins in China. The nonmarine petrolumn-generation doctrine that nonmarine petroleum deposits should be formed in inland tectonic depressions under humid climatic conditions is confirmed. The presence of spores and pollen of various ages pinpoints source and transport directions for potential new petroleum prospects. Their presence in igneous rock reservoirs also provides convincing evidence for the organic petroleum origin theory. The information also shows how the presence of palynomorphs of mixed ages helps to correlate between the many parts of the petroleum system.

Keywords

Palynological data · Nonmarine petroleum generation · Petroleum system · Exploration and development of petroleum resources

For the purposes of exploration and development of petroleum resources, it is necessary to study the generation, migration, accumulation, and preservation of petroleum. These have been reported here for inland and coastal shelf basins in China.

Based on petroleum sporo-pollen assemblages and correlations between spores and pollen in crude oils and those in potential source rocks, it can be concluded that the Triassic, Jurassic, Cretaceous, and Paleogene Systems contain abundant petroleum source rocks in the petroliferous basins of China. The plentiful Mesozoic and Tertiary petroleum in China coincides with world-wide petroleum distribution. This is not a fortuitous phenomena because, at the root, changes in tectonic motion and the ensuing new thermal regimes were favorable for the generation and accumulation of petroleum. At the end of the Late Permian, continental collision was complete, resulting in the formation of folding and subsequent mountain-building. As a result of the Hercynian orogeny, Mesozoic and Cenozoic petroliferous basins formed and developed around the world. Because of geothermal flows rising from the mantle and moving through the crust, the organic matter in sedimentary basins was converted into petroleum. Therefore, many Mesozoic and Cenozoic petroliferous basins were formed under new tectonic regimes in new stages. Consequently Mesozoic and Tertiary petroliferous basins have very bright prospective value.

Furthermore, in the large-sized, superimposed, composite basins such as the Tarim and Junggar basins, the underlying Carboniferous and Permian Systems also contain petroleum source rocks in addition to those in the overlying Mesozoic and Paleogene Systems. Such basins possess the character of multiple-period petroleum sources. Because the Paleozoic basins were superimposed by the Mesozoic and Cenozoic basins, sedimentary and tectonic superimposition caused conditions favorable for the generation and accumulation of petroleum. Therefore, searching for Paleozoic oil fields and gas fields underlying the Mesozoic and Cenozoic petroliferous basins will be new sphere for petroleum exploration. Post-Hercynian prospective petroleum may be expected along with the development and advance of science and technology.

Hou (1959) long ago expressed that paleoclimatic and tectonic factors should be the most important for the generation and accumulation of petroleum in marine and nonmarine sedimentary basins. On the basis of investigations on the formation and distribution of nonmarine petroleum fields in Northwest China, it is suggested that the nonmarine petroleum deposits should be formed in inland tectonic depressions under humid climatic conditions (Lanzhou Institute of Geology, Academia Sinica 1960). This is considered to be a nonmarine petroleum-generating doctrine. Xu (2002) summarized the geoscience framework for nonmarine petroleum-generation by explaining that the petroleum sources were controlled by the environment and the petroleum deposits were controlled by the petroleum sources. He explained that this doctrine emphasized an environmentally-driven function.

Spores and pollen found in crude oils can act as indicators of lacustrine source rocks (Fleet et al. 1988). The spores and pollen that indicate lacustrine source rocks usually also support interpretation for tropical or subtropical wet and hot sedimentary environments, so it may be considered that lacustrine deposition under hot or warm, and wet or humid, climatic conditions should be favorable for the formation of lacustrine petroleum source rocks (Jiang 1988). This conclusion supports the nonmarine petroleum-generating doctrine. Moreover, this conclusion is confirmed by the palynological data from petroleum source rocks in this book. In other words, the nonmarine petroleum-generation doctrine is fully supported by the palynological evidence.

Coal-related oil and gas, and submature to low mature oil and gas, are also consistent with doctrine of nonmarine petroleum-generation, thereby opening up new fields for research on nonmarine petroleum generation (Xu 2002). The Cordaitopsida plants that flourished in the Carboniferous and Permian Periods and the Coniferopsida and Cycadopsida plants that flourished in the Jurassic and Cretaceous Periods were the most important coal-forming plants. Coal-bearing strata of these ages are rich in spores and pollen, so we suggest this may be a future petroleum-related research field. It is concluded that lacustrine and marsh/swamp deposition under hot or warm and wet or humid climatic conditions might possess tremendous potential for generation of oil, coal, and natural gas, while marsh/swamp deposition should be more favorable for the generation of coal alone (Jiang and Wang, 2002). Research on correlations between spores and pollen in coal-related oils and those in coal-bearing strata might be useful for discovering relationships between organic material type, maturity, and the environmental character of linked petroleum and coal origin. Studies on spores and pollen in submature and low mature oils might provide useful information for theories about early hydrocarbon-forming epochs.

New ideas are constantly being developed and applied in oil and gas exploration. For example, in recent years the ideas and methods of the petroleum system are used in the exploration and development of petroleum resources. The essential elements of a petroleum system (Magoon and Dow 1994) are the source rock, reservoir rock, seal rock, and overburden rock; the essential processes in the petroleum system include trap formation and the generation–migration–accumulation of petroleum. These essential elements and processes must occur in time and place so that the organic matter that is included in a source rock can be converted to a petroleum accumulation. The petroleum system has a stratigraphic limit, geographic extent, and an age. The time of hydrocarbon generation for a petroleum system can span considerable time and cover a large area. Diagrams showing burial history are drawn where overburden rock is the thickest, which indicate the time of oil generation (from start to finish), the critical moment, and the times of generation, migration, and accumulation, thereby bracketing the age(s) of the petroleum system. The critical moment is that point in time selected by the investigator that best depicts the generation–migration–accumulation of most hydrocarbons in a particular petroleum system. A map or cross section drawn for depicting the critical moment best shows the geographic and stratigraphic extent of the petroleum system.

Knowing the age of various horizons within the overburden rock is the key to determining when and where a source rock first starts generating petroleum and when and where it ceases to generate petroleum. Palynology can provide information for the age of related stratigraphy. By means of correlation between spores and pollen in crude oil and those in potential source rock, the active petroleum source rock of the petroleum system can be determined, and such petroleum source rock is specific to the petroleum accumulation. In such a petroleum system, an exceptionally good palynological match exists between the petroleum source rock and the petroleum accumulation, so the level of certainty of the petroleum system is known, not hypothetical or speculative (Magoon and Dow 1994). Therefore, palynological data from petroleum sources can provide reliable information to establish a firm foundation for a petroleum system. Obviously, palynological study of petroleum sources is helpful in the application of the ideas and methods of the petroleum system to the exploration and development of petroleum resources.

References

Fleet A. J., Kelts K., & Talbot M. R. (1988). *Lacustrine petroleum source rocks* (pp. 1–391). London: Blackwell Scientific Publications

Hou D. F. (1959). Nonmarine sediments and petroleum geology. *Chinese Journal of Geology (Scientia Geologica Sinica), 8*, 225-227 (in Chinese)

Jiang D. X. (1988). Spores and pollen in oils as indicators of lacustrine source rocks. In A. J. Fleet, K. Kelts, M. R. Talbot (Eds.), *Lacustrine petroleum source rocks* (pp. 159–169). London: Blackwell Scientific Publications

Jiang D. X., & Wang Y. D. (2002). Middle Jurassic palynoflora and its environmental significance of Dongsheng, Inner Mongolia. *Acta Sedimentologica Sinica, 20*(1), 47–54 (in Chinese with English abstract)

Lanzhou Institute of Geology, Academia Sinica. (1960). *Formation and distribution of non-marine petroleum fields in Northwest China* (pp. 1–340). Beijing: Science Press (in Chinese)

Magoon L. B., & Dow W. G. (1994). *The petroleum system—from source to trap*. AAPG Memoir (vol. 60, pp. 1–645). Tulsa, Oklahoma: American Association Petroleum Geologists Press

Xu Y. C. (2002). Non-marine petroleum origin and heated argument. In: National Key Laboratory of Gas Geochemistry, Lanzhou Institute of Geology, Academia Sinica (Eds.), *Collected papers of natural gas geochemistry* (pp. 3–13). Beijing: Geological Publishing House, (in Chinese)

Explanation of Plates and Plates

All specimens are at a magnification of 704× unless otherwise stated. Numbers following species names refer to specific archived slides.

Plate I

1 *Cyathidites australis* Couper No. L-1-146
2–3 *Cyathidites minor* Couper 2 No. L-1-16; 3 No. S-14-161
4–6 *Undulatisporites pflugii* Pocock 4 No. L-1-94; 5 L-1-139; 6 S-14-124
7 *Dictyophyllidites harrisii* Couper No. L-1-32
8 *Gleicheniidites conflexus* (Chlonova) Xu et Zhang No. L-1-138
9 *Gleicheniidites rouseii* Pocock No. L-1-133
10 *Gleicheniidites senonicus* Ross No. L-1-201
11 *Gleicheniidites nilssonii* Pocock No. L-1-169
12 *Cibotiumspora paradoxa* (Maljavkina) Chang No. L-1-59
13, 21 *Osmundacidites wellmanii* Couper 13 No. L-1-5; 21 No. L-1-165
14 *Granulatisporites jurassicus* Pocock No. L-1-26
15 *Apiculatisporis globosus* (Leschik) Playford et Dettmann No. L-1-103
16 *Duplexisporites gyratus* Playford et Dettmann No. L-1-22
17 *Apiculatisporis variabilis* Pocock No. L-1-25
18–19 *Apiculatisporis* spiniger (Leschik) Qu 18 No. S-14-80; 19 No. L-1-6
20 *Duplexisporites anagrammensis* (Kara-Murza) Playford et Dettmann No. L-1-169
22 *Duplexisporites scanicus* (Nilsson) Playford et Dettmann No. L-1-1
23–24 *Duplexisporites amplectiformis* (Kara-Murza) Playford et Dettmann
  23 No. L-1-23; 24 No. L-1-77

Explanation of Plates and Plates

Plate I

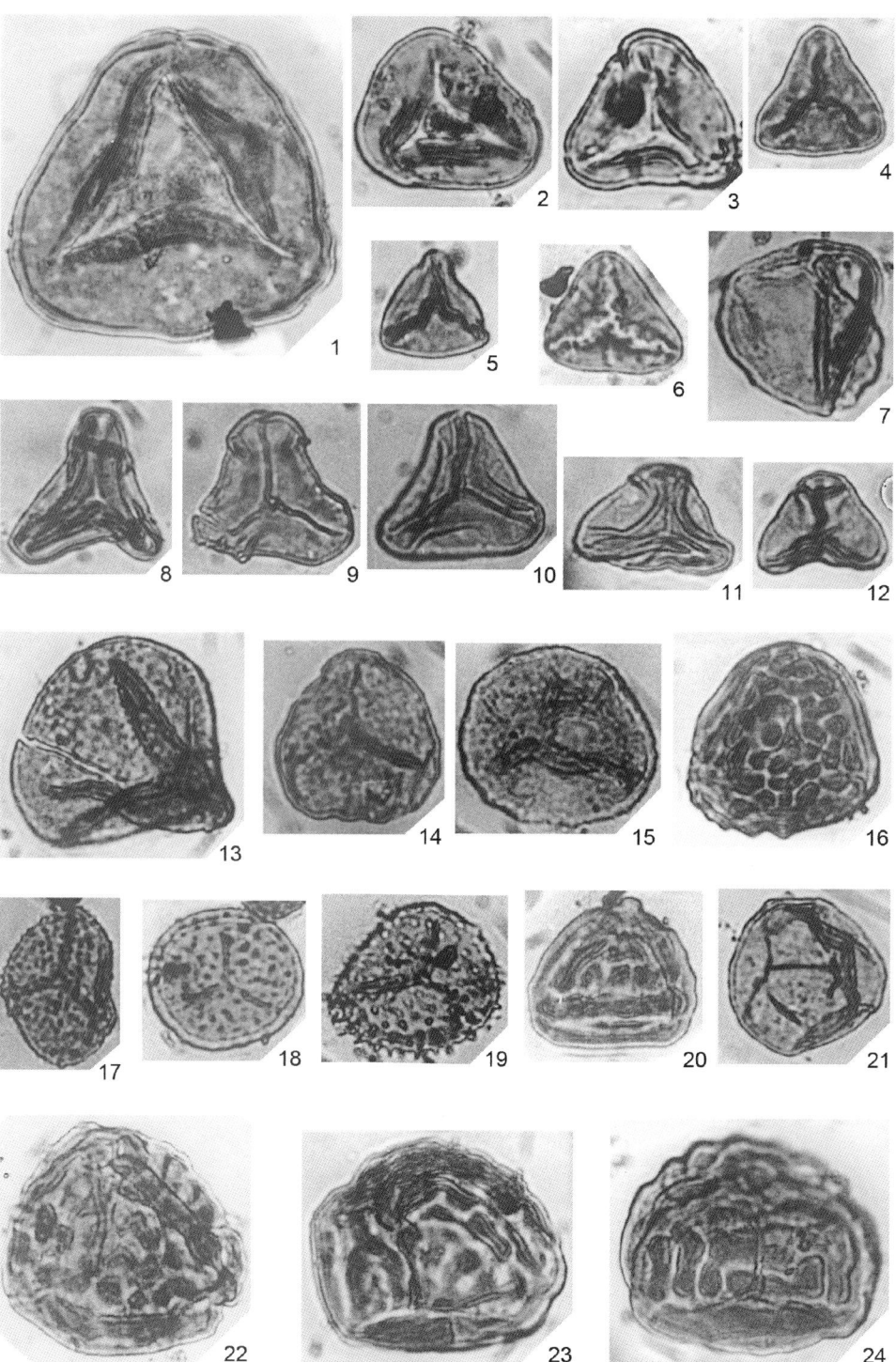

Fossil spores in crude oils from the North Tarim Upheaval of the Tarim Basin

Plate II

1 Lycopodiacidites rhaeticus Schulz No. S-9-21
2 Zebrasporites kahleri Klaus No. S-14-53
3 Tigrisporites halleinis Klaus No. S-14-101
4 Retusotriletes mesozoicus Klaus No. S-9-32
5 Verrucosisporites contactus Clarke No. S-14-40
6 Lundbladispora nejburgii Schulz No. S-9-17
7, 20 Aratrisporites fischeri (Klaus) Playford et Dettmann *7.* No. L-1-165; *20* No. L-1-1
8 Aratrisporites scabratus Klaus No. S-14-61
9 Aratrisporites paenulatus Playford et Dettmann No. L-1-1
10 Aratrisporites paraspinosus Klaus No. L-1-152
11 Aratrisporites granulatus (Klaus) Playford et Dettmann No. S-14-46
12–13 Aratrisporites strigosus Playford *12* No. S-14-144; *13* No. S-14-148
14 Marattisporites scabratus Couper No. S-14-6
15 Aratrisporites tenuispinosus Playford No. L-1-41
16–17 Aratrisporites parvispinosus Leschik *16* No. L-1-92; *17* No. L-1-191
18 Aratrisporites sp. No. S-14-141
19 Verrucosisporites remyanus Mädler No. L-1-111
21 Lycopodiacidites kuepperi Klaus No. S-14-50

Plate II

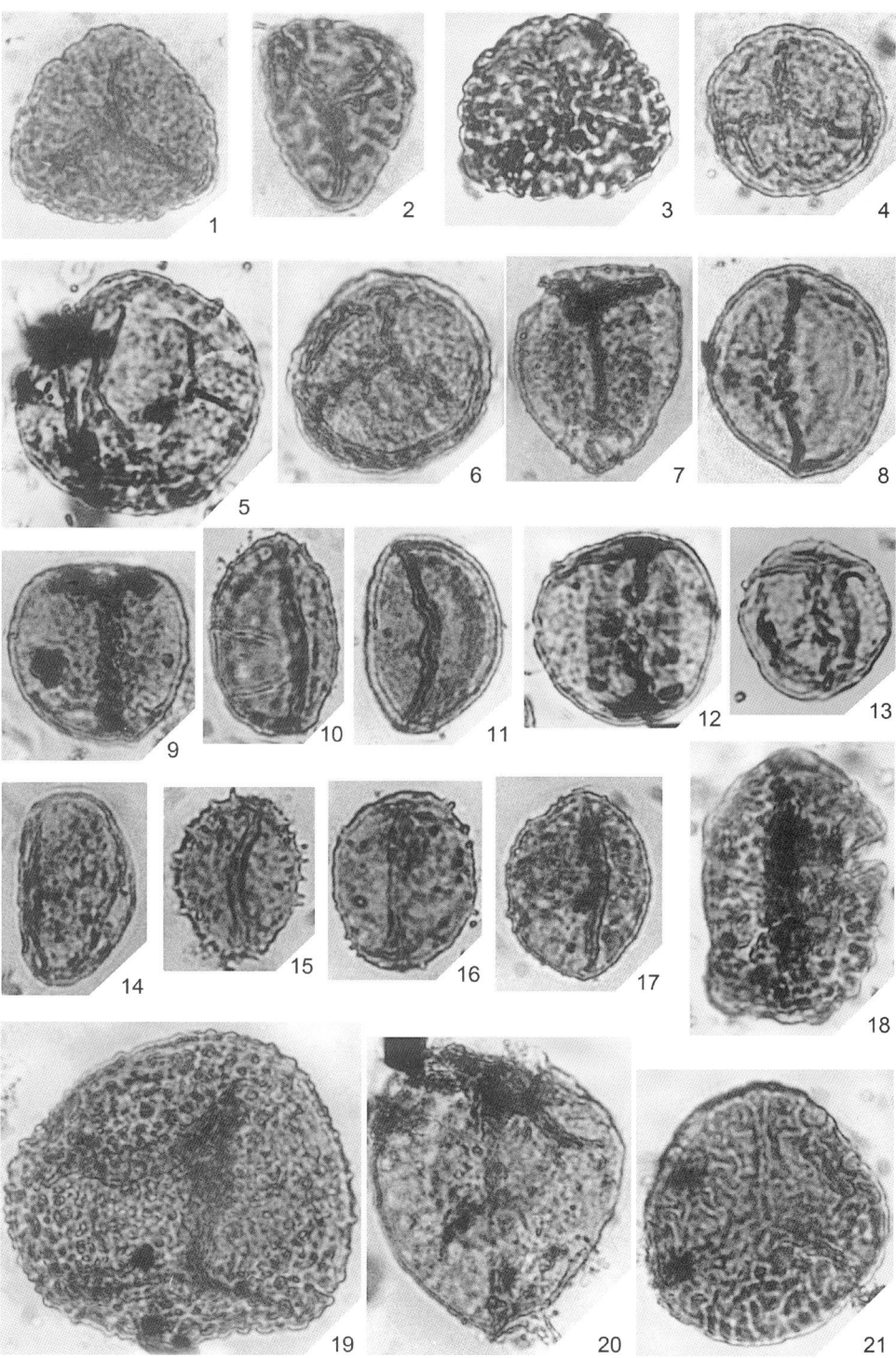

Fossil spores in crude oils from the North Tarim Upheaval of the Tarim Basin (cont'd)

Plate III

1–2 Lophotriletes corrugatus Ouyang et Li *1* No. L-1-29; *2* No. L-1-1
3 Lundbladispora subornata Ouyang et Li No. S-9-20
4 Retusotriletes arcticus Qu et Wang No. L-1-125
5 Camarozonosporites rudis (Leschik) Klaus No. S-9-2
6 Triquitrites desperatus Potonié et Kremp No. L-1-22
7 Limatulasporites parvus Qu et Wang No. S-14-140
8 Limatulasporites dalongkouensis Qu et Wang No. S-14-64
9 Lycopodiumsporites paniculatoides Tralau No. L-1-105
10 Lycopodiumsporites subrotundum (Kara-Murza) Pocock No. L-1-92
11 Calamospora pedata Kosanke No. S-14-114
12 Lundbladispora plicata Bai No. S-14-76
13 Lycopodiacidites kuepperi Klaus No. S-14-40
14 Multinodisporites junctus Ouyang et Li No. L-1-90
15 Chordasporites singulichorda Klaus No. S-14-58
16 Piceites latens Bolchovitina No. L-1-101
17 Lueckisporites triassicus Clarke No. L-1-122
18 Colpectopollis pseudostriatus (Kopytova) Qu et Wang No. L-1-11

Explanation of Plates and Plates

Plate III

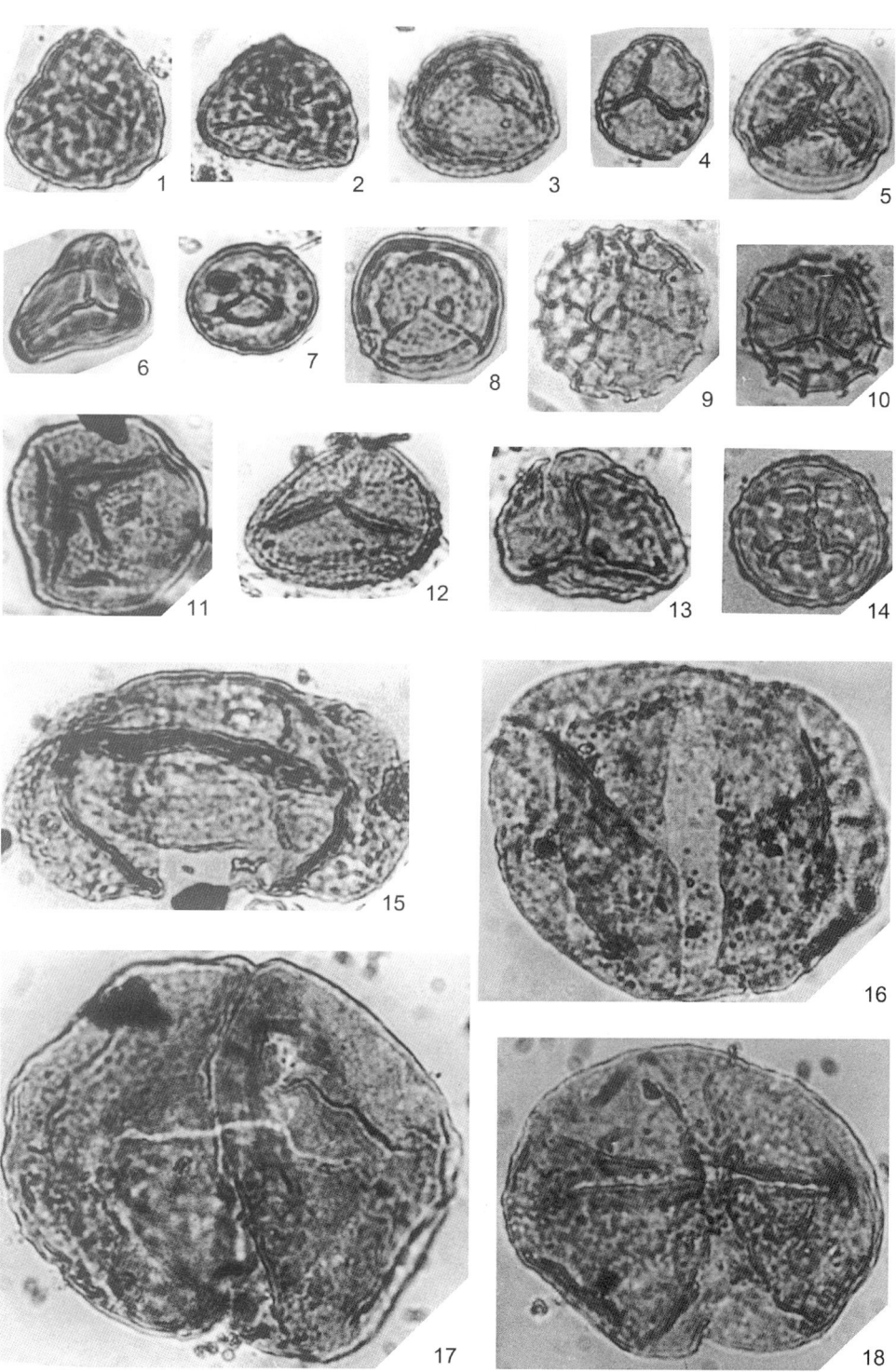

Fossil spores and pollen in crude oils from the North Tarim Upheaval of the Tarim Basin (cont'd)

Plate IV

1 *Calamospora pedata* Kosanke No. S-14-34
2 *Retispora florida* Staplin No. D-1-7
3 *Torispora securis* (Balme) Alpern, Doubinger et Hörst No. S-14-47
4 *Punctatosporites minutus* Ibrahim No. S-14-107
5 *Punctatisporites minutus* Kosanke No. S-14-34
6 *Callialasporites dampieri* (Balme) Sukh Dev No. S-14-122
7 *Callialasporites minus* (Tralau) Guy No. L-8-2
8 *Lycospora pusilla* (Ibrahim) Somers No. D-1-4
9 *Classopollis qiyangensis* Shang No. L-1-1
10 *Chasmatosporites magnolioides* (Erdtman) Nilsson No. S-14-19
11 *Taeniaesporites albertae* Jansonius No. S-14-30
12 *Cedripites priscus* Balme No. S-14-39
13 *Bennettiteaepollenites lucifer* (Thiergart) Potonié No. L-1-20
14 *Alisporites lowoodensis* deJersey No. S-14-149
15 *Vesicaspora wilsonii* Schemel No. S-14-123
16 *Chordasporites impensus* Ouyang et Li No. L-1-81
17 *Cycadopites nitidus* (Balme) Pocock No. S-14-3
18 *Podocarpidites multesimus* (Bolchovitina) Pocock No. L-1-112

Explanation of Plates and Plates

Plate IV

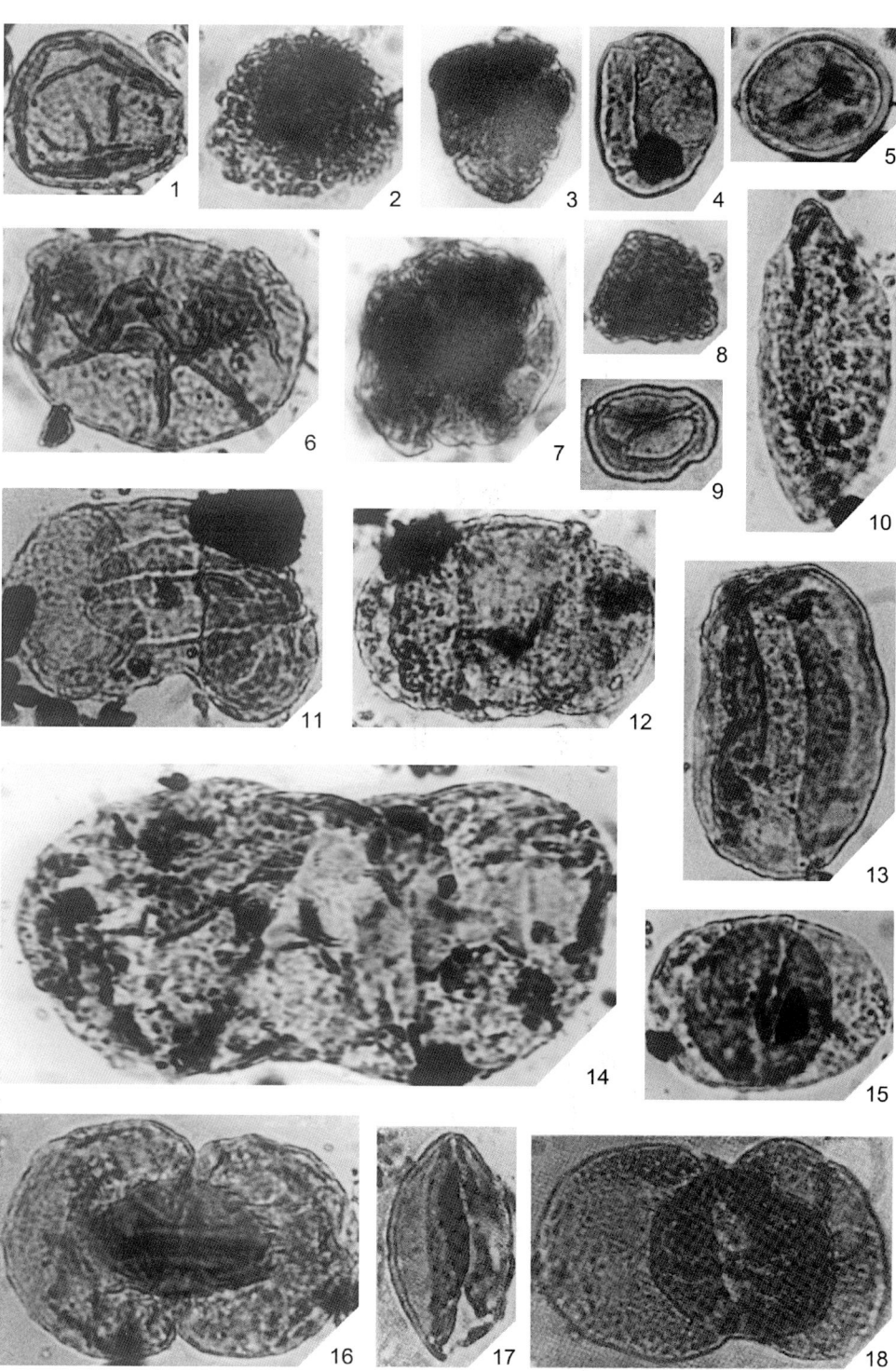

Fossil spores and pollen in crude oils from the North Tarim Upheaval of the Tarim Basin (cont'd)

Plate V

1 *Aratrisporites coryliseminis* Klaus No. S-14-38
2 *Pityosporites parvisaccatus* de Jersey No. L-1-16
3 *Alisporites australis* de Jersey No. L-1-5
4 *Araucariacites australis* Cookson No. L-1-55
5 *Platysaccus undulates* Ouyang et Li No. L-1-31
6 *Parcisporites rarus* Ouyang et Li No. S-14-13
7 *Cedripites minor* Pocock No. 2-1-39
8, 11 *Granosaccus ornatus* (Pautsch) Pautsch 8 No. L-1-104; 11 No. L-1-16
9 *Piceites latens* Bolchovitina No. L-1-45
10 *Bennettiteaepollenites lucifer* (Thiergart) Potonié No. L-1-1
12 *Alisporites parvus* de Jersey No. L-1-38
13 *Alisporites fusiformis* Ouyang et Li No. L-1-86
14 *Cycadopites carpentieri* (Delcourt et Sprumont) Singh No. S-9-5
15 *Alisporites aequalis* Mädler No. L-1-9

Plate V

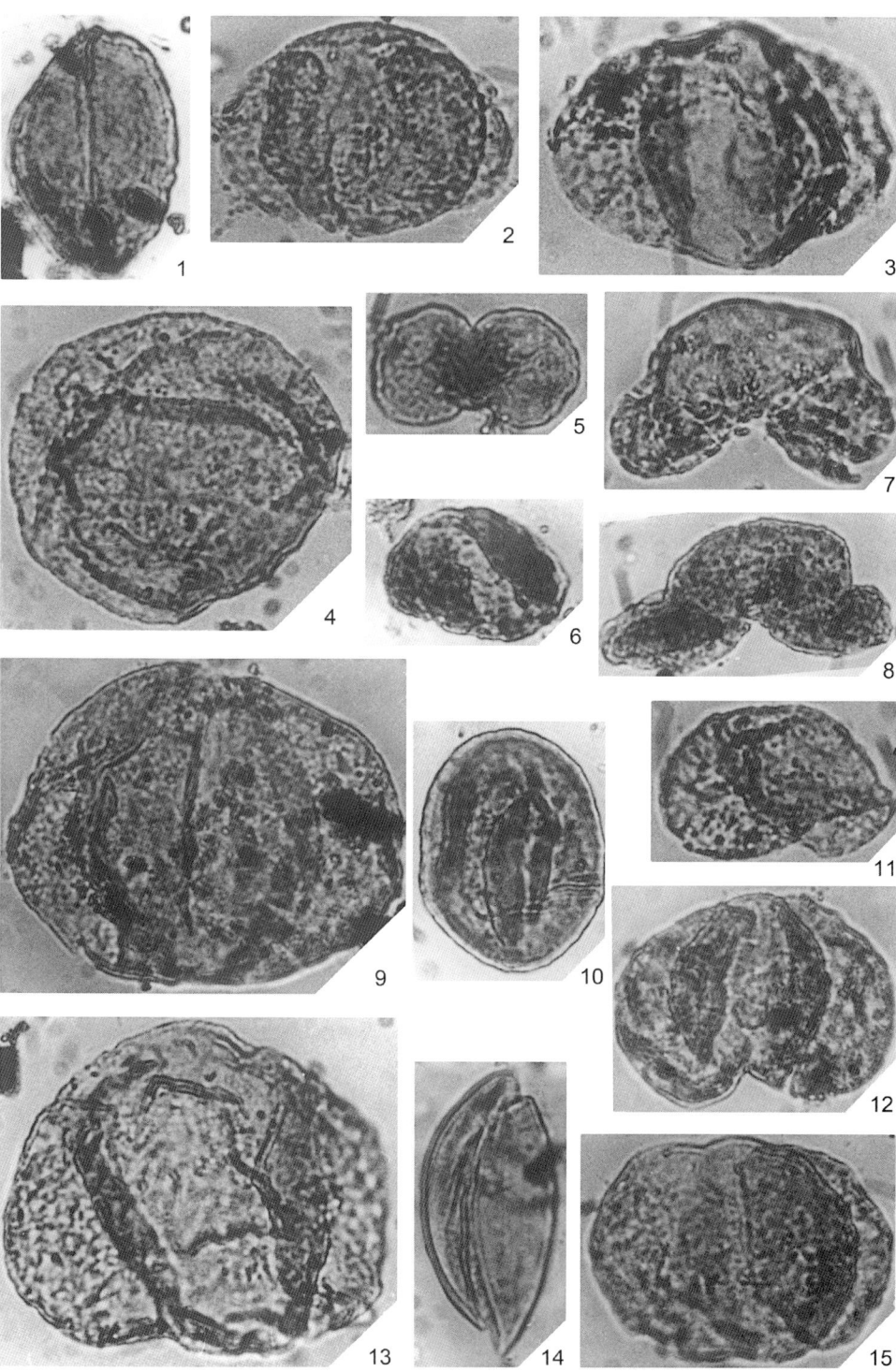

Fossil spores and pollen in crude oils from the North Tarim Upheaval of the Tarim Basin (cont'd)

Plate VI

1, 3 Taeniaesporites pellucidus (Goubin) Balme
  1 No. S-9-9; *3* No. S-14-50
2 Lueckisporites tattooensis Jansonius No. S-9-27
4 Taeniaesporites rhaeticus Schulz No. S-9-26
5 Protohaploxypinus microcorpus (Schaarschmidt) Clarke No. S-9-8
6 Taeniaesporites divisus Qu No. S-9-29
7 Lueckisporites virkkiae Potoié et Klaus No. S-9-11
8 Calamospora tener (Leschik) Mädler No. S-9-7

Plate VI

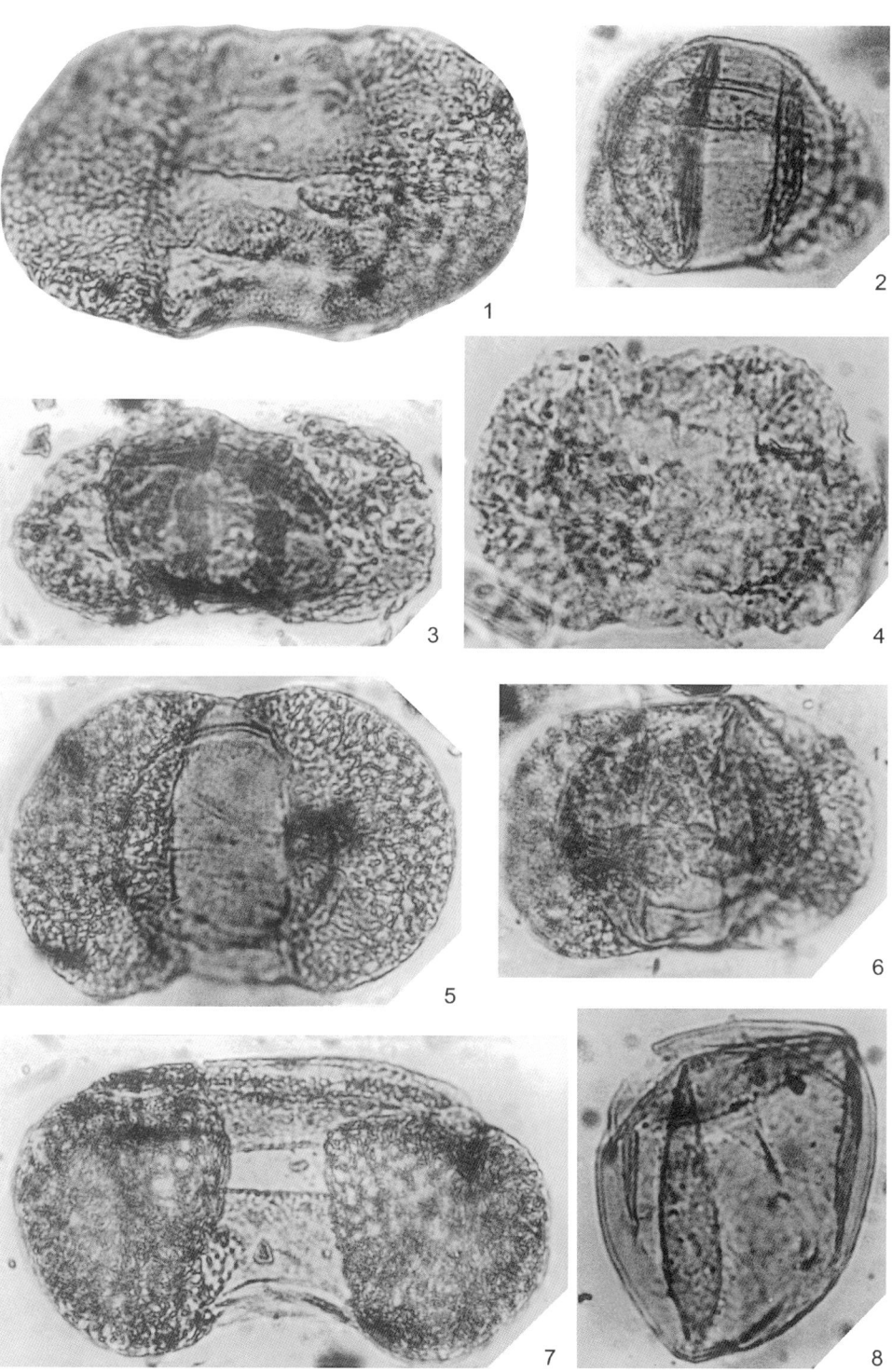

Fossil spores and pollen in crude oils from the North Tarim Upheaval of the Tarim Basin (cont'd)

Plate VII

1, 4 Rugubivesiculites lepidus Bai *1.* No. L-1-162; *4* No. L-1-16

2, 6 Pinuspollenites divulgatus (Bolchovitina) Qu *2* No. L-1-31; *6* No. S-14-140

3 Cedripites parvisaccus Quyang et Li No. L-1-91

5 Schizosporis bilobatus (Faddeeva) Qu No. S-14-167

7 Striatopodocarpites cf. *tojmensis* Sedova No. S-14-31

8 Torispora securis (Balme) Alpern, Doubinger et Hödrst No. S-14-43

9 Klausipollenites schaubergeri (Potonié et Klaus) Jansonius No. S-14-124

10 Protohaploxypinus cf. *samoilovichii* (Jansonius) Hart No. S-9-26

11 Pinuspollenites normalis Qu et Wang No. L-1-12

12 Lueckisporites triassicus Clarke No. S-14-145

13 Enzonalasporites tenuis Leschik No. S-14-42

Explanation of Plates and Plates

Plate VII

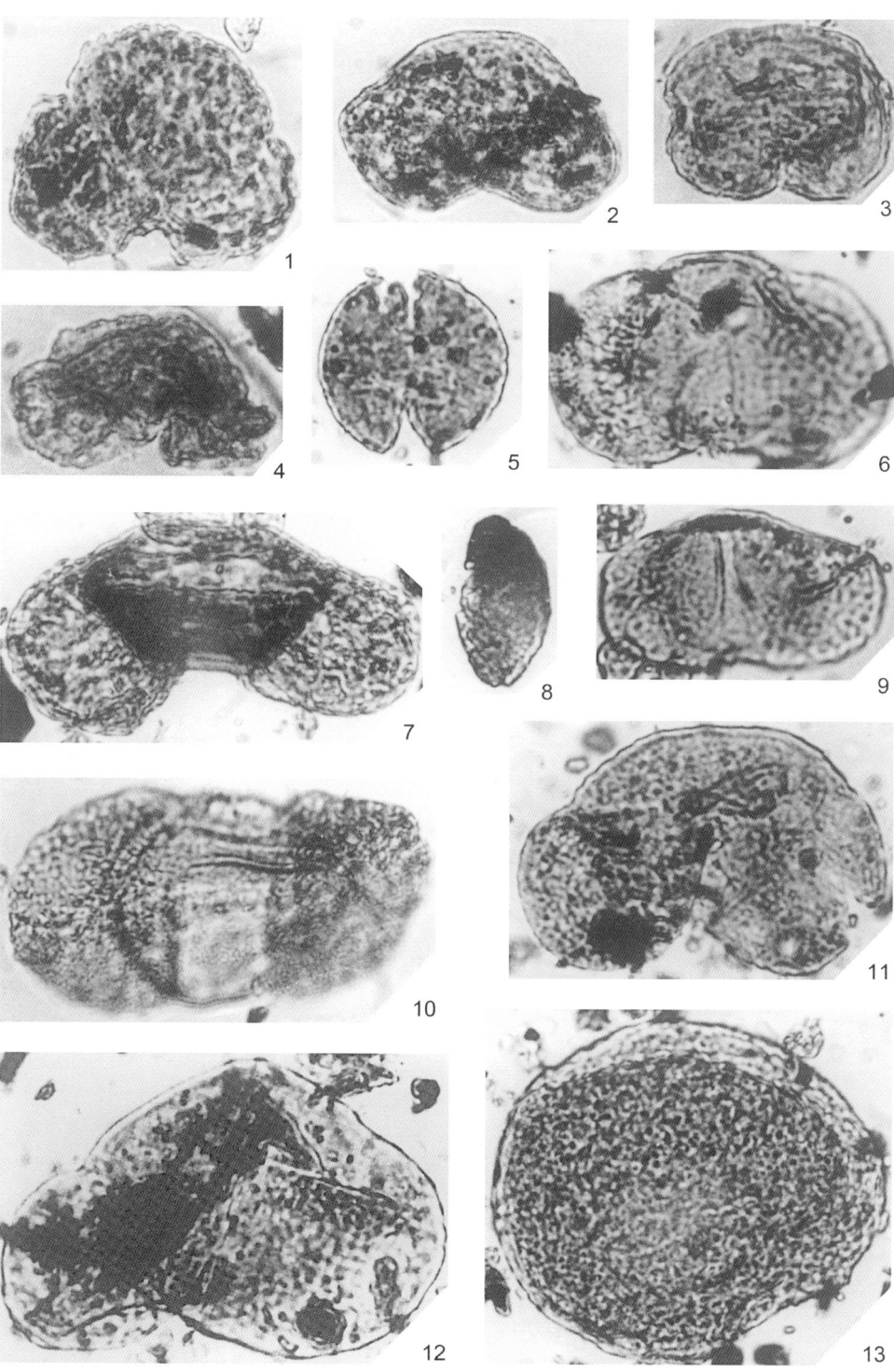

Fossil spores and pollen in crude oils from the North Tarim Upheaval of the Tarim Basin (cont'd)

Plate VIII

1 Colpectopollis pseudostriatus (Kopytova) Qu et Wang No. L-1-91

2 Colpectopollis scitulus (Qu et Pu) Qu et Wang No. L-1-11

3 Taeniaesporites cf. *pellucidus* (Goubin) Balme No. L-1-1

4 Striatoabietites duivenii (Jansonius) Hart No. L-1-11

5 Striatoabietites richteri (Klaus) Hart No. S-14-85

6 Piceites podocarpoides Bolchovitina No. L-1-29

7 Protohaploxypinus samoilovichii (Jansonius) Hart No. S-9-36

8 Piceites pseudorotundiformis (Maljavkina) Pocock No. L-1-15

Plate VIII

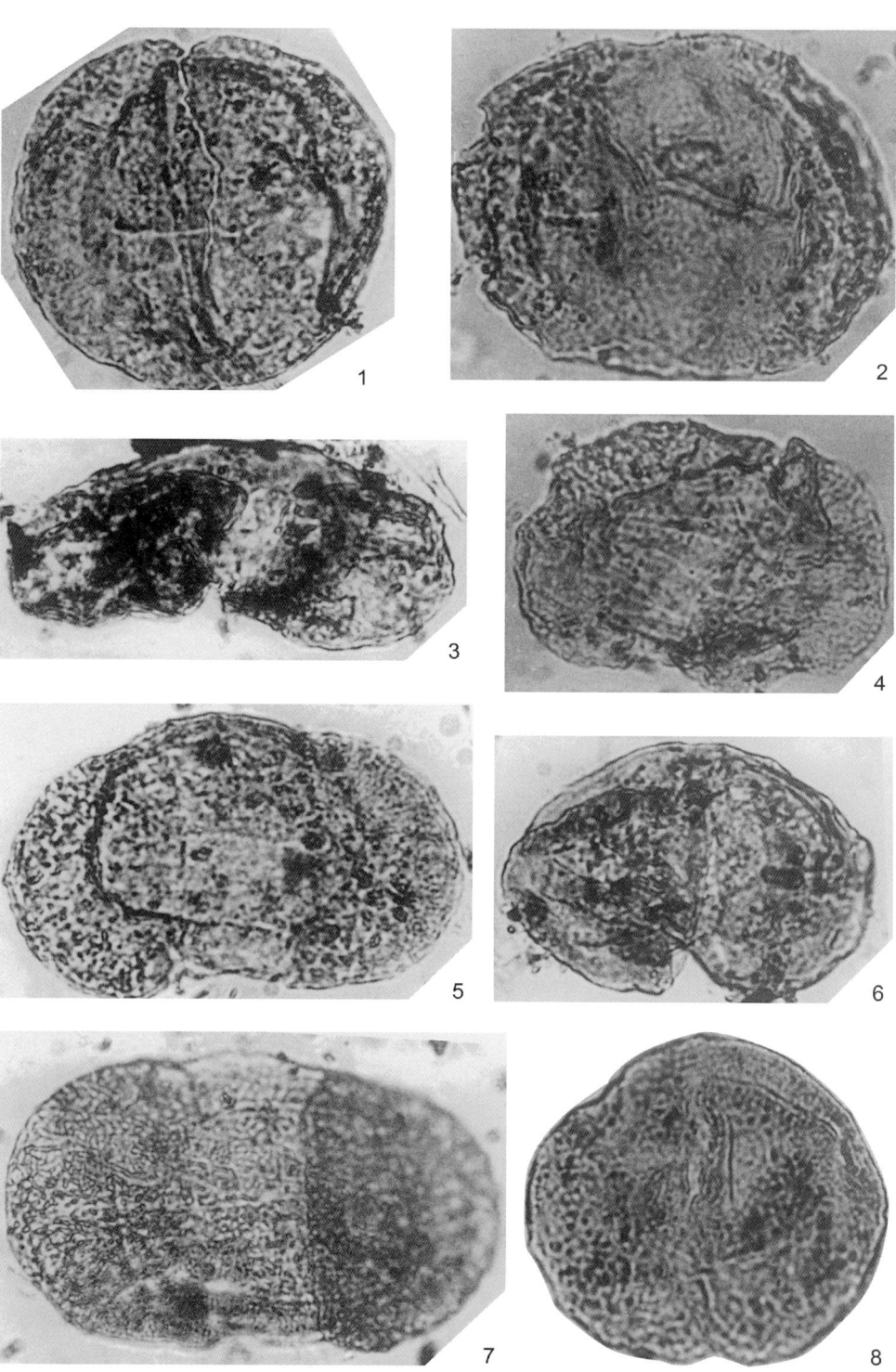

Fossil pollen in crude oils from the North Tarim Upheaval of the Tarim Basin (cont'd)

Plate IX

1 *Deltoidospora gradata* (Maljavkina) Pocock No. H-1-4
2 *Concavisporites toralis* (Leschik) Nilsson No. H-1-8
3–3a *Duplexisporites scanicus* (Nilsson) Playford et Dettmann No. H-1-3
4 *Osmundacidites wellmanii* Couper No. Y-61-2
5 *Deltoidospora perpusilla* (Bolchovitina) Pocock No. Y-30-2
6 *Cibotiumspora paradoxa* (Maljavkina) Chang No. Y-61-4
7 *Chenopodipollis multiporatus* (Pflug et Thomson) Zhou No. K-1-5
8 *Artemisiaesporites sellularis* Nagy No. K-1-2
9 *Cycadopites nitidus* (Balme) Pocock No. Y-30-1
10 *Alisporites bilateralis* Rouse No. H-1-7
11 *Ephedripites tertiarus* Krutzsch No. K-1-12
12 *Podocarpidites multesimus* (Bolchovitina) Pocock No. J-1-9
13 *Cycadopites typicus* (Maljavkina) Pocock No. Y-61-5
14 *Pinuspollenites labdacus* (Potonié) Raatz No. H-1-13
15 *Protopicea exilioides* (Bolchovitina) Pocock No. H-1-6
16 *Parvisaccites enigmatus* Couper No. J-1-11
17 *Abietineaepollenites minimus* Couper No. Y-420-1

Plate IX

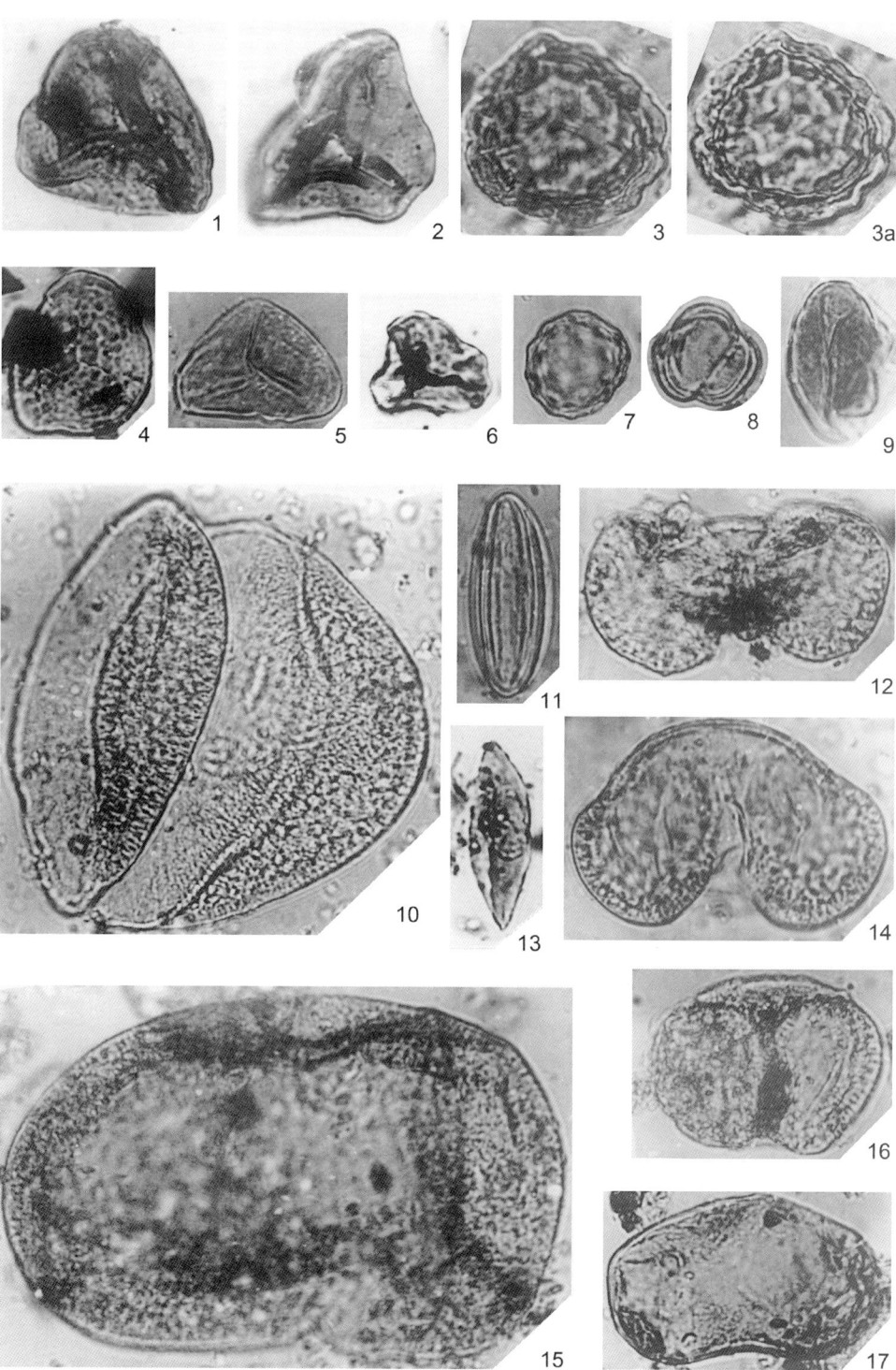

Fossil spores and pollen in crude oils from the Kuqa Depression of the Tarim Basin

Plate X

1 *Quadraeculina limbata* Maljavkina No. Y-1-9
2 *Cerebropollenites carlylensis* Pocock No. Y-1-11
3 *Podocarpidites multesimus* (Bolchovitina) Pocock No. K-2-22
4 *Paleoconiferus asaccatus* Bolchovitina No. K-2-42
5 *Protopinus scanicus* Nilsson No. K-2-34
6 *Classopollis classoides* (Pflug) Pocock et Jansonius No. Y-1-8
7 *Cycadopites minimus* (Cookson) Pocock No. K-2-5
8 *Protopicea exilioides* (Bolchovitina) Pocock No. Y-1-18
9 *Podocarpidites florinii* Pocock No. K-2-4
10 *Piceites latens* Bolchovitina No. K-2-4

Explanation of Plates and Plates

Plate X

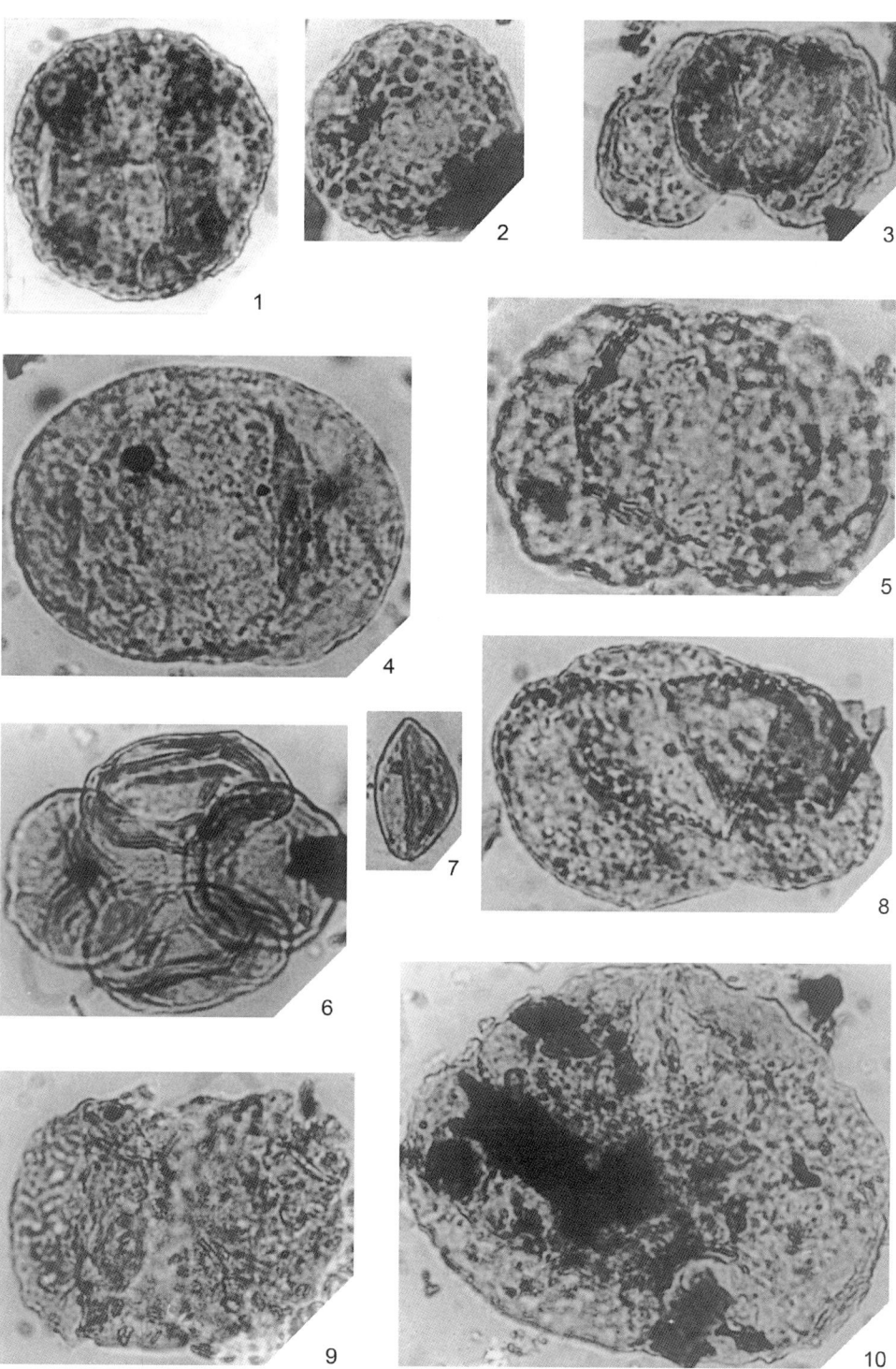

Fossil pollen in crude oils from the Kashi Sag in the Southwest Tarim Depression of the Tarim Basin

Plate XI

1–2 Cyathidites australis Couper *1* No. K-6-37; *2* No. K-30

3 Apiculatisporis ovalis (Nilsson) Norris No. K-6-33

4–5 Chasmatosporites elegans Nilsson *4* No. K-6-15; *5* No. K-6-9

6–7 Eucommiidites troedssonii Erdtman *6* No. K-6-19; *7* No. K-6-7

8 Cycadopites nitidus (Balme) Pocock No. K-6-22

9 Cycadopites typicus (Maljavkina) Pocock No. K-6-18

10 Pteruchipollenites thomasii Couper No. K-6-14

11 Alisporites bilateralis Rouse No. K-6-25

12 Podocarpidites multicinus (Bolchovitina) Pocock No. K-6-33

13 Piceites podocarpoides Bolchovitina No. K-6-35

Explanation of Plates and Plates

Plate XI

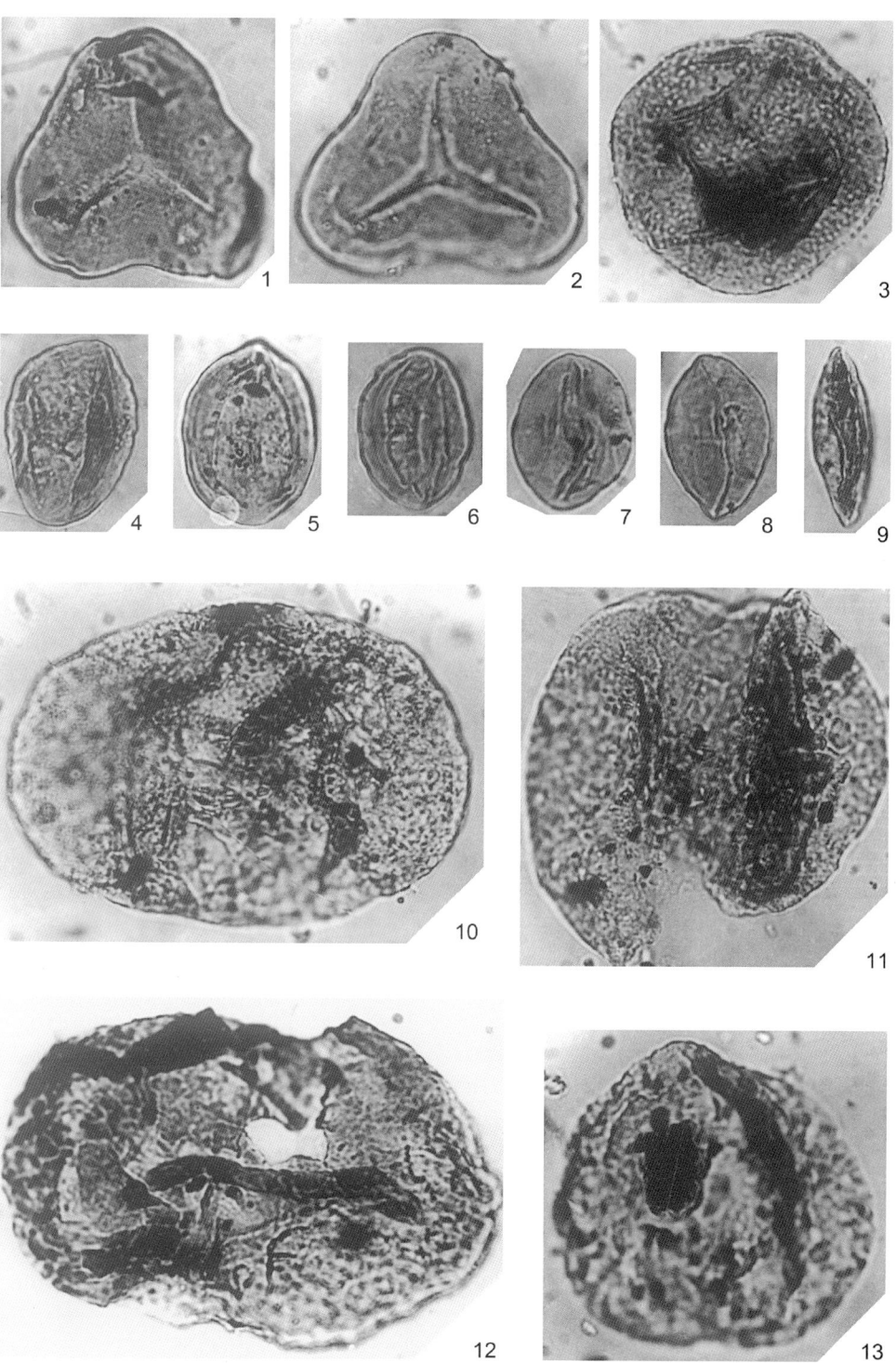

Fossil spores and pollen in crude oils from the Yecheng Sag in the Southwest Tarim Depression of the Tarim Basin

Plate XII

1 *Deltoidospora gradata* (Maljavkina) Pocock No. K-6-18
2 *Leptolepidites major* Couper No. K-6-11
3 *Caytonipollenites pallidus* (Reissinger) Couper No. K-6-32
4 *Bennettiteaepollenites lucifer* (Thiergart) Pocock No. K-6-15
5 *Chasmatosporites major* Nilsson No. K-6-25
6 *Biretisporites potoniaei* Delcourt et Sprumont No. K-6-10
7 *Undulatisporites concavus* Kedves No. K-6-22
8 *Lycopodiumsporites clavatoides* Couper No. K-6-10
9 *Leptolepidites verrucatus* Couper No. K-6-23
10 *Murospora minor* Pocock No. K-6-5
11 *Rugubivesiculites* sp. No. K-6-25
12 *Podocarpidites multicinus* (Bolchovitina) Pocock No. K-6-35
13 *Cedripites minor* Pocock No. K-6-10
14–15 *Quadraeculina limbata* Maljavkina 14 No. K-6-27; 15 No. K-6-38
16 *Podocarpidites wapellensis* Pocock No. K-6-25
17 *Parvisaccites enigmatus* Couper No. K-6-70
18 *Protopicea exilioides* (Bolchovitina) Pocock No. K-6-35
19 *Ephedripites tertiarus* Krutzsch No. K-6-47

Explanation of Plates and Plates

Plate XII

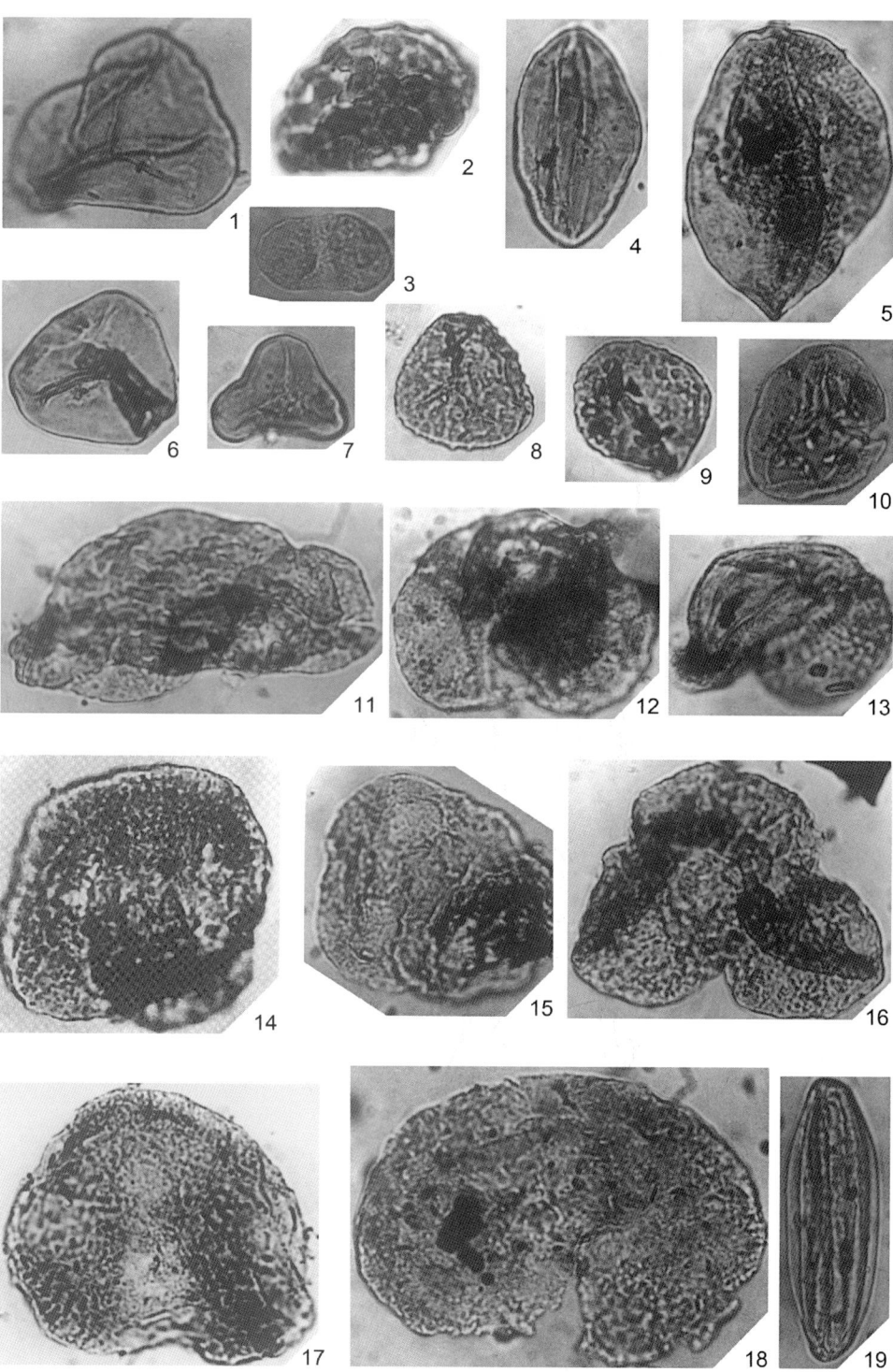

Fossil spores and pollen in crude oils from the Yecheng Sag in the Southwest Tarim Depression of the Tarim Basin (cont'd)

Plate XIII

1 Calamospora breviradiata Kosanke No. B-5-35
2 Discisporites psilatus de Jersey No. B-5-57
3 Limatulasporites dalongkouensis Qu et Wang No. B-5-68
4 Limatulasporites limatulus (Playford) Helby et Foster No. B-5-4
5 Granulatisporites parvus (Ibrahim) Potonié et Kremp No. B-5-35
6 Leiotriletes ornatus Ischenko No. B-5-65
7 Lophotriletes humilus Hou et Wang No. B-5-35
8 Apiculatisporis decorus Singh No. B-5-58
9 Cyclogranisporites leopoldii (Kremp) Potonié et Kremp No. B-5-34
10 Dictyophyllidites discretus Ouyang No. B-5-45
11 Nevesisporites sp. No. B-5-50
12 Triangulisaccus primitivus Hou et Wang No. B-5-56
13 Calamospora nathorstii (Halle) Klaus No. B-5-34
14 Duplexisporites gyratus Playford et Dettmann No. B-5-10
15 Aratrisporites strigosus Playford No. B-5-9
16 Laevigatosporites medius Kosanke No. B-5-16
17 Calamospora densirugosus Hou et Wang No. B-5-41
18 Punctatisporites parasolidus Ouyang No. B-5-1
19 Punctatosporites sp. No. B-5-68
20 Aratrisporites fischeri (Klaus) Playford et Dettmann No. B-5-12
21 Torispora securis (Balme) Alpern, Doubinger et Hörst No. B-5-57

Plate XIII

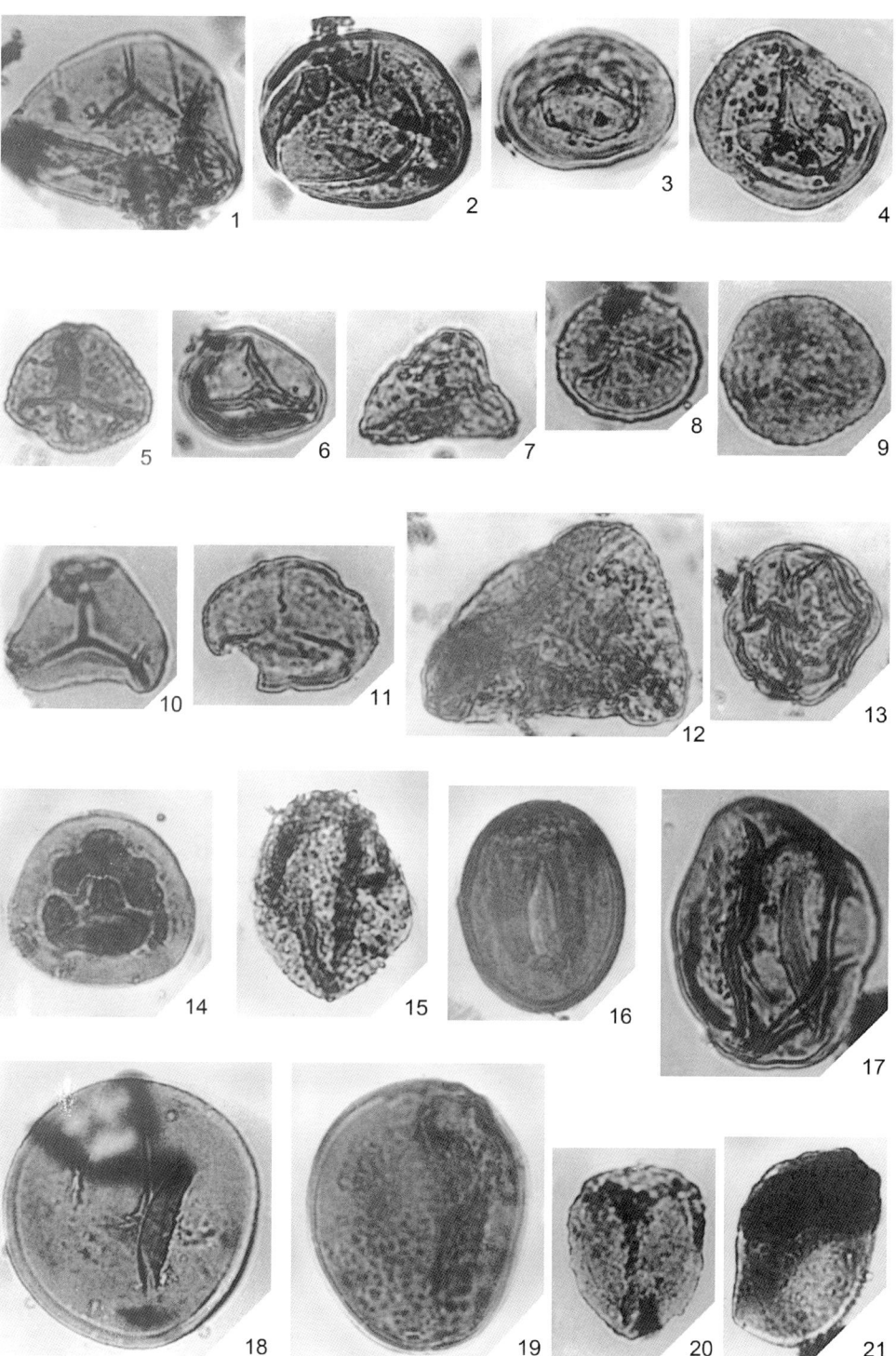

Fossil spores and pollen in crude oils from the Beisantai oil field in the East Junggar Depression of the Junggar Basin

Plate XIV

1 *Kraeuselisporites argutus* Hou et Wang No. B-5-31
2–3 *Crassispora orientalis* Ouyang et Li 2 No. B-5-10; *3* No. B-5-2
4 *Chordasporites orientalis* Ouyang et Li No. B-5-2
5–6 *Wilsonia vesicata* Kosanke 5 No. B-5-43; 6 No. B-5-55
7 *Endosporites ornatus* Wilson et Coe No. B-5-13
8 *Taeniaesporites pellucidus* (Goubin) Balme No. B-5-76
9 *Entylissa spinosus* (Samoilovich) Du No. B-5-7
10 *Cordaitina uralensis* (Luber) Samoilovich No. B-5-75

Plate XIV

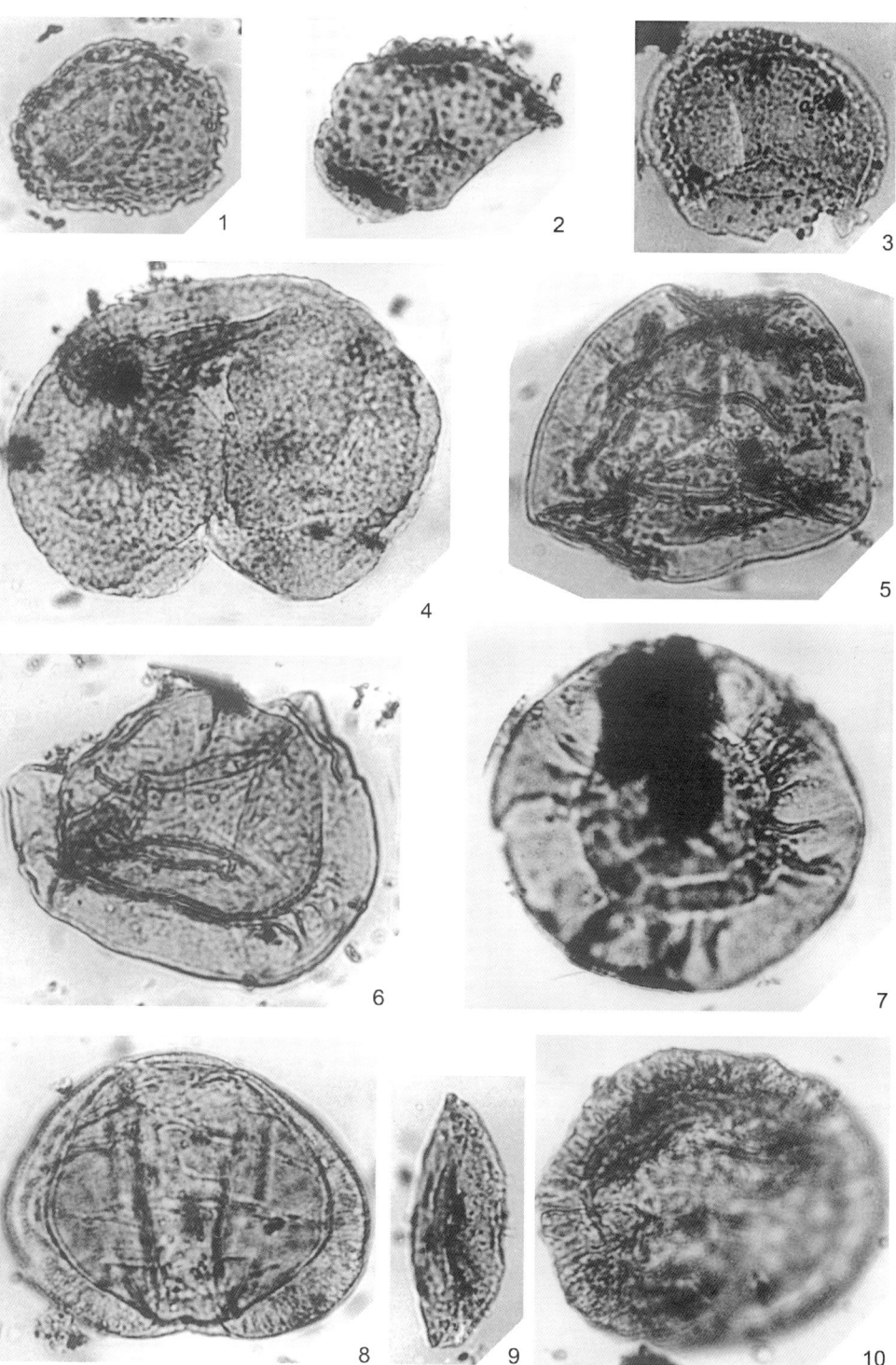

Fossil spores and pollen in crude oils from the Beisantai oil field in the East Junggar Depression of the Junggar Basin (cont'd)

Plate XV

1–2 Striatopodocarpites tojmensis (Sedowa) Hart *1* No. B-5-2; *2* No. B-5-9
3 Striatopodocarpites crassus Singh No. B-5-70
4 Limitisporites rhombicorpus Zhou No. B-5-70
5 Vitreisporites signatus Leschik No. B-5-11
6 Alisporites parvus de Jersey No. B-5-40
7, 9 Minutosaccus parcus Qu et Wang *7* No. B-5-19; *9* No. B-5-72
8 Chordasporites sp. No. B-5-6
10 Protohaploxypinus samoilovichii (Jansonius) Hart No. B-5-74
11 Cordaitina duralimita Hou et Wang No. B-5-33

Plate XV

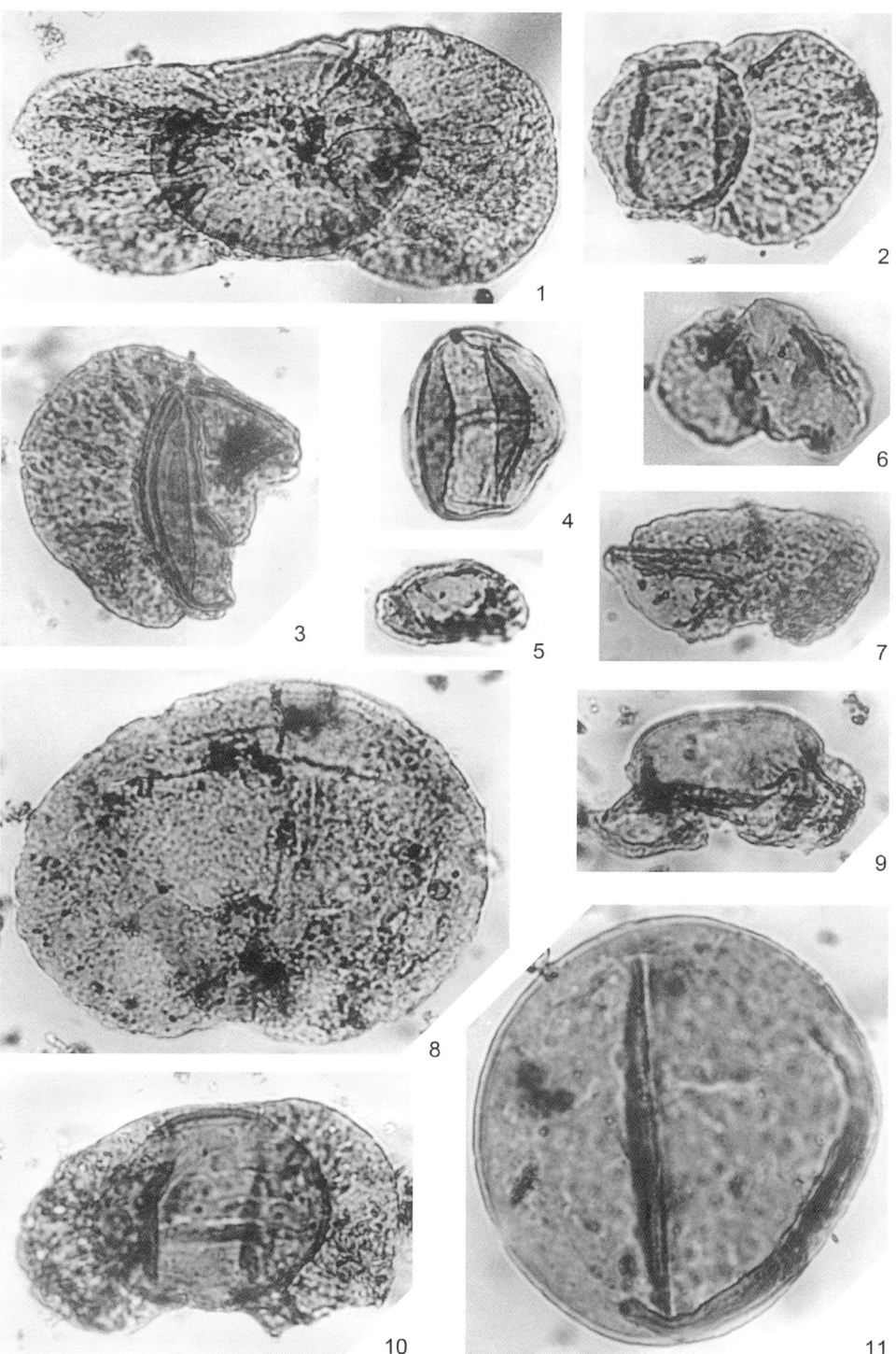

Fossil pollen in crude oils from the Beisantai oil field in the East Junggar Depression of the Junggar Basin (cont'd)

Plate XVI

1 Taeniaesporites kraeuseli Leschik No. B-5-1
2 Entylissa spinosus (Samoilovich) Du No. B-5-54
3 Striatopodocarpites crassus Singh No. B-5-75
4, 9 Endosporites multirugulatus Hou et Wang *4* No. B-5-45; *9* No. B-5-46
5 Cordaitina uralensis (Luber) Samoilovich No. B-5-21
6 Alisporites parvus de Jersey No. B-5-45
7 Chordasporites impensus Ouyang et Li No. B-5-51
8 Alisporites australis de Jersey No. B-5-58
10 Endosporites cf. *ornatus* Wilson et Coe No. B-5-67
11 Protohaploxypinus dvinensis (Sedova) Hart No. B-5-68

Plate XVI

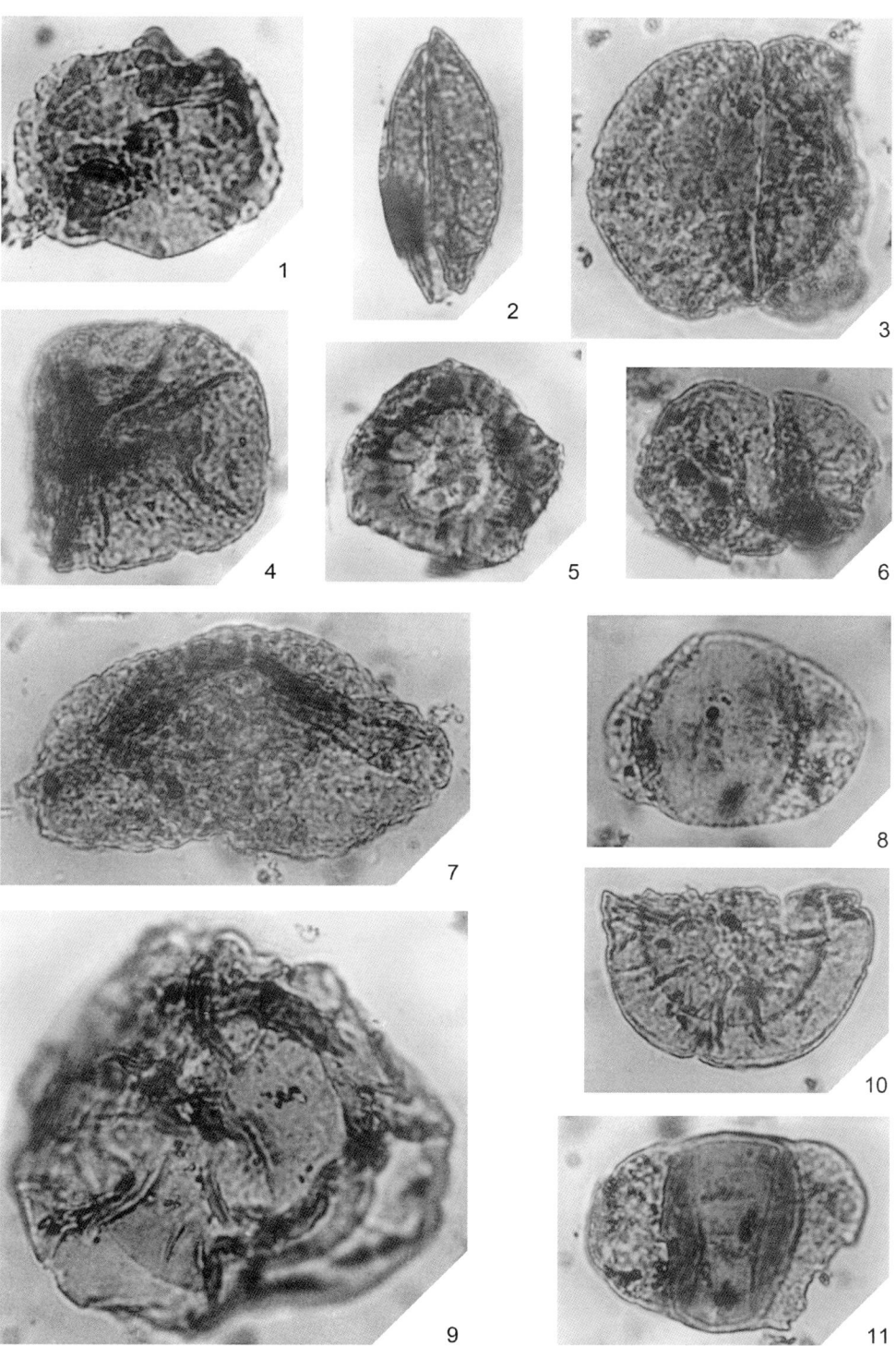

Fossil spores and pollen in crude oils from the Beisantai oil field in the East Junggar Depression of the Junggar Basin (cont'd)

Plate XVII

1 Leiotriletes adnatus (Kosanke) Potonié et Kremp No. H-1-49

2 Punctatisporites triassicus Schulz No. H-1-18

3–4 Dictyophyllidites mortoni (de Jersey) Playford et Dettmann
 3 No. H-1-11; *4* No. H-1-27

5 Apiculatisporis setaceformis Hou et Wang No. H-1-5

6 Lophotriletes delicatus Ouyang et Li No. H-1-30

7, 9 Lophotriletes pseudaculeatus Potonié et Kremp *7* No. H-1-17;
 9 No. H-1-46

8 Lophotriletes sp. No. H-1-55

10 Lophotriletes flavus Qu et Wang No. H-1-22

11 Acanthotriletes microspinosus (Ibrahim) Potonié et Kremp No. H-1-7

12 Apiculatisporis xiaolongkouensis Hou et Wang No. H-1-1

13 Apiculatisporis spiniger (Leschik) Qu No. H-1-14

14 Kraeuselisporites argutus Hou et Wang No. H-1-22

15 Lophotriletes corrugatus Ouyang et Li No. H-1-8

16 Lophotriletes humilus Hou et Wang No. H-1-10

17 Triquitrites attenuatus Ouyang No. H-1-12

18 Kraeuselisporites sp. No. H-1-6

19 Convolutispora asiatica Ouyang et Li No. H-1-1

20 Convolutispora triangularis Ouyang et Li No. H-1-30

21 Converrucosisporites mictus Ouyang No. H-1-42

Plate XVII

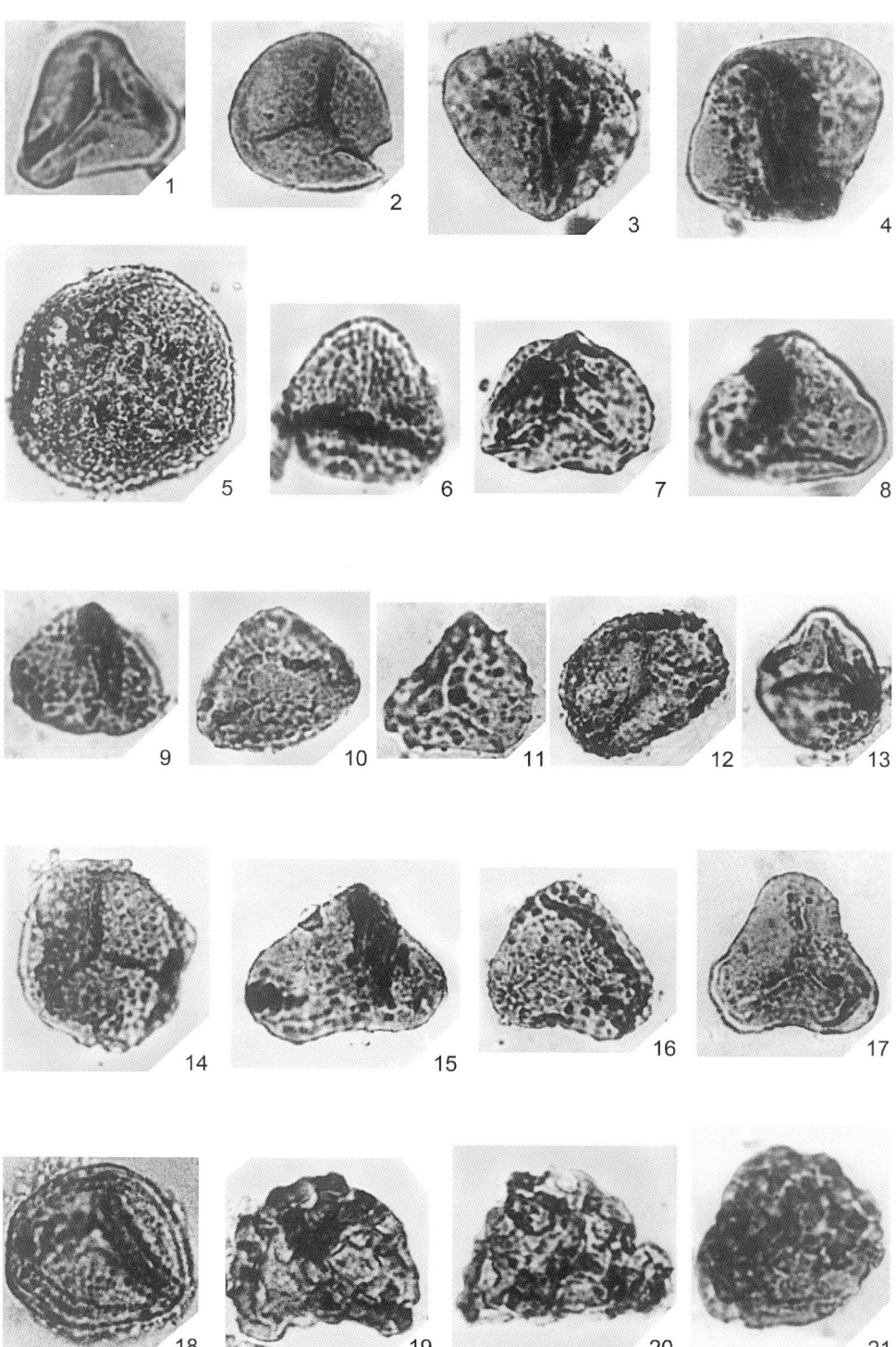

Fossil spores in crude oils from the Huonan oil field in the East Junggar Depression of the Junggar Basin

Plate XVIII

1 *Tuberculatosporites homotubercularis* Hou et Wang No. H-1-25
2–3 *Cordaitina radialis* Ouyang et Li 2. No. H-1-22; 3. No. H-1-2
4 *Limitisporites* sp. No. H-1-10
5–6 *Cordaitina rotata* (Luber) Samoilovich 5 No. H-1-40; 6 No. H-1-44
7 *Vittatina subsaccata* Samoilovich No. H-1-35
8 *Cedripites parvisaccus* Ouyang et Li No. H-1-22
9 *Gardenasporites xinjiangensis* Hou et Wang No. H-1-13
10 *Hamiapollenites obliquus* Zhan No. H-1-34
11 *Punctatosporites minutus* Ibrahim No. H-1-27
12 *Klausipollenites decipiens* Jansonius No. H-1-33
13 *Lueckisporites virkkiae* Potonié et Klaus No. H-1-26
14 *Gardenasporites latisectus* Hou et Wang No. H-1-43

Plate XVIII

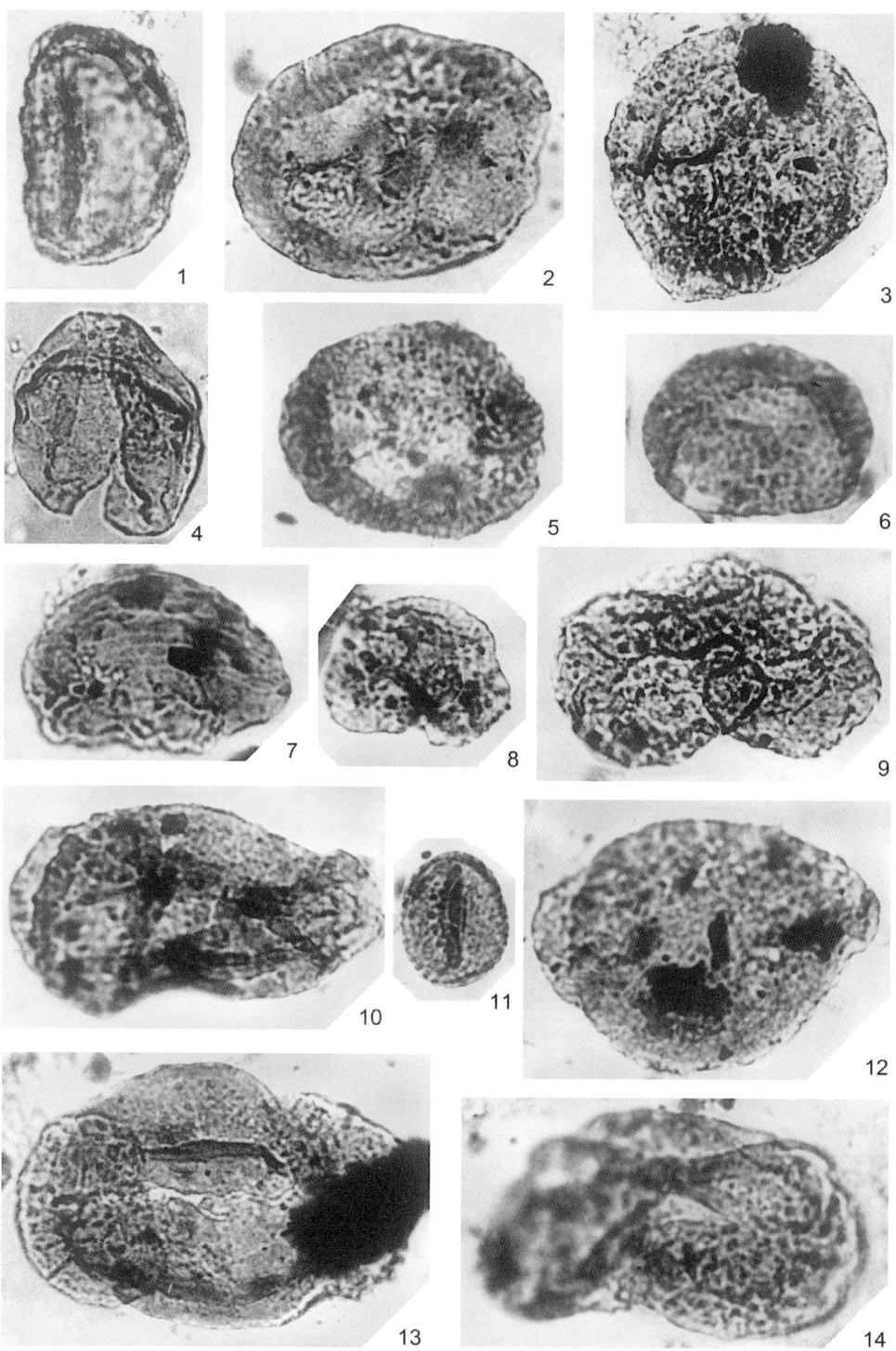

Fossil spores and pollen in crude oils from the Huonan oil field in the East Junggar Depression of the Junggar Basin (cont'd)

Plate XIX

1–2 Verrucorpipollis archaicus Ouyang *1, 2* No. H-1-15

3 Cedripites priscus Balme No. H-1-46

4–6 Granosaccus ornatus (Pautsch) Pautsch *4* No. H-1-32; *5* No. H-1-24; *6* No. H-1-2

7 Striatoabietites duivenii (Jansonius) Hart No. H-1-46

8 Welwitschipollenites clarus (Qu et Wang) Ouyang No. H-1-2

9 Hamiapollenites bullaeformis (Samoilovich) Jansonius No. H-1-4

10 Klausipollenites caperatus Ouyang No. H-1-1

11–12 Entylissa spinosus (Samoilovich) Du *11* No. H-1-16; *12* No. H-1-3

13 Ricciisporites tuberculatus Lundblad No. H-1-8

14 Parvisaccites sp. No. H-1-1

15 Cycadopites caperatus (Luber et Valts) Hart No. H-1-20

16 Acritarch No. H-1-22

17 Taeniaesporites quadratus Qu et Wang No. H-1-8

Plate XIX

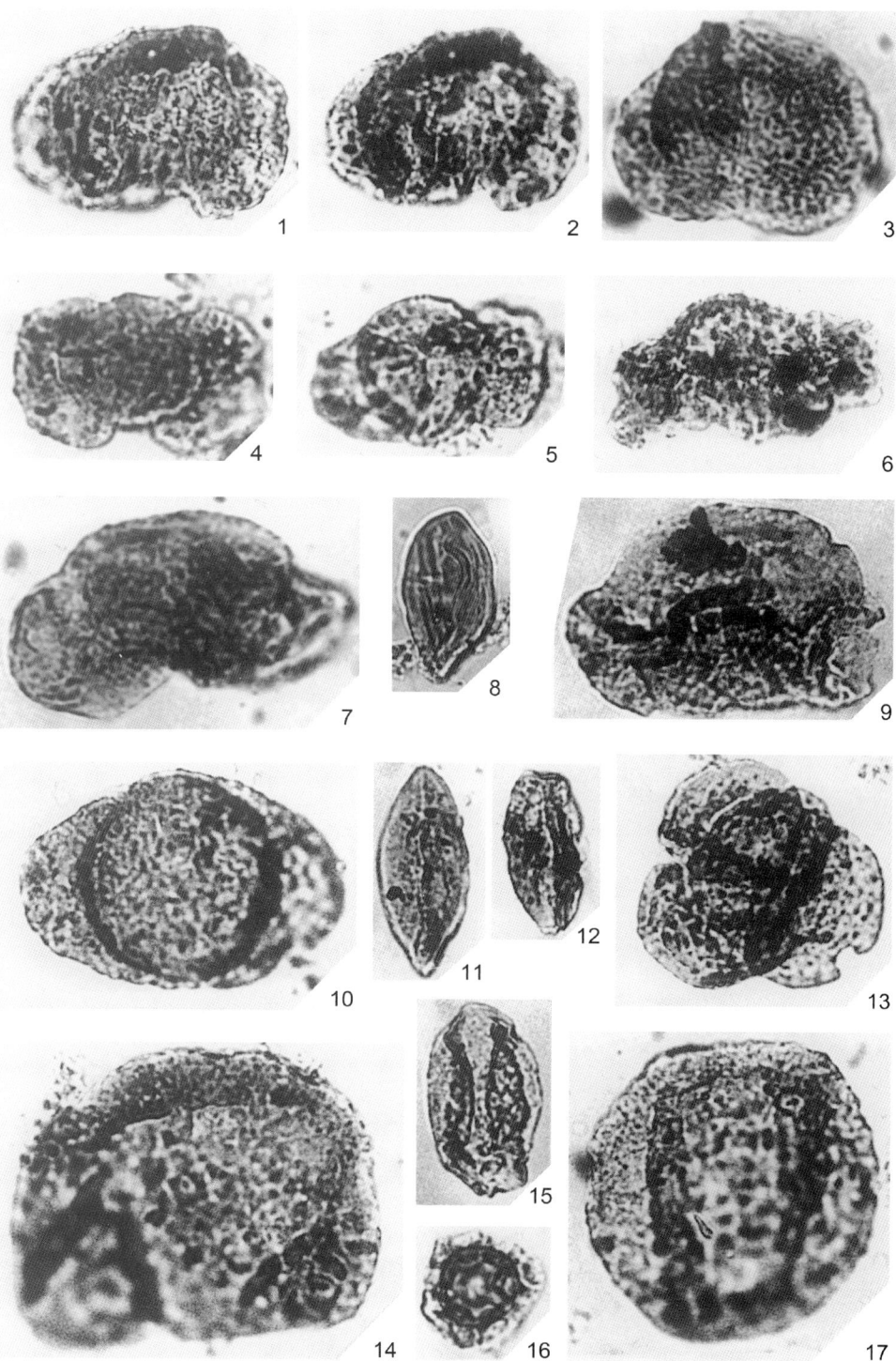

Fossil pollen and acritarchs in crude oils from the Huonan oil field in the East Junggar Depression of the Junggar Basin (cont'd)

Plate XX

1 Cirratriradites sp. No. H-1-5
2–3, 11 Cycadopites caperatus (Luber et Valts) Hart *2* No. H-1-28;
 3 No. H-1-17; *11* No. H-1-40
4 Triangulisaccus primitivus Hou et Wang No. H-1-17
5 Vittatina cryptosaccata Wang No. H-1-10
6 Pteruchipollenites reticorpus Ouyang et Li No. H-1-24
7 Vittatina specialis Wang No. H-1-16
8 Striatites sp No. H-1-18
9 Leiotriletes microtriangulus Artüz No. H-1-20
10 Cycadopites granulatus Ouyang No. H-1-11
12 Triangulatisporites junggarensis Yang et Sun No. H-1-23

(*12* at a magnification of 176×)

Plate XX

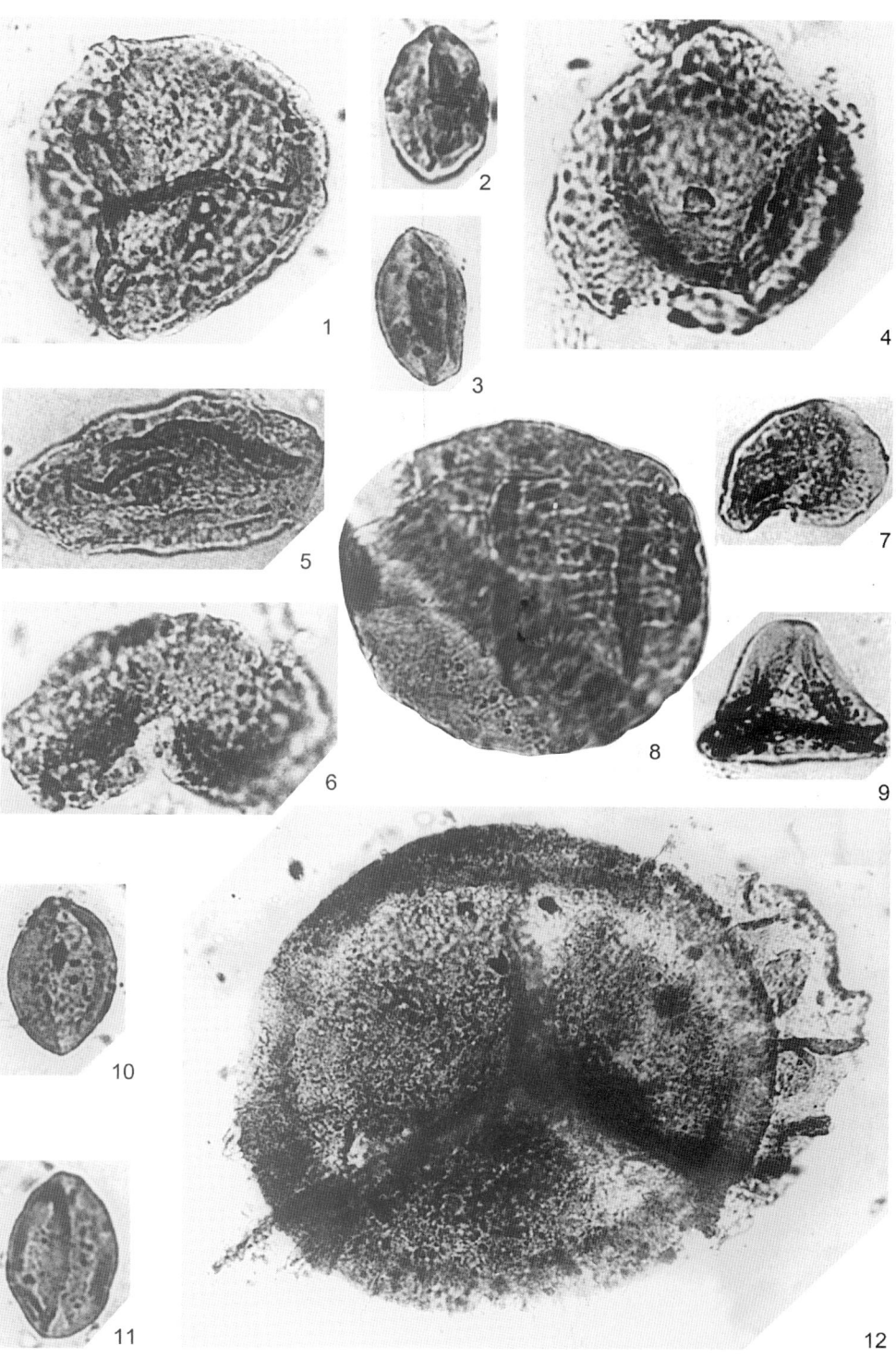

Fossil spores and pollen in crude oils from the Huonan oil field in the East Junggar Depression of the Junggar Basin (cont'd)

Plate XXI

1 *Cyathidites minor* Couper No. Q-34-8
2 *Dictyophyllidites harrisii* Couper No. Q-34-37
3 *Cibotiumspora paradoxa* (Maljavkina) Chang No. Q-34-18
4 *Cibotiumspora jurienensis* (Balme) Filatoff No. Q-34-21
5 *Granulatisporites minor* de Jersey No. Q-34-38
6 *Concavisporites* sp. No. Q-34-25
7 *Gleicheniidites conflexus* (Chlonova) Xu et Zhang No. Q-34-44
8 *Gleicheniidites rouseii* Pocock No. Q-34-49
9 *Gleicheniidites nilssonii* Pocock No. Q-34-38
10–11 *Granulatisporites jurassicus* Pocock 10, 11 No. Q-34-50
12 *Ceratosporites jurassicus* Pocock No. Q-34-1
13 *Ceratosporites varispinosus* Pocock No. Q-34-33
14–15 *Converrucosisporites venitus* Batten 14 No. Q-34-22; 15 No. Q-34-38
16 *Converrucosisporites minor* Pocock No. Q-34-28
17 *Concavissimisporites southeyensis* Pocock No. Q-34-25
18 *Concavissimisporites subgranulosus* (Couper) Pocock No. Q-34-37
19 *Apiculatisporis ovalis* (Nilsson) Norris No. Q-34-47
20 *Concavissimisporites delcourtii* Pocock No. Q-34-6

Plate XXI

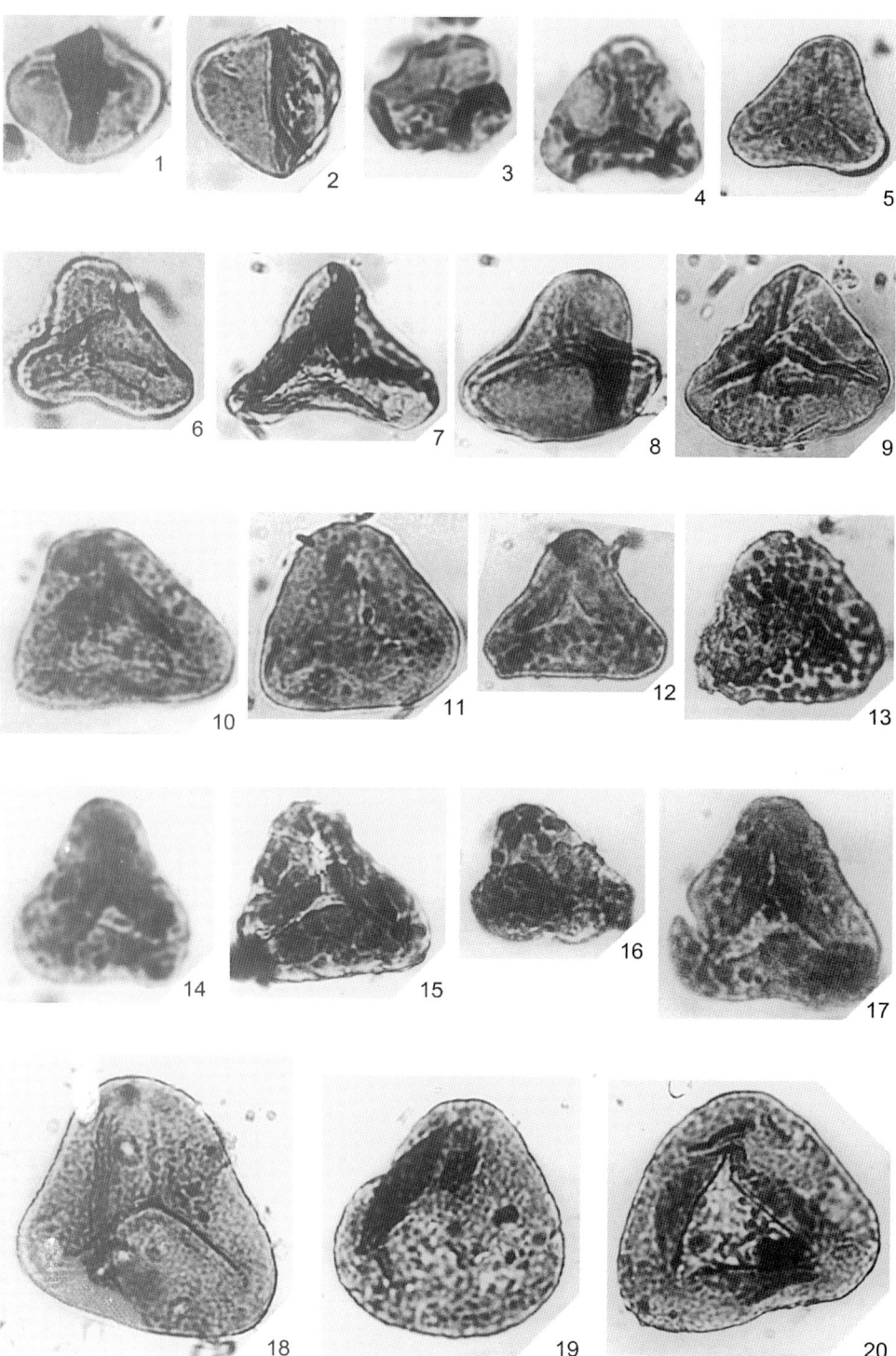

Fossil spores in crude oils from the Qigu oil field in the South Junggar Depression of the Junggar Basin

Plate XXII

1 *Dictyophyllidites harrisii* Couper No. Q-34-41
2 *Dictyophyllidites* sp No. Q-34-28
3–4 *Converrucosisporites elegans* Bai *3* No. Q-34-17; *4* No. Q-34-16
5 *Granulatisporites jurassicus* Pocock No. Q-34-9
6 *Rouseisporites* cf. *reticulatus* Pocock No. Q-34-21
7 *Podocarpidites multesimus* (Bolchovitina) Pocock No. Q-34-42
8 *Pteruchipollenites thomasii* Couper No. Q-34-14
9 *Protopicea exilioides* (Bolchovitina) Pocock No. Q-34-21
10 *Cedripites minor* Pocock No. Q-34-45
11 *Parvisaccites* sp. No. Q-34-6
12 *Piceites latens* Bolchovitina No. Q-34-33

Plate XXII

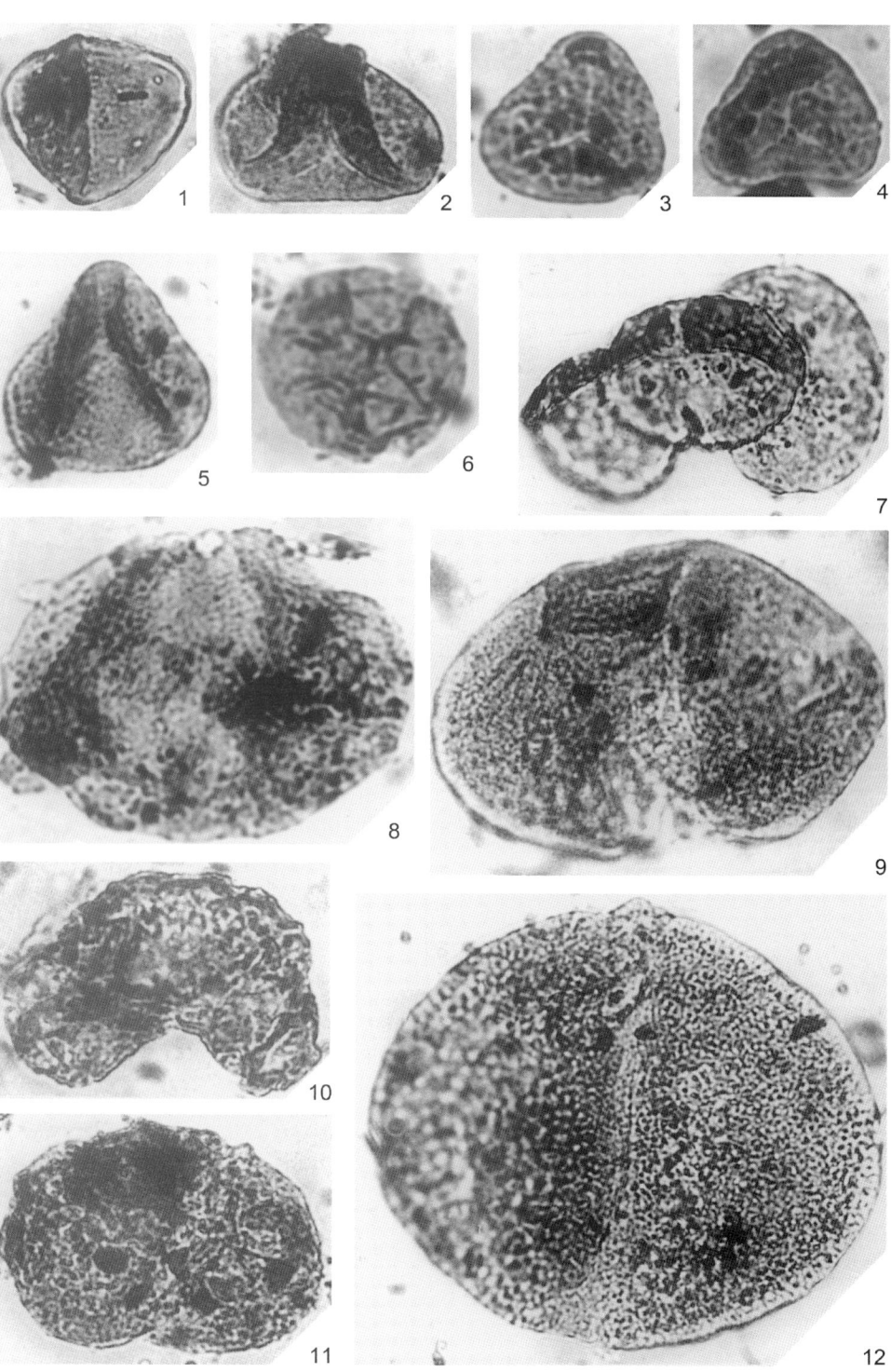

Fossil spores and pollen in crude oils from the Qigu oil field in the South Junggar Depression of the Junggar Basin (cont'd)

Plate XXIII

1-2 Leiotriletes adriensis (Potonié et Gelletich) Krutzsch *1* No. D-68-54; *2* No. D-58-1
3 Taxodiaceaepollenites hiatus (Potonié) Kremp No. D-68-59
4 Tsugaepollenites viridifluminipites (Wodehouse) Potonié No. D-68-60
5 Paleoconiferus asaccatus Bolchovitina No. D-68-4
6 Ephedripites fusiformis (Shakhmundes) Krutzsch No. D-68-12
7 Ephedripites eocenipites (Wodehouse) Krutzsch No. D-68-39
8 Ephedripites scabridus Song et Zheng No. D-68-34
9-9a Juglanspollenites verus Raatz No. D-68-49
10 Piceaepollenites alatus (Potonié) Potonié No. D-68-43
11 Liquidambarpollenites stigmosus (Potonié) Raatz No. D-68-38
12 Chenopodipollis minor Song No. D-68-66
13 Eucommiidites troedssonii Erdtman No. D-68-13
14 Piceaepollenites sp. No. D-68-19
15 Salixipollenites discoloripites (Wodehouse) Srivastava No. D-68-23
16 Caryapollenites simplex (Potonié) Raatz No. D-68-68
17-18 Tubulifloridites macroechinatus (Trevisan) Song et Zhu *17* No. D-68-36; *18* No. D-68-42

Explanation of Plates and Plates

Plate XXIII

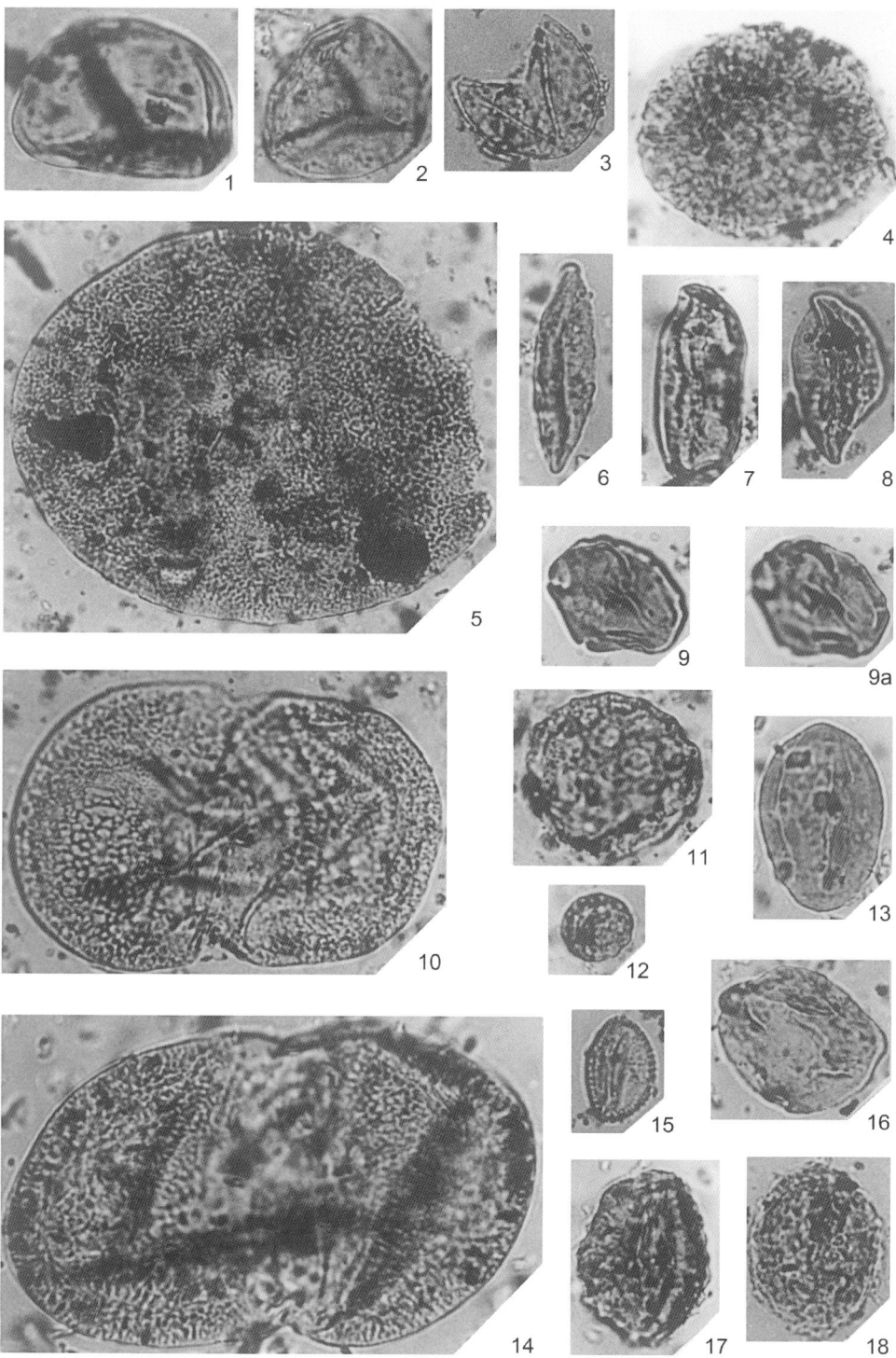

Fossil spores and pollen in crude oils from the Dushanzi oil field in the South Junggar Depression of the Junggar Basin

Plate XXIV

1 Quercoidites asper (Thomson et Pflug) Sung et Zheng No. D-68-36
2 Quercoidites henrici (Potonié) Potonié, Thomson et Thiergart No. D-68-41
3 Quercoidites microhenrici (Potonié) Potonié No. D-68-53
4–5 Protopicea exilioides (Bolchovitina) Pocock *4* No. D-68-27; *5* No. D-68-10
6 Abietineaepollenites minimus Couper No. D-68-41
7 Paleoconiferus asaccatus Bolchovitina No. D-68-67
8 Piceites pseudorotundiformis (Maljavkina) Pocock No. D-68-14
9 Alisporites grandis (Cookson) Dettmann No. D-201-1
10 Araucariacites australis Cookson No. D-68-6

Plate XXIV

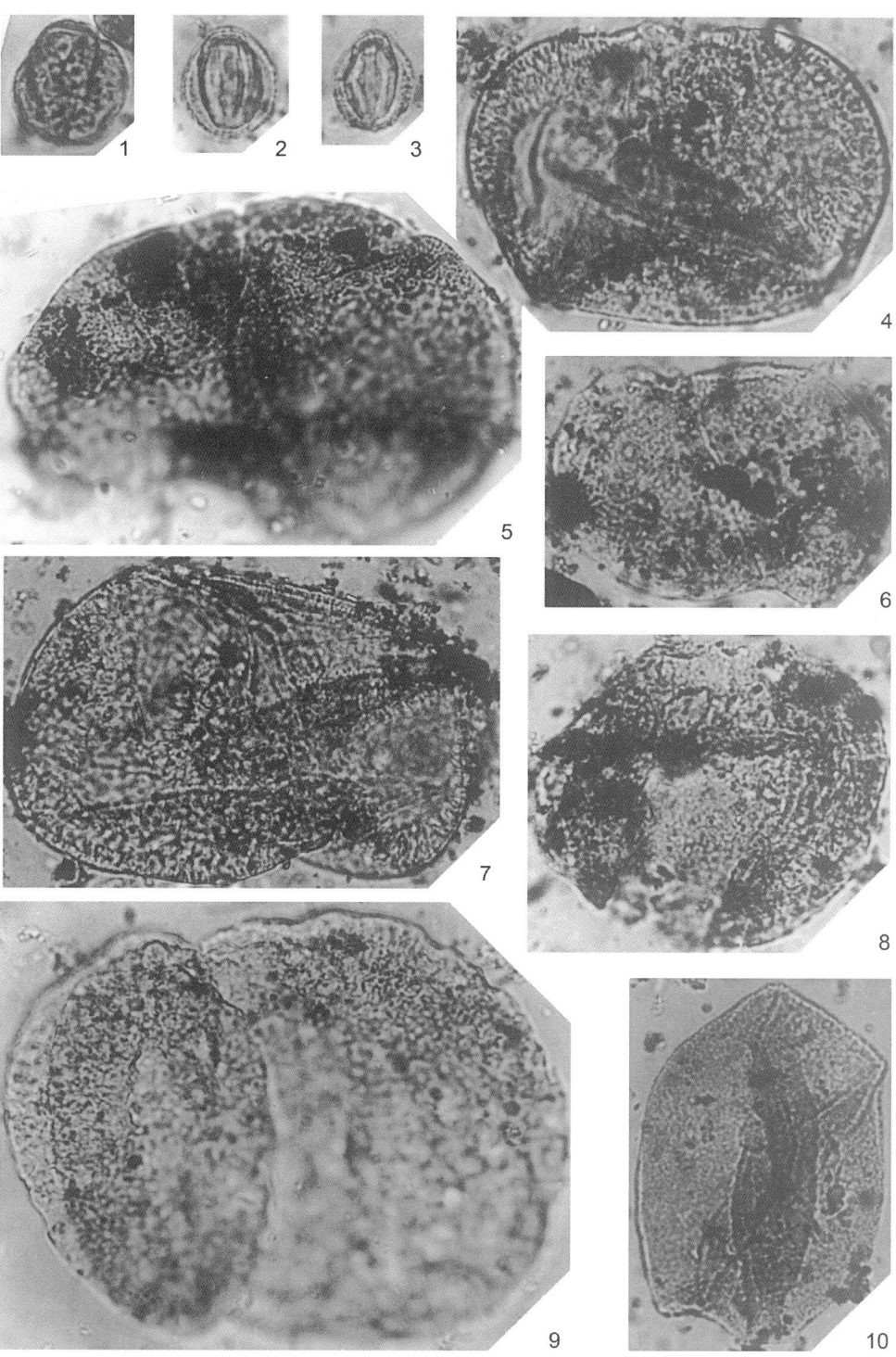

Fossil pollen in crude oils from the Dushanzi oil field in the South Junggar Depression of the Junggar Basin (cont'd)

Plate XXV

1–2 Cyathidites minor Couper *1* No. Q-1-65; *2* No. Q-1-37
3 Cibotiumspora jurienensis (Balme) Filatoff No. Q-1-10
4 Dictyophyllidites harrisii Couper No. Q-1-42
5 Divisisporites undulatus Huang No. Q-1-91
6 Tripartina variabilis Maljavkina No. Q-1-24
7 Gleicheniidites conflexus (Chlonova) Xu et Zhang No. Q-1-6
8 Todisporites minor Couper No. Q-1-1
9 Deltoidospora magna (de Jersey) Norris No. Q-1-2
10 Cyathidites australis Couper No. Q-1-89
11 Verrucosisporites variabilis Pocock No. Q-1-63
12 Piceites expositus Bolchovitina No. Q-1-74
13 Piceites latens Bolchovitina No. Q-1-87
14 Callialasporites radius Xu et Zhang No. Q-1-63
15 Osmundacidites wellmanii Couper No. Q-1-8
16 Classopollis classoides (Pflug) Pocock No. Q-1-59
17 Vitreisporites jansonii Pocock No. Q-1-9

Plate XXV

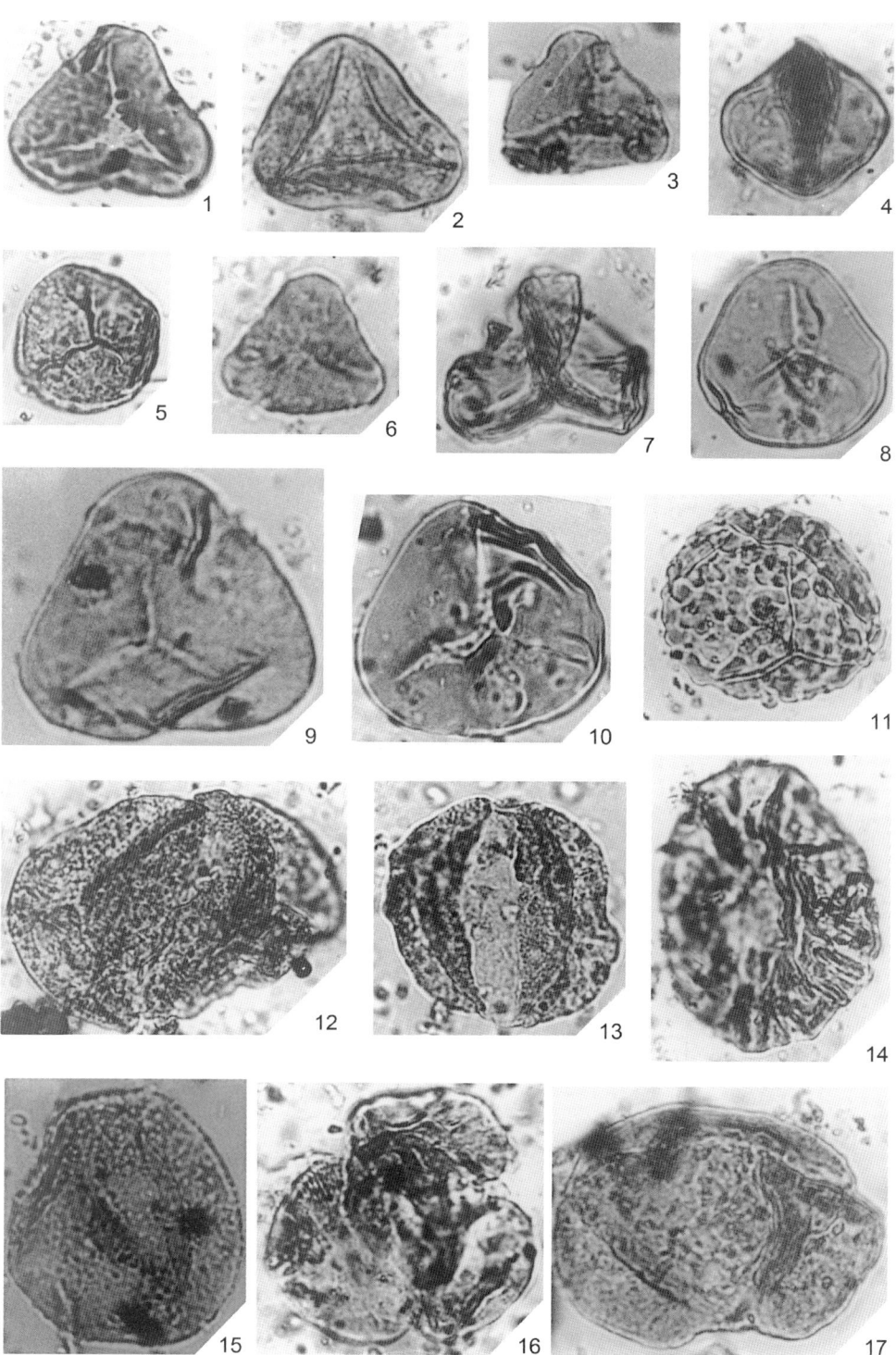

Fossil spores and pollen in crude oils from the Qiktim oil field of the Turpan Basin

Plate XXVI

1 Piceaepollenites omoriciformis (Bolchovitina) Xu et Zhang No. Q-1-88
2 Podocarpidites multicinus (Bolchovitina) Pocock No. Q-1-11
3 Paleoconiferus asaccatus Bolchovitina No. Q-1-10
4 Pteruchipollenites thomasii Couper No. Q-1-4
5 Protopicea exilioides (Bolchovitina) Pocock No. Q-1-24
6–7 Classopollis annulatus (Verbitzkaya) Li 6 No. Q-1-88; 7. No. Q-1-84
8 Platysaccus lopsinensis (Maljavkina) Pocock No. Q-1-4
9 Abietineaepollenites dunrobinensis Couper No. Q-1-24
10 Cedripites minor Pocock No. Q-1-66
11 Cycadopites typicus (Maljavkina) Pocock No. Q-1-19
12 Parvisaccites enigmatus Couper No. Q-1-17

Explanation of Plates and Plates

Plate XXVI

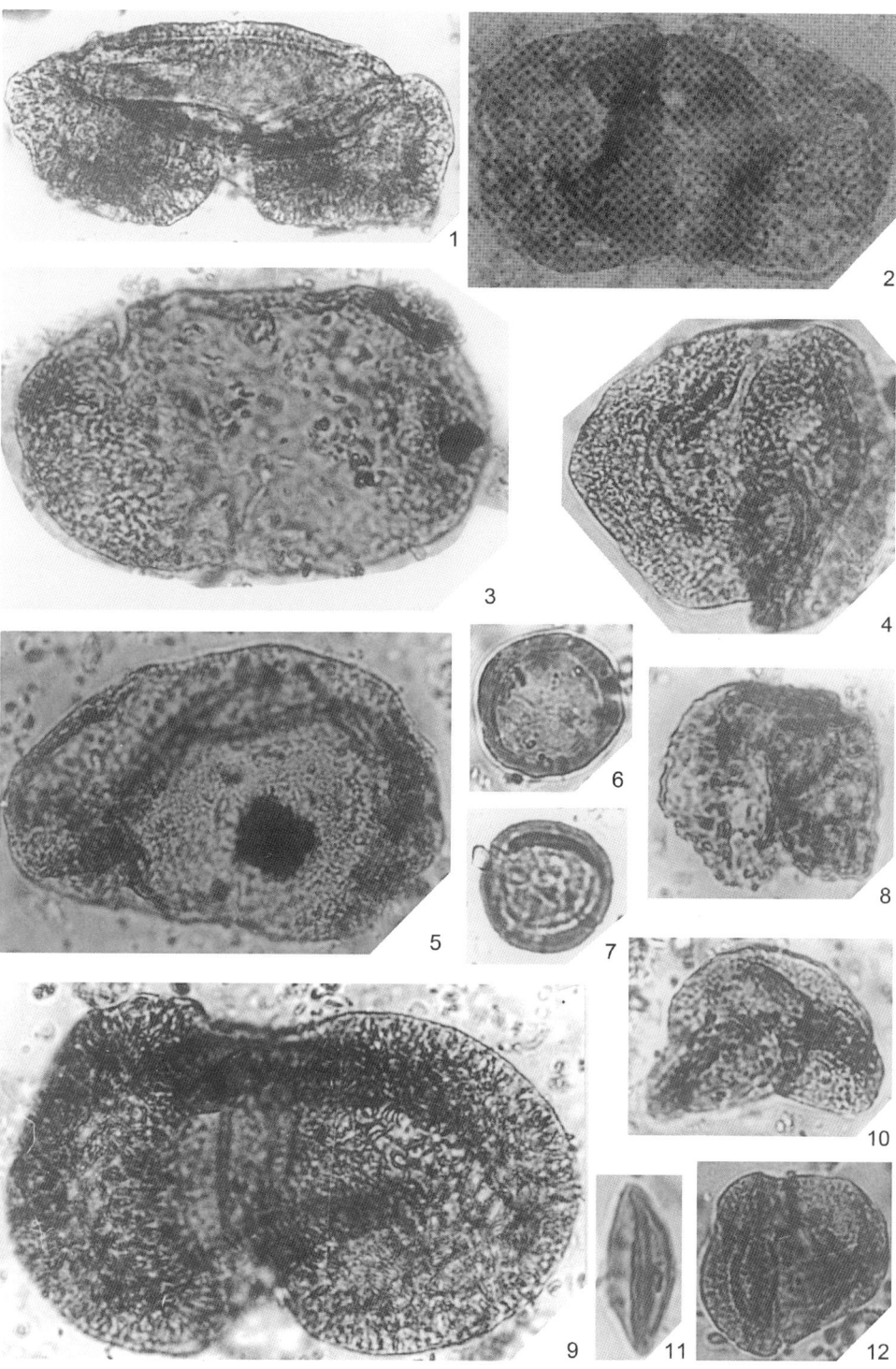

Fossil pollen in crude oils from the Qiktim oil fiedl of the Turpan Basin (cont'd)

Plate XXVII

1 Cyathidites minor Couper No. L-7411-24
2 Gleicheniidites conflexus (Chlonova) Xu et Zhang No. L-7411-14
3–4 Duplexisporites scanicus (Nilsson) Playford et Dettmann
3 No. L-7411-11; *4* No. L-7411-12
5–6 Osmundacidites wellmanii Couper *5* No. L-7411-21; *6* No. L-7411-14
7 Osmundacidites elegans (Verbitzkaya) Xu et Zhang No. L-7411-25
8 Duplexisporites amplectiformis (Kara-Murza) Playford et Dettmann No. L-7411-10
9–10 Apiculatisporis variabilis Pocock *9* No. L-7411-7; *10* No. L-7411-20
11 Classopollis annulatus (Verbitzkaya) Li No. L-7411-25
12 Acanthotriletes midwayensis Pocock No. L-7411-10
13 Quadraeculina limbata Maljavkina No. L-7411-8
14 Callialasporites radius Xu et Zhang No. L-7411-20
15 Duplexisporites gyratus Playford et Dettmann No. L-7411-11
16 Protopodocarpus mollis Bolchovitina No. L-7411-19

Plate XXVII

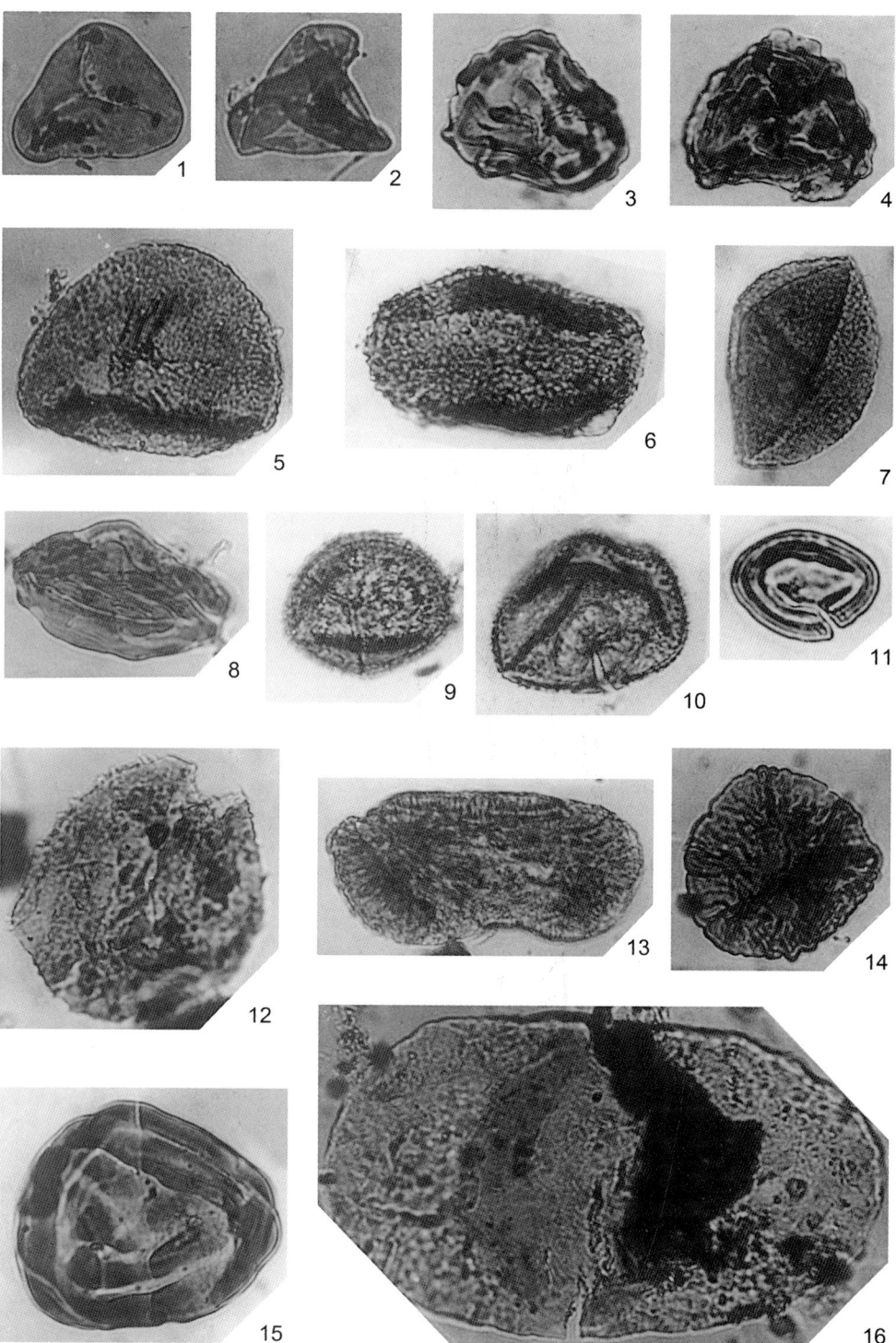

Fossil spores and pollen in crude oils from the Lenghu oil field in the North Border Block-fault Zone of the Qaidam Basin

Plate XXVIII

1, 9 Podocarpidites paulus (Bolchovitina) Xu et Zhang *1* No. L-7411-1;
 9 No. L-7411-4
2–3 Protopicea vilujensis Bolchovitina *2* No. L-7411-10; *3* No. L-7411-2
4 Chasmatosporites major Nilsson No. L-7411-8
5 Cycadopites carpentieri (Delcourt et Sprumont) Singh No. L-7411-2
6 Protopicea minutereticulata Bolchovitina No. L-7411-17
7 Cycadopites subgranulosus (Couper) Clarke No. L-7411-23
8 Parvisaccites enigmatus Couper No. L-7411-7
10 Piceites expositus Bolchovitina No. L-7411-12
11 Piceites latens Bolchovitina No. L-7411-24

Plate XXVIII

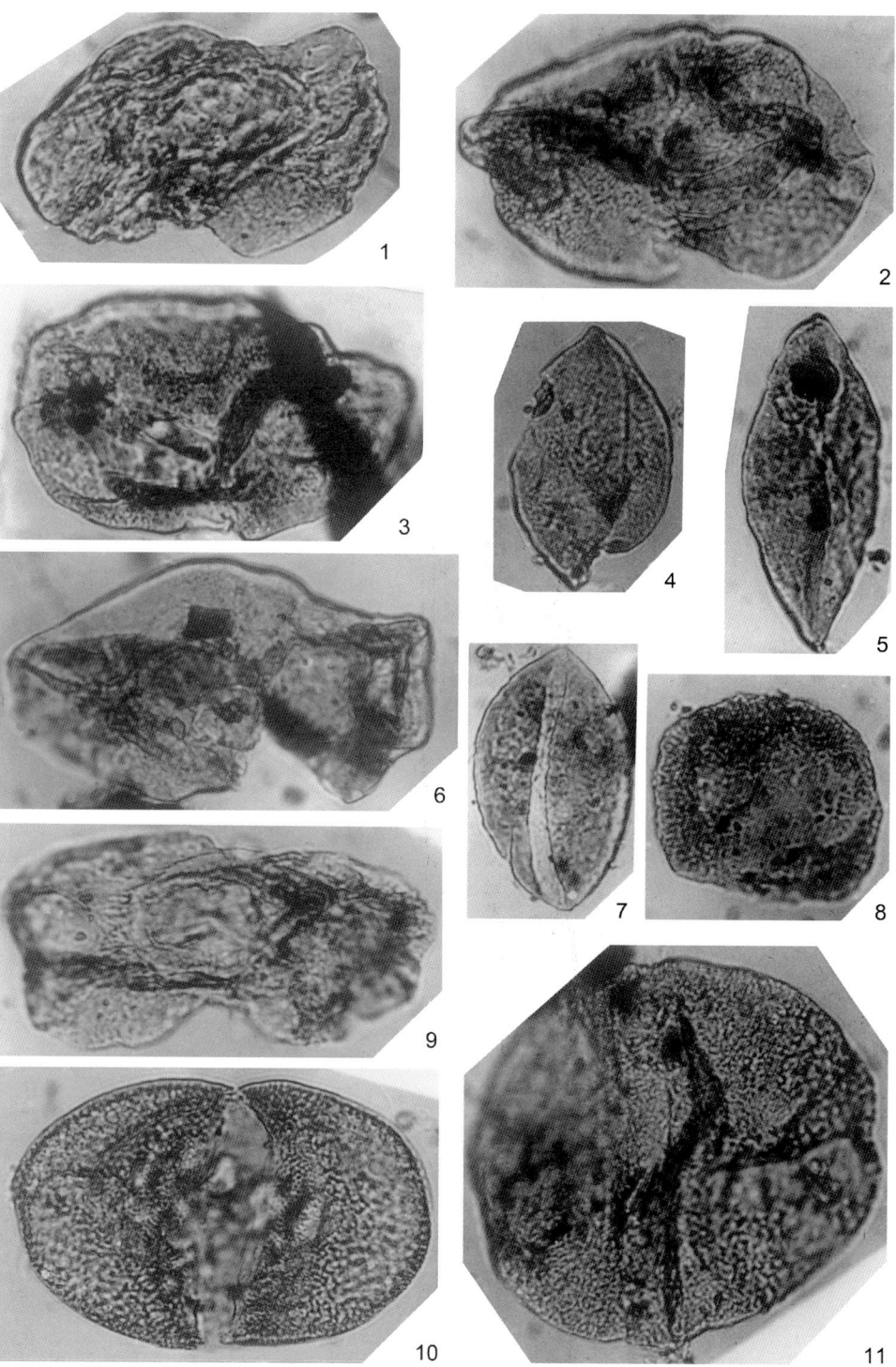

Fossil pollen in crude oils from the Lenghu oil field in the North Border Block-fault Zone of the Qaidam Basin (cont'd)

Plate XXIX

1 *Deltoidospora regularis* (Pflug) Song et Zheng No. Y-1-12
2 *Polypodiaceaesporites haardti* (Potonié et Venitz) Thiergart No. Y-1-27
3 *Polypodiisporites favus* (Potonié) Potonié No. Y-1-16
4–6 *Podocarpidites verrucorpus* Wu 4 No. Y-1-14; 5 No. Y-1-21; 6 No. Y-1-11
7 *Podocarpidites qigequanensis* Wu No. Y-1-15
8 *Ephedripites fusiformis* (Shakhmundes) Krutzsch No. Y-1-10
9 *Keteleeriaepollenites dubius* (Chlonova) Li No. Y-1-15
10 *Piceaepollenites planoides* (Krutzsch) Sun et Li No. Y-1-29
11 *Piceaepollenites alatus* (Potonié) Potonié No. Y-1-21

Explanation of Plates and Plates

Plate XXIX

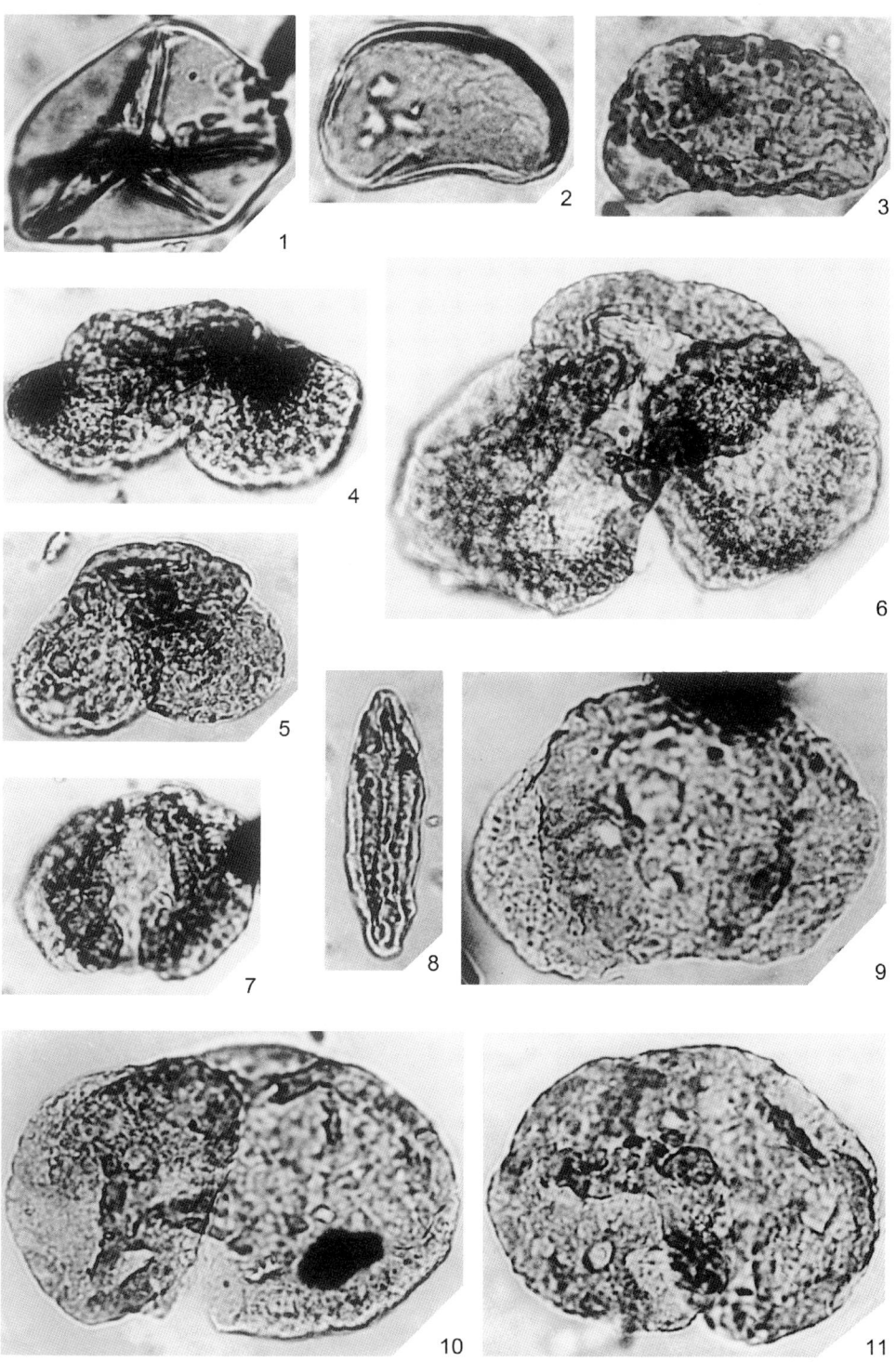

Fossil spores and pollen in crude oils from the Youquanzi oil field in the Mangnai Depression of the Qaidam Basin

Plate XXX

1 *Pinuspollenites labdacus maximus* (Potonié) Potonié No. Y-1-9
2 *Abietineaepollenites microalatus major* (Potonié) Potonié No. Y-1-25
3 *Pinuspollenites labdacus minor* (Potonié) Potonié No. Y-1-15
4 *Deltoidospora regularis* (Pflug) Song et Zheng No. Y-1-28
5–6 *Cedripites deodariformis* (Zauer) Krutzsch 5 No. Y-1-7; 6 No. Y-1-10
7 *Ephedripites eocenipites* (Wodehouse) Krutzsch No. Y-1-24
8 *Podocarpidites paranageiaformis* Ke et Shi No. Y-1-27
9 *Chenopodipollis multiplex* (Weyland et Pflug) Krutzsch No. Y-1-6
10–11 *Cedripites pachydermus* (Zauer) Krutzsch 10 No. Y-1-5; 11 No. Y-1-9
12 *Cedripites microsaccoides* Song et Zheng No. Y-1-4

Plate XXX

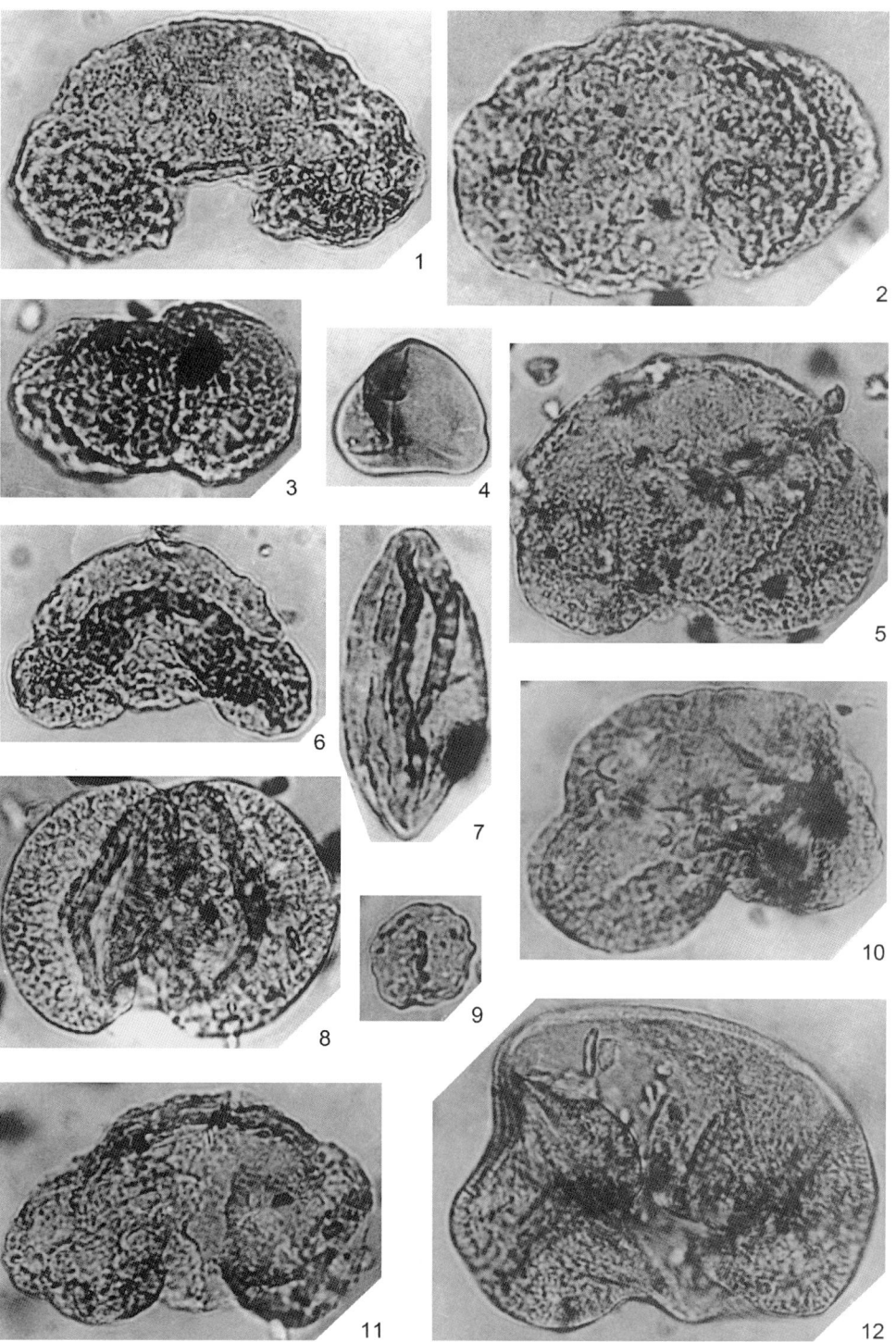

Fossil spores and pollen in crude oils from the Youquanzi oil field in the Mangnai Depression of the Qaidam Basin (cont'd)

Plate XXXI

1 Keteleeriaepollenites dubius (Chlonova) Li No. Y-1-14

2 Ephedripites eocenipites (Wodehouse) Krutzsch No. Y-1-19

3 Keteleeriaepollenites mangnaiensis Zhu No. Y-1-22

4 Abietineaepollenites cembraeformis (Zaklinskaja) Ke et Shi No. Y-1-23

5 Abietineaepollenites microalatus major (Potonié) Potonié No. Y-1-2

6 Tsugaepollenites igniculus major (Potonié) Potonié No. Y-1-21

7 Tsugaepollenites igniculus minor (Potonié) Potonié No. Y-1-24

8 Quercoidites henrici (Potonié) Potonié, Thomson et Thiergart No. Y-1-14

9 Chenopodipollis multiporatus (Pflug et Thomson) Zhou No. Y-1-17

10 Tsugaepollenites spinulosus (Krutzsch) Ke et shi No. Y-1-5

11 Sparganiaceaepollenites sparganioides (Meyer) Krutzsch No. Y-1-4

12 Meliaceoidites rhomboiporus Wang No. Y-1-6

13 Nitrariadites subrotundus Zhu et Xi No. Y-1-4

14 Pinuspollenites labdacus maximus (Potonié) Potonié No. Y-1-25

15 Tubulifloridites macroechinatus (Trevisan) Song et Zhu No. Y-1-3

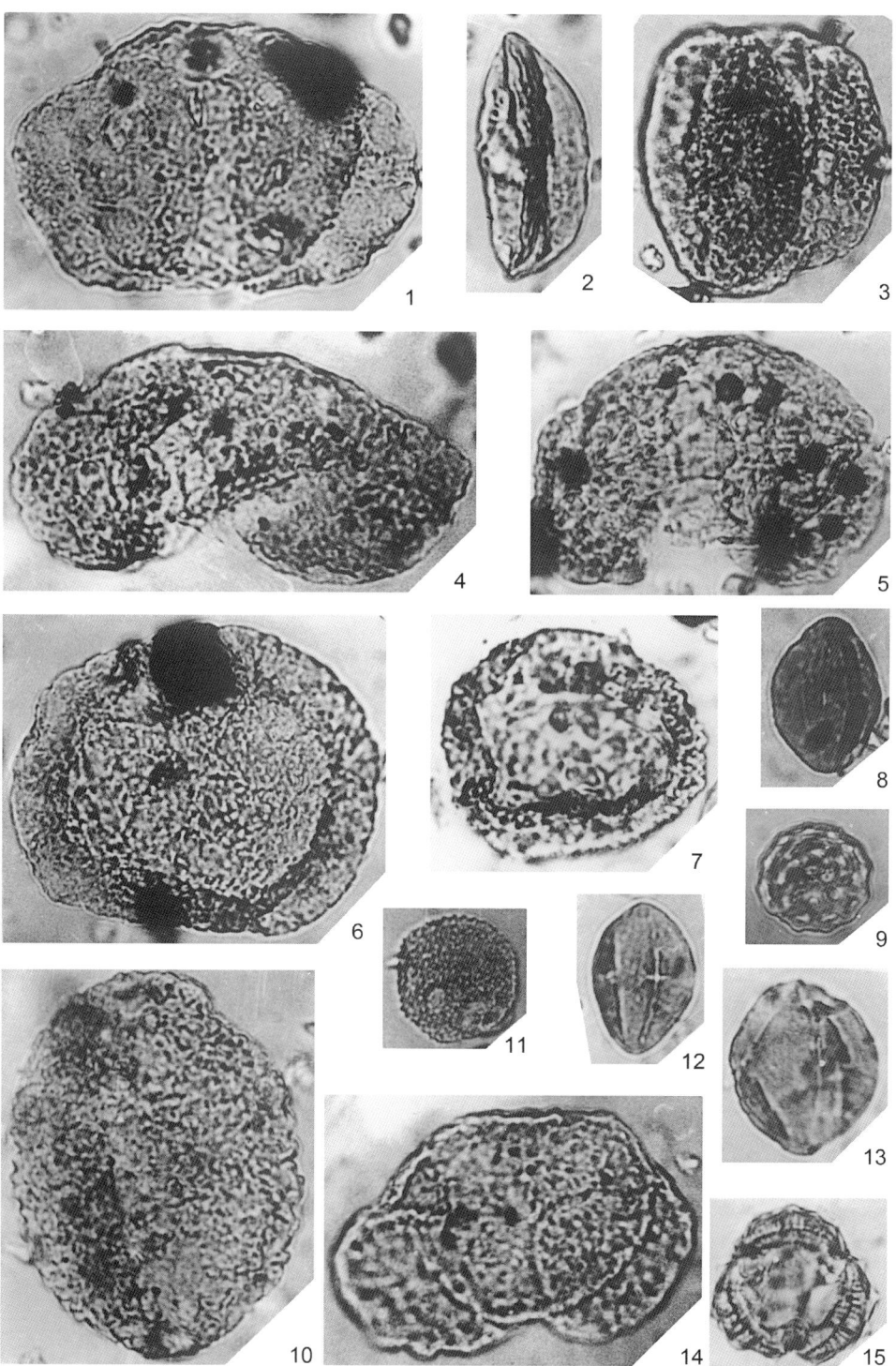

Fossil pollen in crude oils from the Youquanzi oil field in the Mangnai Depression of the Qaidam Basin (cont'd)

Plate XXXII

1 *Granulatisporites pteridiumoides* Zhang No. X-2-7
2 *Polypodiaceaesporites haardti* (Potonié et Venitz) Thiergart No. X-2-14
3 *Pinuspollenites labdacus minor* (Potonié) Potonié No. X-2-6
4 *Cedripites deodariformis* (Zauer) Krutzsch No. X-2-15
5 *Ephedripites mangnaiensis* Zhu et Wu No. X-2-8
6 *Pinuspollenites labdacus maximus* (Potonié) Potonié No. X-2-16
7 *Piceaepollenites quadracorpus* Zhu et Xi No. X-2-17
8 *Tsugaepollenites igniculus major* (Potonié) Potonié No. X-2-22
9 *Keteleeriaepollenites megasaccus* Zhu No. X-2-5
10 *Ephedripites neogenicus* Zhu et Wu No. X-2-15
11 *Nitrariadites subrotundus* Zhu et Xi No. X-2-10
12 *Artemisiaepollenites sellularis* Nagy No. X-2-6
13 *Nitrariadites communis* Zhu et Xi No. X-2-13

Plate XXXII

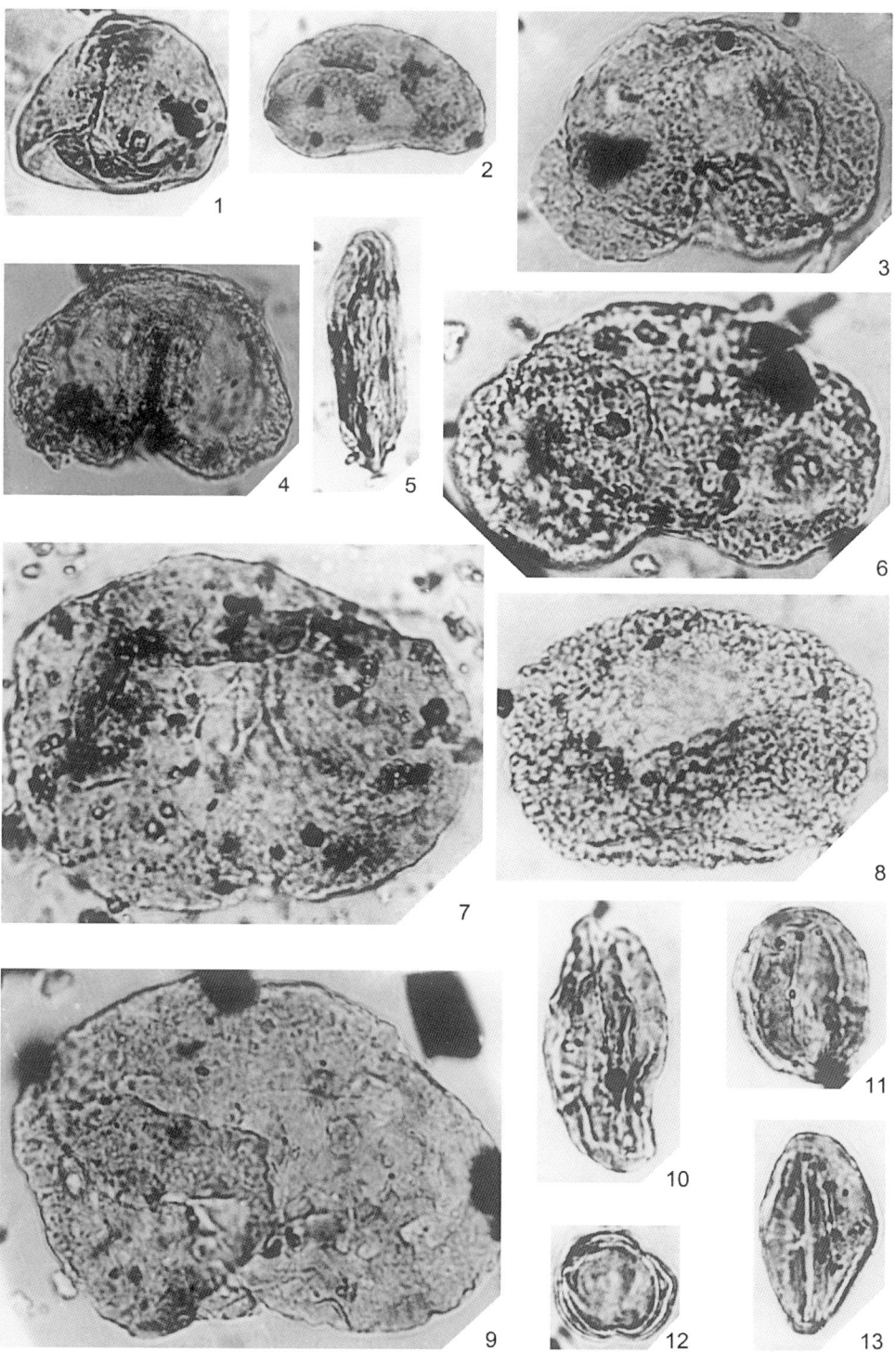

Fossil spores and pollen in crude oils from the Xianshuiquan oil field in the Mangnai Depression of the Qaidam Basin

Plate XXXIII

1 Cingulatisporites ruginosus Hsü,Chiang et Young No. L-716-23
2 Cibotiumspora juncta (Kara-Murza) Xu et Zhang No. L-716-17
3–4, 17 Classopollis annulatus (Verbitzkaya) Li
 3 No. L-716-1;
 4 No. L-716-5; *17* No. L-215-3
5–8 Schizaeoisporites zizyphinus Hsü; Chiang et Young *5* No. L-215-12;
 6 No. B-94-12; *7* No. B-94-1; *8* No. B-94-17
9 Cyperaceaepollis sp. No. Y-115-11
10 Psophosphaera grandis Bolchovitina No. L-215-18
11–13 Cycadopites minimus (Cookson) Pocock
 11 No. L-7062-7;
 12 No. Y-135-10; *13* No. Y-503-18
14 Eutrema sp. No. B-117-9
15 Achillea sp. No. B-94-8
16 Artemisia sp. No. S-173-1
18 Nymphaea sp. No. L-604-5
19 Solidago sp. No. B-94-19
20–21 Chenopodium sp. *20* No. L-604-4; *21* No. B-94-1
22 Pinus sp. No. B-117-10
23 Graminidites sp. No. Y-115-5

(*10* at a magnification of 440×, others at a magnification of 880×)

Plate XXXIII

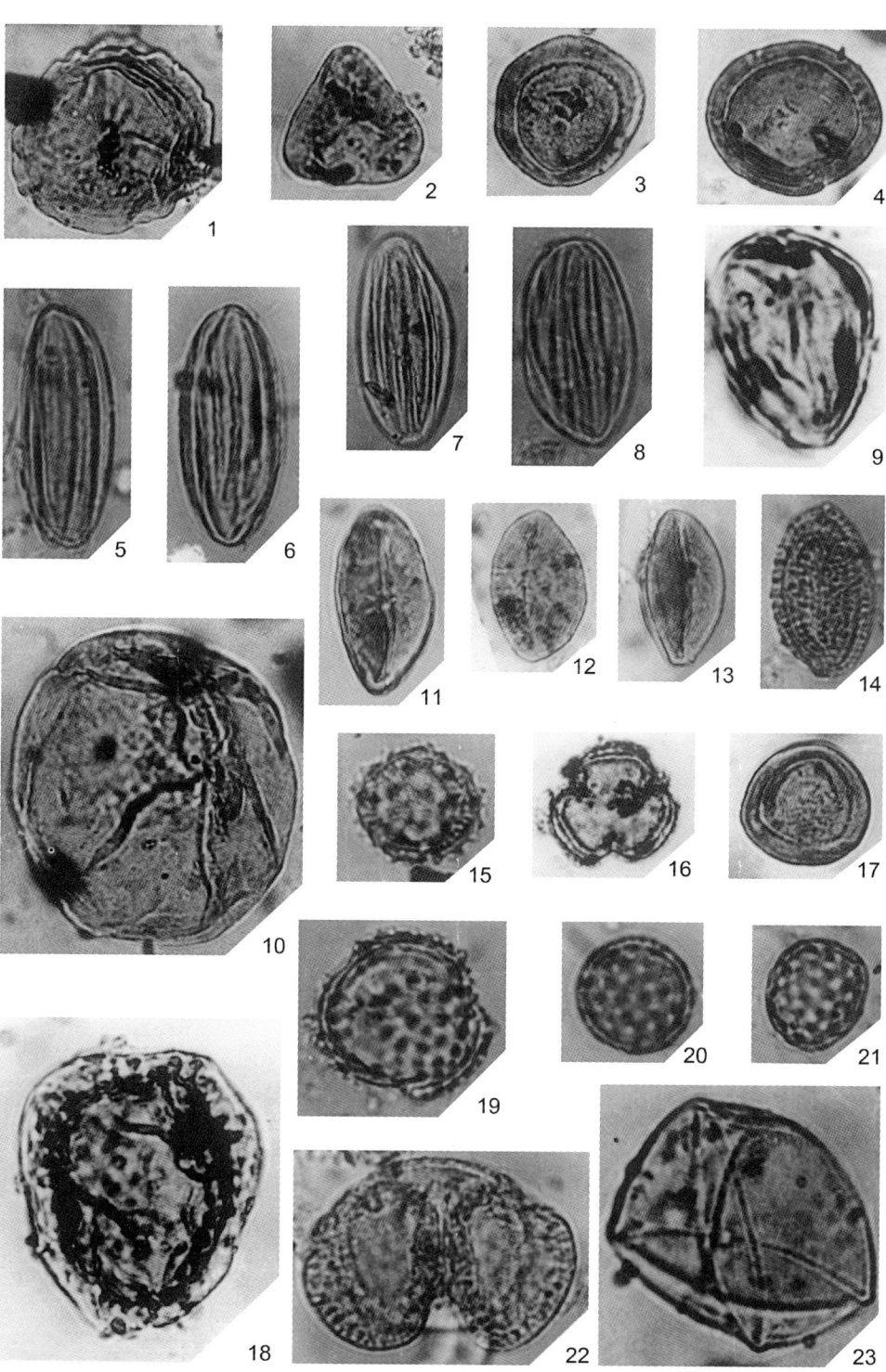

Fossil spores and pollen in crude oils from the Laojunmiao oil field of the West Jiuquan Basin

Plate XXXIV

1–2 Pterisisporites undulatus Sung et Zheng *1* No. 8-11-10; *2* No. 8-11-13

3 Plicifera decora (Chlonova) Bolchovitina No. 213-6

4 Abietineaepollenites cembraeformis (Zaklinskaja) Ke et Shi No. 8-11-18

5 Pinuspollenites strobipites (Wodehouse) Sun et Li No. 8-11-6

6 Cedripites diversus Ke et Shi No. 8-11-12

7 Pinuspollenites labdacus minor (Potonié) Potonié No. 8-11-11

8 Abietineaepollenites microsibiricus (Zaklinskaja) Ke et Shi No. 8-11-17

9 Cedripites pachydermus (Zauer) Krutzsch No. 213-30

10 Keteleeriaepollenites dubius (Chlonova) Li No. 8-11-14

11 Quercoidites asper (Thomson et Pflug) Sung et Zheng No. 8-11-14

12–13 Sparganiaceaepollenites sparganioides (Meyer) Krutzsch
 12 No. 213-10; *13* No. 213-22

14 Graminidites sp. No. 213-11

15–16 Cyperaceaepollis sp. *15* No. 8-11-14; *16* No. 213-5

17 Chenopodipollis microporatus (Nakoman) Liu No. 213-17

18, 20 Artemisiaepollenites sp. *18* No. 213-23; *20* No. 213-3

19 Chenopodipollis multiporatus (Pflug et Thomson) Zhou No. 213-10

21 Compositoipollenites sp. No. 8-11-1

22 Liliacidites sp. No. 213-13

Explanation of Plates and Plates

Plate XXXIV

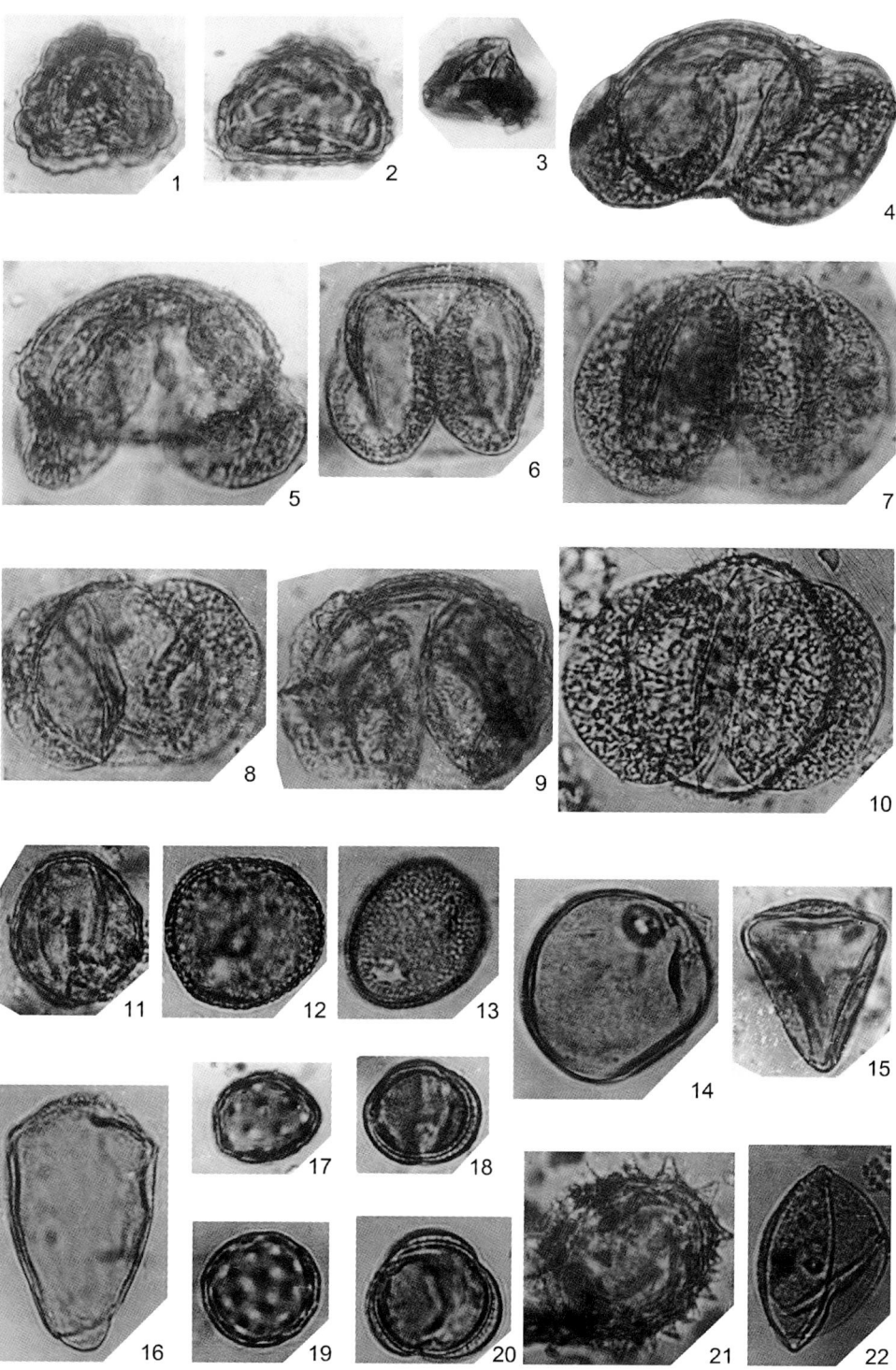

Fossil spores and pollen in crude oils from the Xinglongtai oil field of the Liaohe Basin

Plate XXXV

1 Leiotriletes adriensis (Potonié et Gelletich) Krutzsch No. W10-3-4-15

2 Crassoretitriletes nanhaiensis Zhang et Li No. W10-3-4-21

3 Polypodiisporites afavus (Krutzsch) Sun et Li No. W10-3-4-28

4 Podocarpidites andiniformis (Zaklinskaja) Takahashi No. W10-3-4-15

5, 8 Pinuspollenites strobipites (Wodehouse) Sun et Li

 5 No. W10-3-4-24; *8* No. W10-3-4-1

6 Cedripites cedroides (Thomson et Pflug) Sun et Li No. W10-3-4-11

7 Cedripites eocenicus Wodehouse No. W10-3-4-27

9 Retimultiporopollenites liushaensis Li et Sun No. W10-3-4-25

10 Piceaepollenites alatus (Potonié) Potonié No. W10-3-2-12

11 Momipites coryloides Wodehouse No. W10-3-2-14

12 Cupuliferoipollenites pusillus (Potonié) Potonié No. W10-3-4-15

13–14 Quercoidites microhenrici (Potonié) Potonié *13* No. W10-3-4-22;

 14 No. W10-3-4-1

15 Salixipollenites discoloripites (Wodehouse) Srivastava No. W10-3-4-21

16 Verrutricolporites pachydermus Sun, Kong et Li No. W10-3-4-14

17 Corylopsis princeps Lubomirova No. W10-3-4-23

Plate XXXV

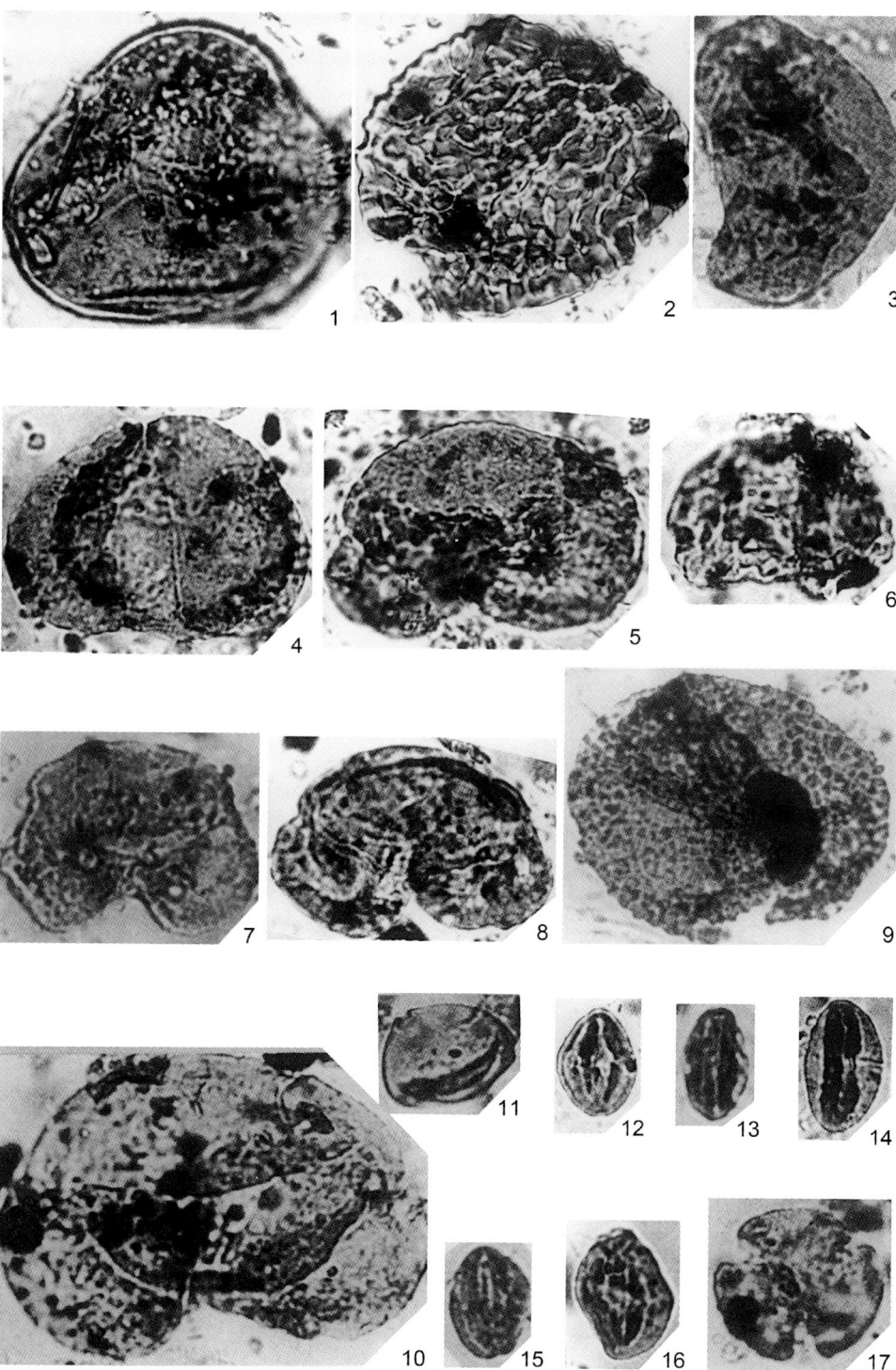

Fossil spores and pollen in crude oils from the Weizhou oil field of the Beibu Gulf Basin

Plate XXXVI

1 Polypodiaceoisporites vitiosus Krutzsch No. Z5-3-7

2, 6 Pinuspollenites strobipites (Wodehouse) Sun et Li *2* No. Z5-3-9; *6* No. Z5-3-14

3–4 Cedripites eocenicus Wodehouse *3* No. Z5-3-17; *4* No. Z5-3-22

5 Crassoretitriletes nanhaiensis Zhang et Li No. Z5-3-5

7–9 Abietineaepollenites microalatus minor (Potonié) Potonié *7* No. Z5-3-3; *8* No. Z5-3-7; *9* No. Z5-3-16

10 Momipites coryloides Wodehouse No. Z5-3-18

11 Cupuliferoipollenites pusillus (Potonié) Potonié No. Z5-3-15

12–13 Pinuspollenites labdacus minor (Potonié) Potonié *12* No. Z5-3-6; *13* No. Z5-3-19

14–15 Piceaepollenites alatus (Potonié) Potonié *14* No. Z5-3-3; *15* No. Z5-3-10

16 Cedripites cedroides (Thomson et Pflug) Sun et Li No. Z5-3-6

Plate XXXVI

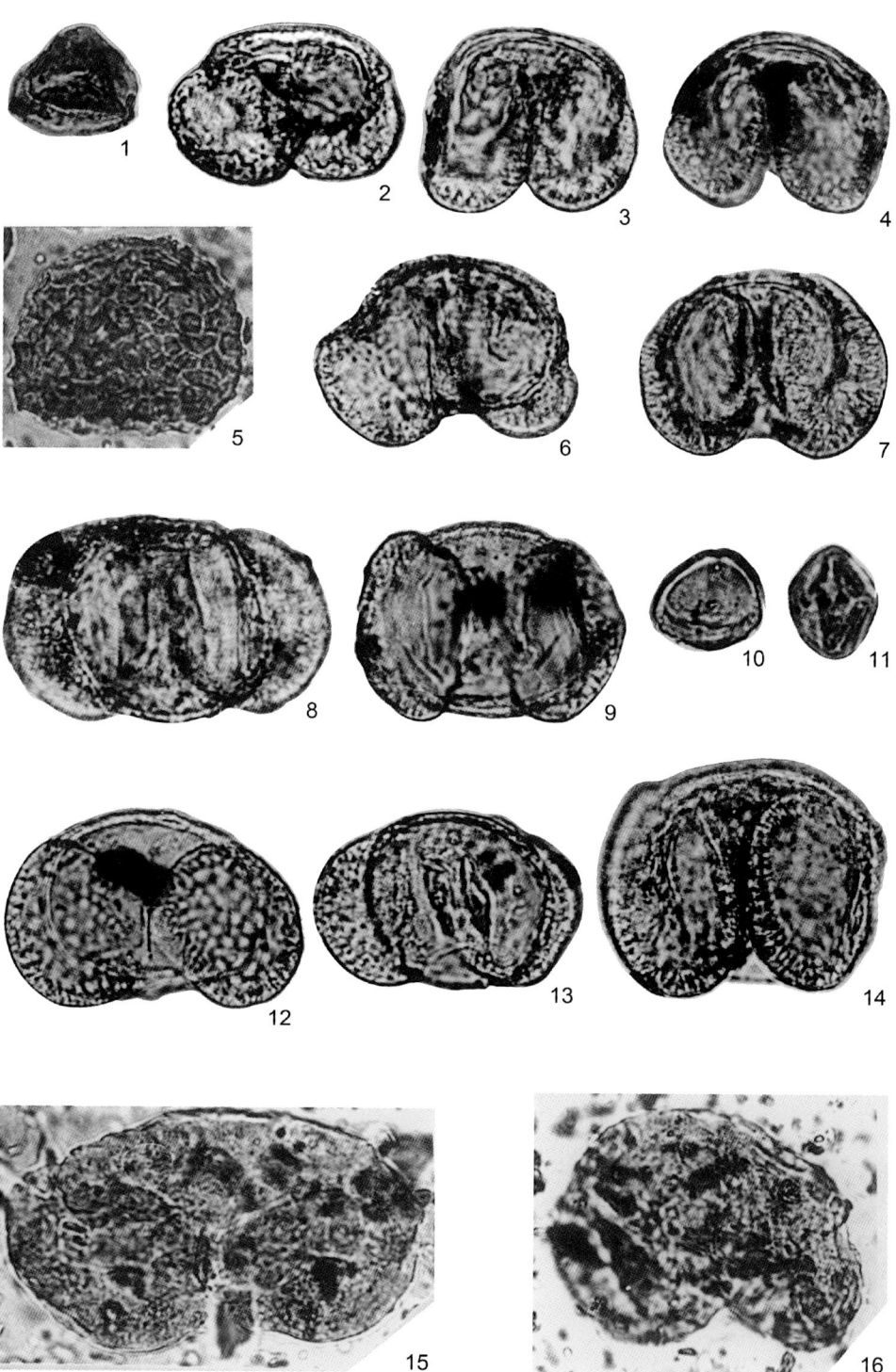

Fossil spores and pollen in crude oils from the Zhuhai oil field of the Zhujiang Mouth Basin

Plate XXXVII

1 Cyathidites australis Couper No. L-1-146
2 Duplexisporites amplectiformis (Kara-Murza) Playford et Dettmann No. L-1-35
3 Lycopodiacidites kuepperi Klaus No. S-14-50
4 Duplexisporites gyratus Playford et Dettmann No. L-1-97
5–6 Duplexisporites scanicus (Nilsson) Playford et Dettmann *5* No. L-1-1; *6* No. L-1-55

(All at a magnification of 1056×)

Explanation of Plates and Plates

Plate XXXVII

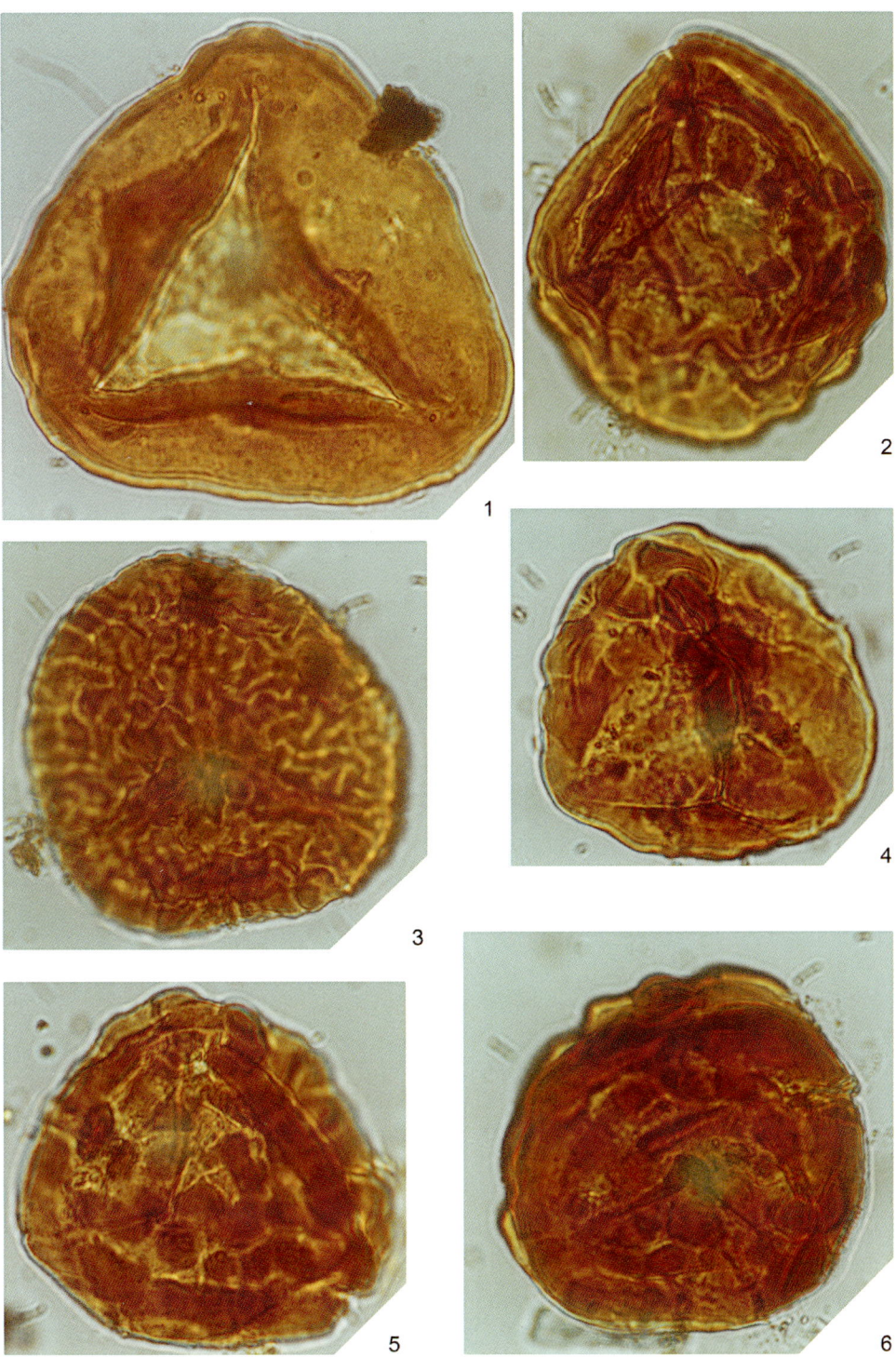

Fossil spores in crude oils of the Tarim Basin

Plate XXXVIII

1 *Aratrisporites fischeri* (Klaus) Playford et Dettmann No. L-1-1
2 *Verrucosisporites remyanus* Mädler No. L-1-111
3 *Celpectopollis pseudostriatus* (Kopytova) Qu et Wang No. L-1-11
4–5 *Lophotriletes corrugatus* Ouyang et Li 4 No. L-1-1; 5 No. L-1-29
6 *Rugubivesiculites* sp. No. L-1-112
7 *Apiculatisporis spiniger* (Leschik) Qu No. L-1-6

(All at a magnification of 1056×)

Plate XXXVIII

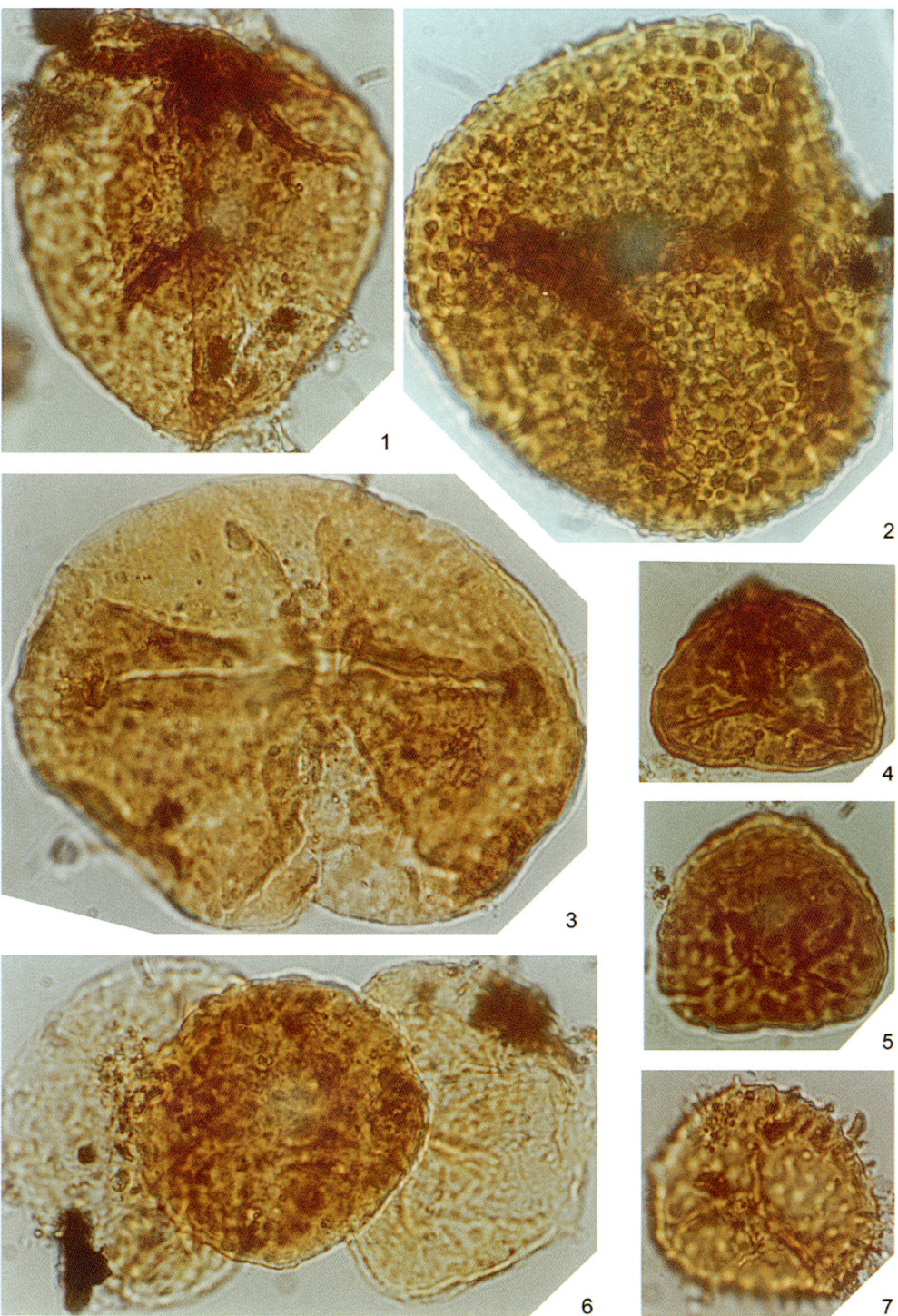

Fossil spores and pollen in crude oils of the Tarim Basin (cont'd)

Plate XXXIX

1 *Taeniaesporites pellucidus* (Goubin) Balme No. S-9-9
2 *Enzonalasporites tenuis* Leschik No. S-14-42
3, 6 *Chordasporites singulichorda* Klaus 3 No. L-1-125; 6 No. L-1-118
4 *Cedripites priscus* Balme No. S-14-39
5 *Taeniaesporites albertae* Jansonius No. S-14-30
7 *Camarozonosporites rudis* (Leschik) Klaus No. S-9-2
8 *Lundbladispora subornata* Ouyang et Li No. S-9-20
9 *Lundbladispora nejburgii* Schulz No. S-9-17

(*3, 6* at a magnification of 1056×, others at a magnification of 528×)

Explanation of Plates and Plates

Plate XXXIX

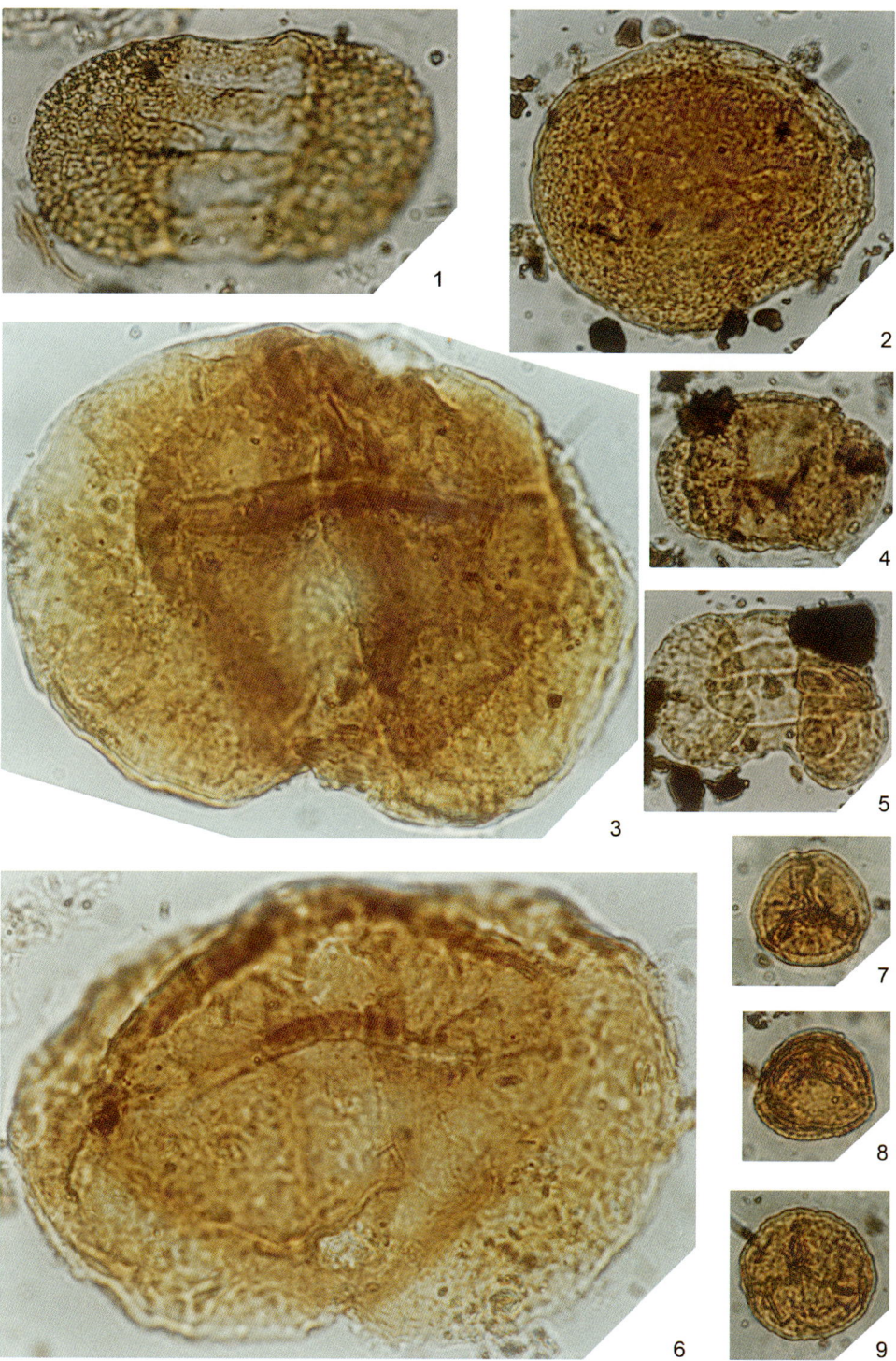

Fossil spores and pollen in crude oils of the Tarim Basin (cont'd)

Plate XL

1–2 Cyathidites australis Couper *1* No. K-6-30; *2* No. K-6-37

3 Apiculatisporis ovalis (Nilsson) Norris No. K-6-33

4 Leptolepidites major Couper No. K-6-11

5 Biretisporites potoniaei Delcourt et Sprumont No. K-6-10

6 Chasmatosporites elegans Nilsson No. K-6-9

7 Cycadopites minimus (Cookson) Pocock No. K-6-28

8 Cycadopites nitidus (Balme) Pocock No. K-6-23

9 Pteruchipollenites thomasii Couper No. K-6-14

10 Cycadopites typicus (Maljavkina) Pocock No. K-6-15

11 Chasmatosporites major Nilsson No. K-6-25

12 Classopollis annulatus (Verbitzkaya) Li No. K-6-35

13 Deltoidospora perpusilla (Bolchovitina) Pocock No. K-6-12

14 Podocarpidites multicinus (Bolchovitina) Pocock No. K-6-33

15 Rugubivesiculites sp. No. K-6-25

Plate XL

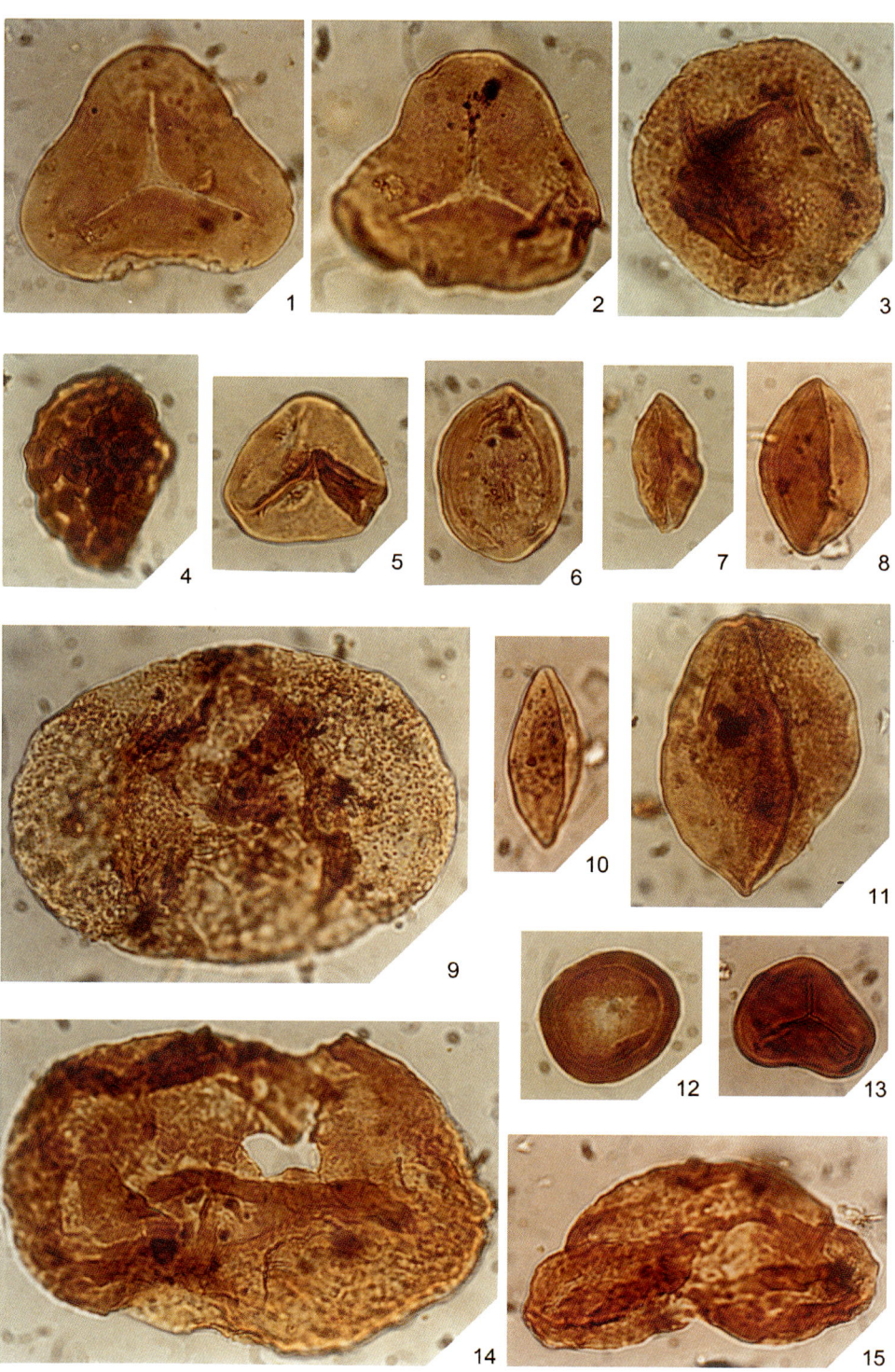

Fossil spores and pollen in crude oils of the Tarim Basin (cont'd)

Plate XLI

1 *Parvisaccites* sp. No. H-1-1
2 *Verrucorpipollis archaicus* Ouyang No. H-1-15
3 *Laevigatosporites medius* Kosanke No. B-5-16
4 *Klausipollenites decipiens* Jansonius No. H-1-33
5 *Chordasporites orientalis* Ouyang et Li No. B-5-2
6 *Protohaploxypinus samoilovichii* (Jansonius) Hart No. B-5-74
7 *Cycadopites caperatus* (Luber et Valts) Hart No. H-1-40
8 *Cedripites priscus* Balme No. H-1-46
9 *Cordaitina rotata* (Luber) Samoilovich No. H-1-40
10 *Striatopodocarpites tojmensis* (Sedowa) Hart No. B-5-2

(*1* at a magnification of 528×)

Plate XLI

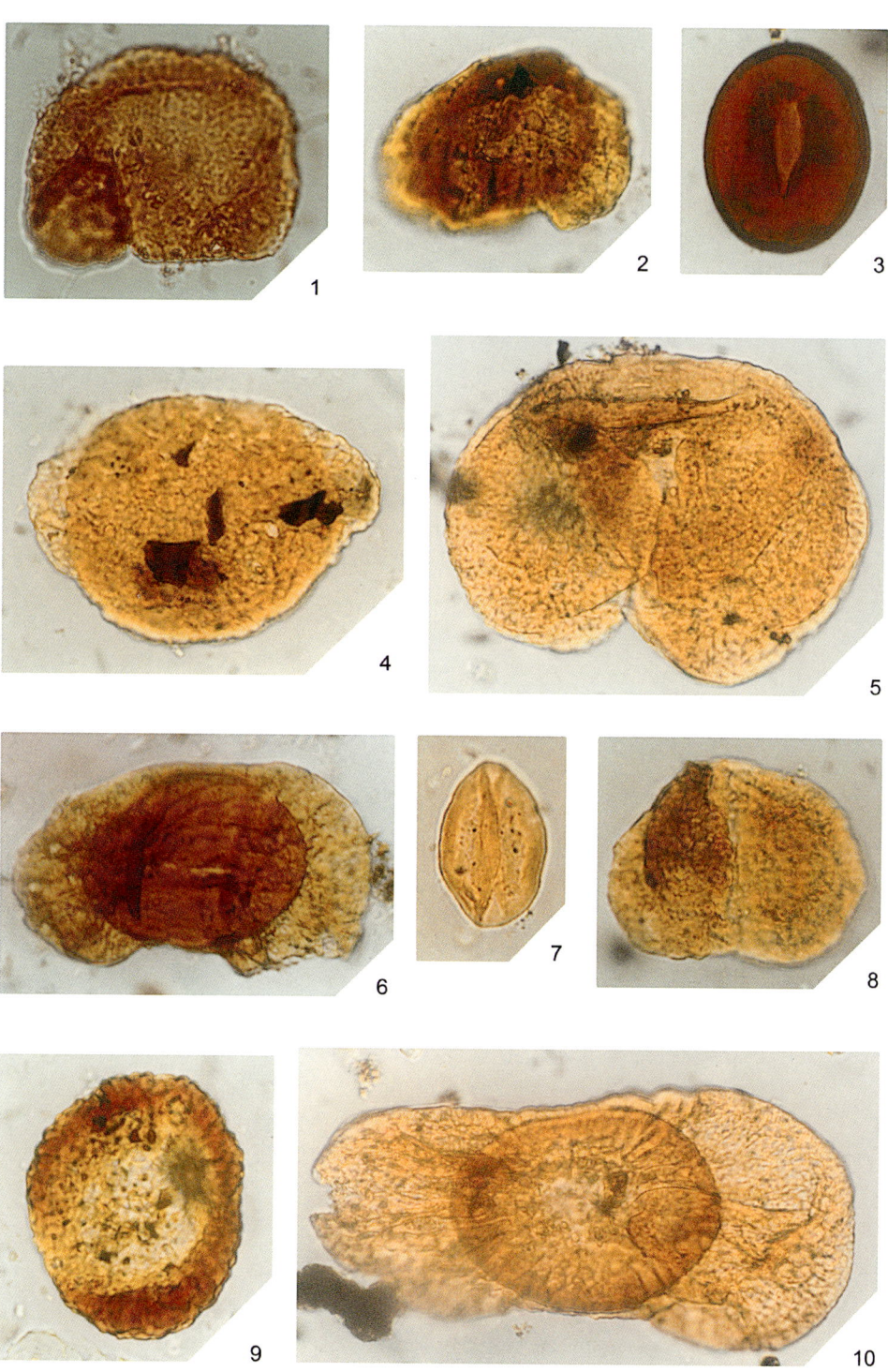

Fossil spores and pollen in crude oils from the East Junggar Depression of the Junggar Basin

Plate XLII

1 *Lophotriletes delicatus* Ouyang et Li No. H-1-30
2 *Lophotriletes pseudaculeatus* Potonié et Kremp No. H-1-46
3 *Punctatosporites minutus* Ibrahim No. H-1-27
4 *Dictyophyllidites mortoni* (de Jersey) Playford et Dettmann No. H-1-27
5 *Kraeuselisporites argutus* Hou et Wang No. H-1-22
6 *Converrucosisporites mictus* Ouyang No. H-1-42
7 *Striatoabietites duivenii* (Jansonius) Hart No. No. H-1-46
8 *Hamiapollenites obliquus* Zhan No. H-1-34
9 *Lueckisporites virkkiae* Potonié et Klaus No. H-1-26
10 *Vittatina subsaccata* Samoilovich No. H-1-35
11 *Klausipollenites caperatus* Ouyang No. H-1-1
12 *Triangulatisporites junggarensis* Yang et Sun No. H-1-23

(*11* at a magnification of 528×, 12. at a magnification of 123×)

Plate XLII

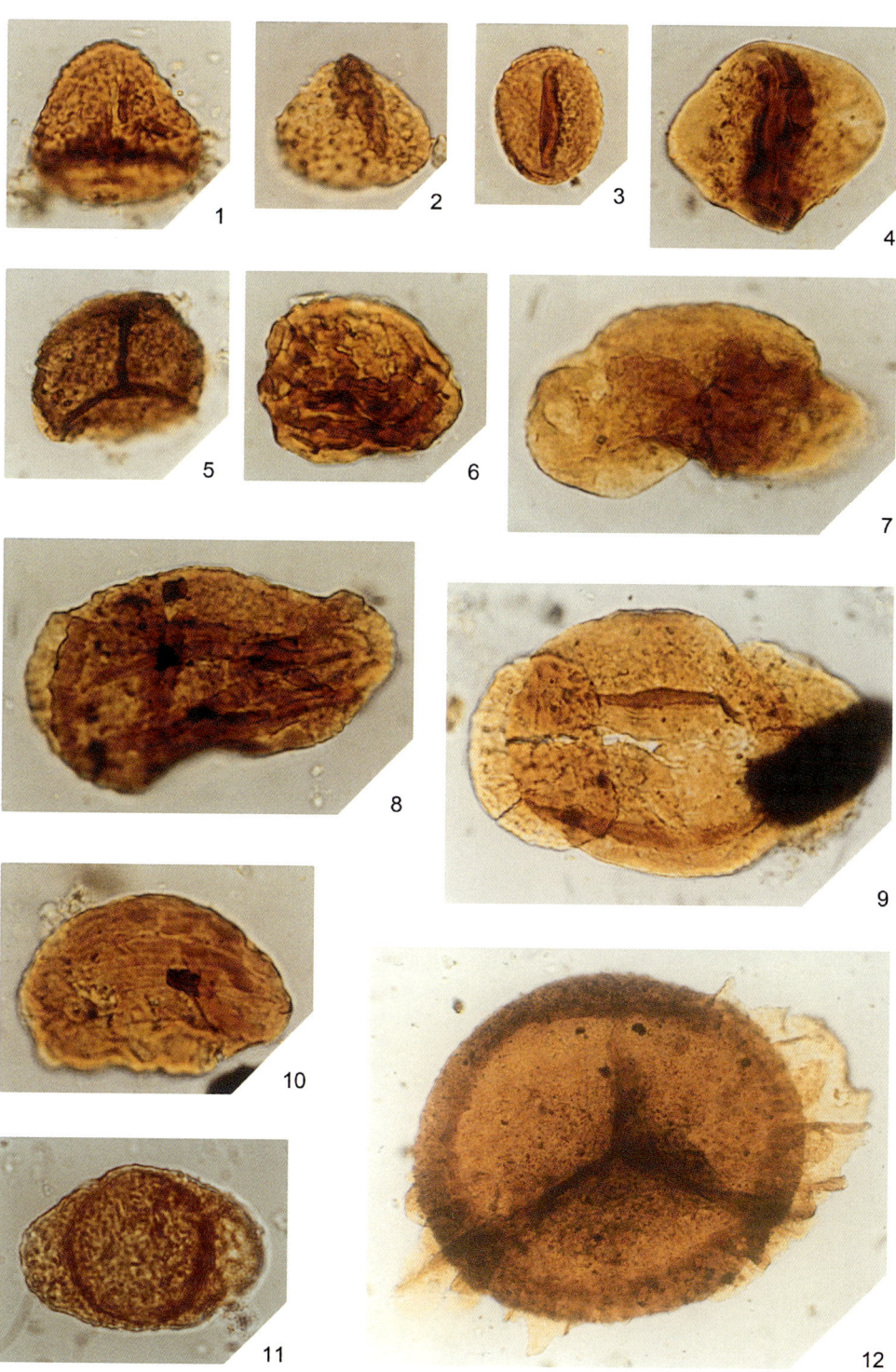

Fossil spores and pollen in crude oils from the East Junggar Depression of the Junggar Basin (cont'd)

Plate XLIII

1 *Granulatisporites minor* de Jersey No. Q-34-38
2 *Cibotiumspora jurienensis* (Balme) Filatoff No. Q-34-21
3 *Ceratosporites jurassicus* Pocock No. Q-34-1
4 *Cibotiumspora paradoxa* (Maljavkina) Chang No. Q-34-18
5 *Piceites expositus* Bolchovitina No. Q-34-25
6 *Piceites latens* Bolchovitina No. Q-34-35
7–8 *Converrucosisporites venitus* Batten 7 No. Q-34-38; 8 No. Q-34-22
9 *Podocarpidites multesimus* (Bolchovitina) Pocock No. Q-34-42
10 *Protopicea exilioides* (Bolchovitina) Pocock No. Q-34-44
11 *Granulatisporites jurassicus* Pocock No. Q-34-9
12 *Concavissimisporites delcourtii* Pocock No. Q-34-6

Plate XLIII

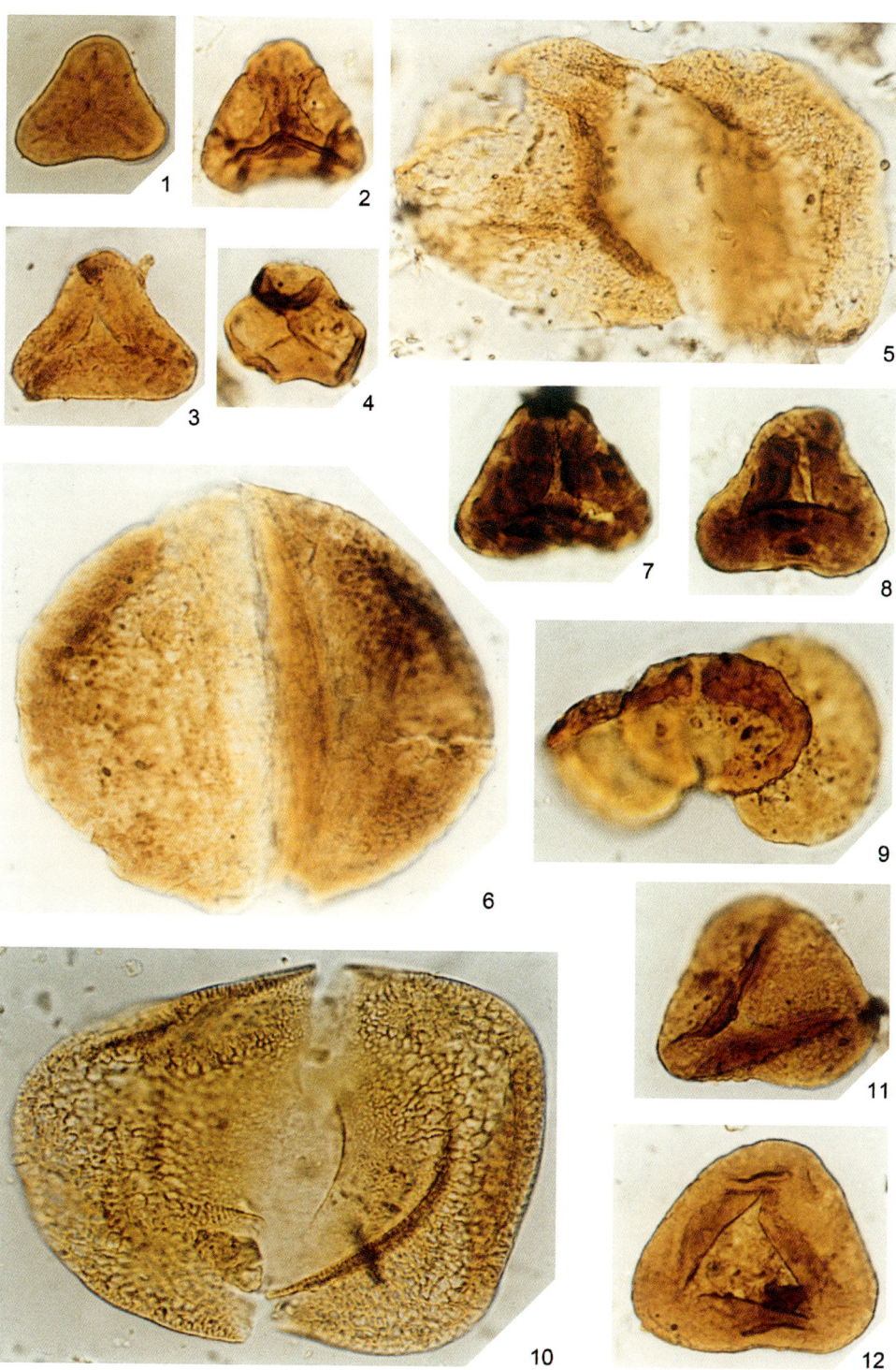

Fossil spores and pollen in crude oils from the South Junggar Depression of the Junggar Basin

Plate XLIV

1 Cyathidites minor Couper No. Q-1-37
2 Verrucosisporites variabilis Pocock No. Q-1-63
3 Cyathidites australis Couper No. Q-1-89
4 Dictyophyllidites harrisii Couper No. Q-1-42
5 Gleicheniidites rouseii Pocock No. Q-1-90
6 Cibotiumspora jurienensis (Balme) Filatoff No. Q-1-4
7 Divisisporites undulatus Huang No. Q-1-91
8 Classopollis classoides (Pflug) Pocock No. Q-1-59
9 Piceites latens Bolchovitina No. Q-1-87
10, 12 Callialasporites radius Xu et Zhang No. Q-1-63
11 Abietineaepollenites dunrobinensis Couper No. Q-1-24

(*10* at a magnification of 1056×)

Plate XLIV

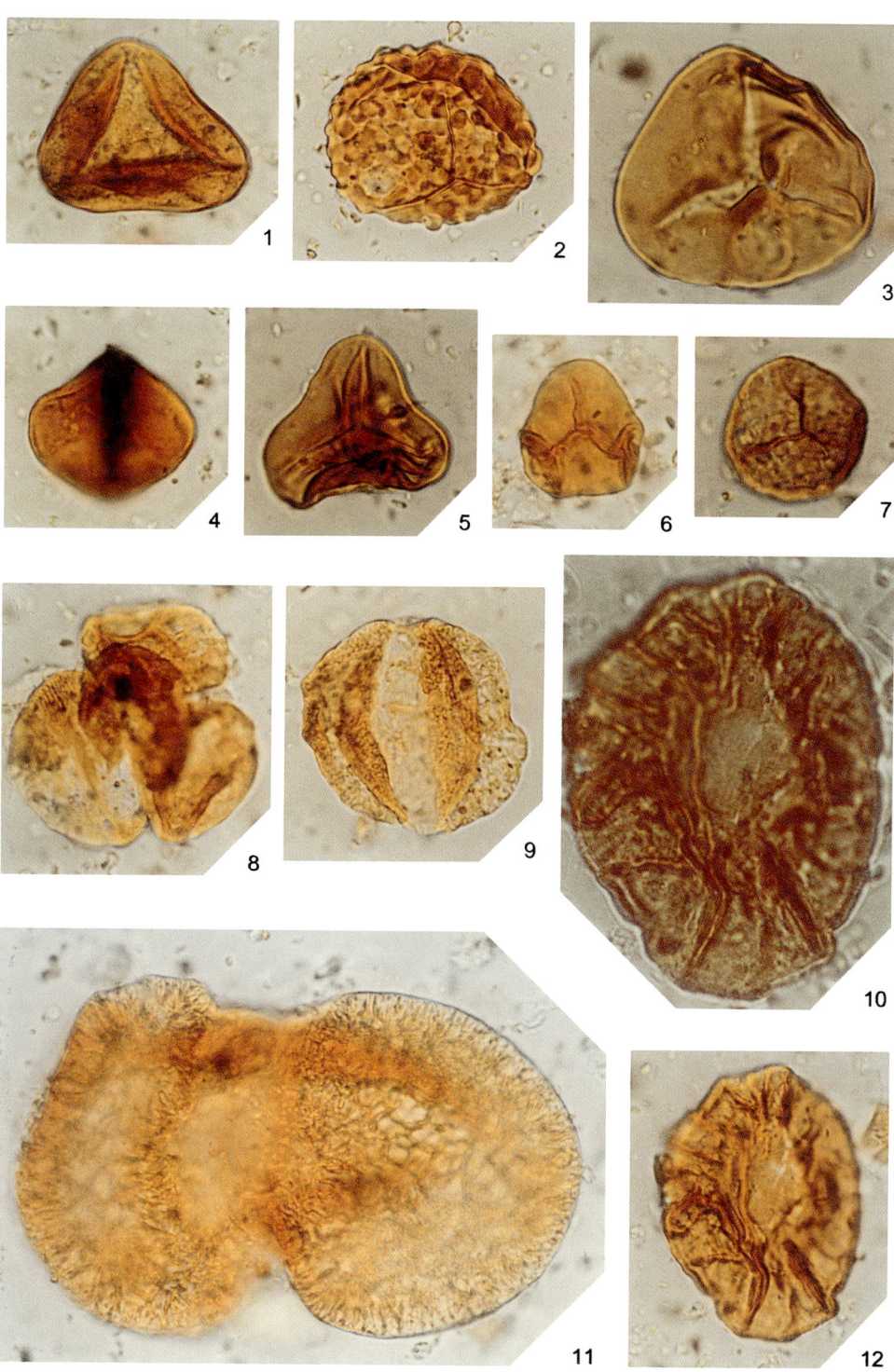

Fossil spores and pollen in crude oils of the Turpan Basin

Plate XLV

1 *Apiculatisporis variabilis* Pocock No. L-7411-7
2 *Duplexisporites amplectiformis* (Kara-Murza) Playford et Dettmann
No. L-7411-10
3 *Duplexisporties gyratus* Playford et Dettmann No. L-7411-11
4 *Duplexisporites scanicus* (Nilsson) Playford et Dettmann No. L-7411-12
5 *Callialasporites radius* Xu et Zhang No. L-7411-20
6 *Osmundacidites elegans* (Verbitzkaya) Xu et Zhang No. L-7411-25
7 *Quadraeculina limbata* Maljavkina No. L-7411-8
8 *Cyathidites minor* Couper No. L-7411-24
9 *Podocarpidites paulus* (Bolchovitina) Xu et Zhang No. L-7411-4
10 *Cycadopites subgranulosus* (Couper) Clarke No. L-7411-23
11 *Cycadopites carpentieri* (Delcourt et Sprumont) Singh No. L-7411-2
12 *Protopicea vilujensis* Bolchovitina No. L-7411-10
13 *Piceites expositus* Bolchovitina No. L-7411-12

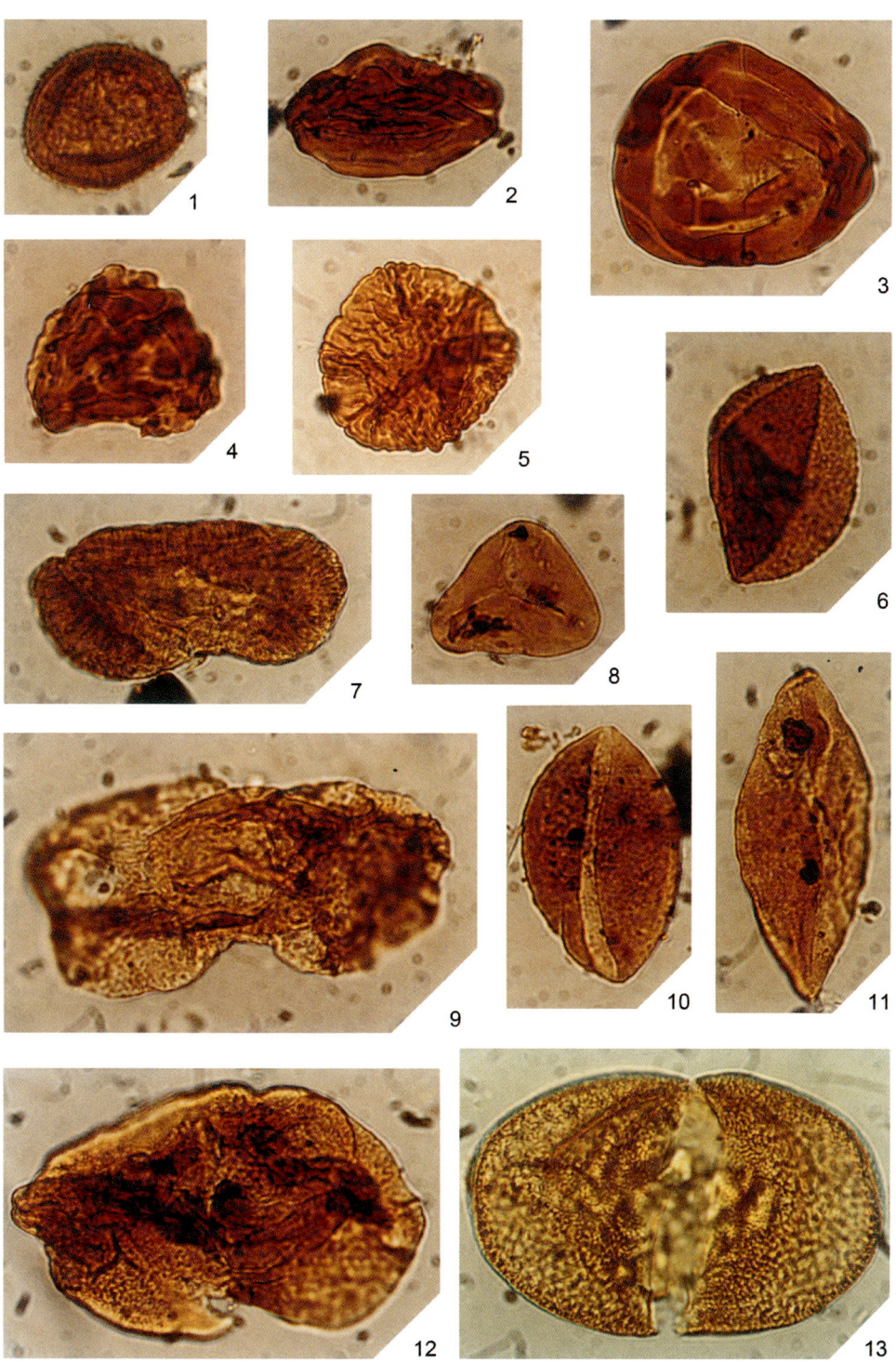

Plate XLV

Fossil spores and pollen in crude oils of the Qaidam Basin

Plate XLVI

1 *Pinuspollenites labdacus minor* (Potonié) Potonié No. Y-1-15
2, 9 *Deltoidospora regularis* (Pflug) Song et Zheng 2 No. Y-1-28; 9 No. Y-1-12
3 *Polypodiisporites favus* (Potonié) Potonié No. Y-1-16
4 *Cedripites deodariformis* (Zauer) Krutzsch No. Y-1-7
5 *Tsugaepollenites spinulosus* (Krutzsch) Ke et Shi No. Y-1-5
6 *Tsugaepollenites igniculus major* (Potonié) Potonié No. Y-1-21
7 *Cedripites microsaccoides* Song et Zheng No. Y-1-4
8 *Piceaepollenites quadracorpus* Zhu et Xi No. Y-1-13
10 *Tsugaepollenites igniculus minor* (Potonié) Potonié No. Y-1-24

Explanation of Plates and Plates 259

Plate XLVI

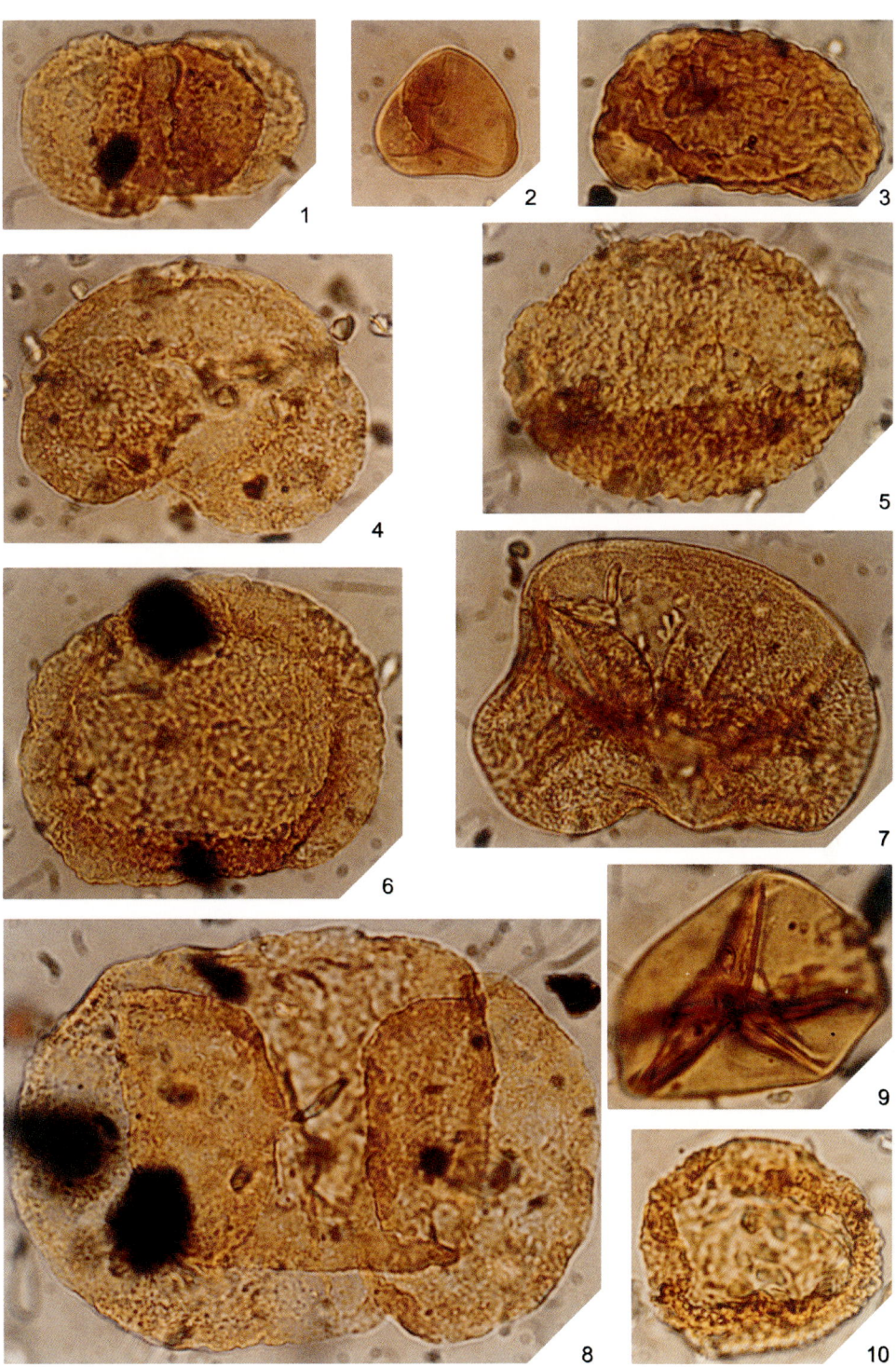

Fossil spores and pollen in crude oils of the Qaidam Basin (cont'd)

Plate XLVII

1–2 *Podocarpidites verrucorpus* Wu *1* No. Y-1-14; *2* No. Y-1-11
3 *Meliaceoidites rhomboiporus* Wang No. Y-1-21
4 *Keteleeriaepollenites dubius* (Chlonova) Li No. Y-1-7
5 *Cedripites deodariformis* (Zauer) Krutzsch No. Y-1-10
6 *Piceaepollenites planoides* (Krutzsch) Sun et Li No. Y-1-29
7–8 *Cedripites pachydermus* (Zauer) Krutzsch *7* No. Y-1-5; *8* No. Y-1-9
9 *Piceaepollenites alatus* (Potonié) Potonié No. Y-1-21
10 *Pinuspollenites labdacus maximus* (Potonié) Potonié No. Y-1-9

Explanation of Plates and Plates

Plate XLVII

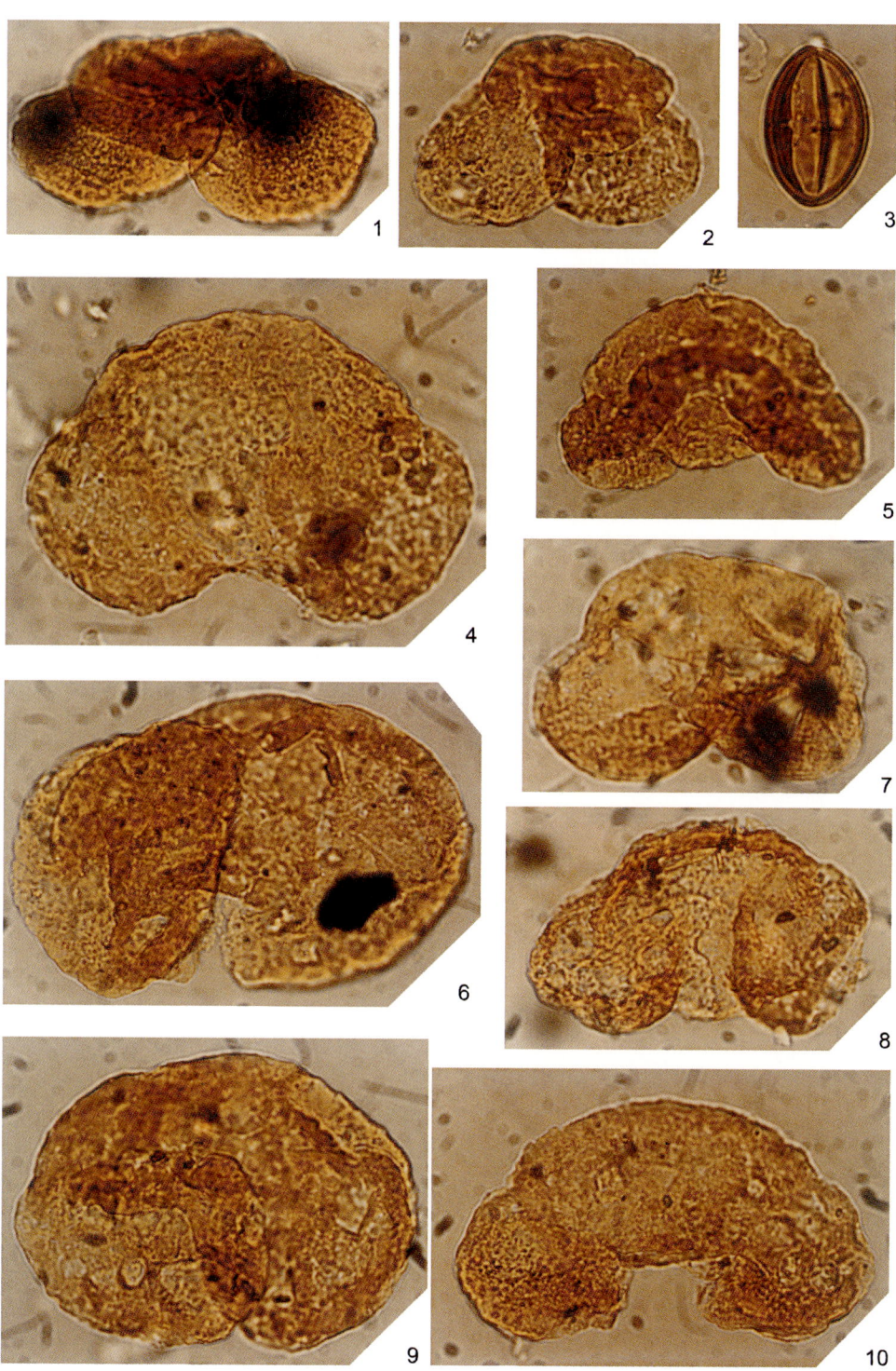

Fossil pollen in crude oils of the Qaidam Basin (cont'd)

Plate XLVIII

1–2 Crassoretitriletes nanhaiensis Zhang et Li *1* No. W10-3-4-21; *2* No. W10-3-4-28

3 Polypodiaceaesporites ovatus (Wilson et Webster) Sun et Li No. W10-3-4-18

4, 7 Pinuspollenites strobipites (Wodehouse) Sun et Li

4 No. W10-3-4-1; *7* No. W10-3-4-24

5 Podocarpidites andiniformis (Zaklinskaja) Takahashi No. W10-3-4-15

6 Osmundacidites primarius (Wolff) Sun et Li No. W10-3-4-17

8 Pinuspollenites minutus (Zaklinskaja) Song et Zheng No. W10-3-4-1

9, 10 Pinuspollenites labdacus minor (Potonié) Potonié

9 No. W10-3-4-25; *10* No. W10-3-4-21

11 Tricolpites tenuicolpus Sun, Kong et Li No. W10-3-4-23

12 Cedripites cf. *eocenicus* Wodehouse No. W10-3-4-17

13 Piceaepollenites alatus (Potonié) Potonié No. W10-3-4-26

14 Cedripites pachydermus (Zauer) Krutzsch No. W10-3-4-25

Plate XLVIII

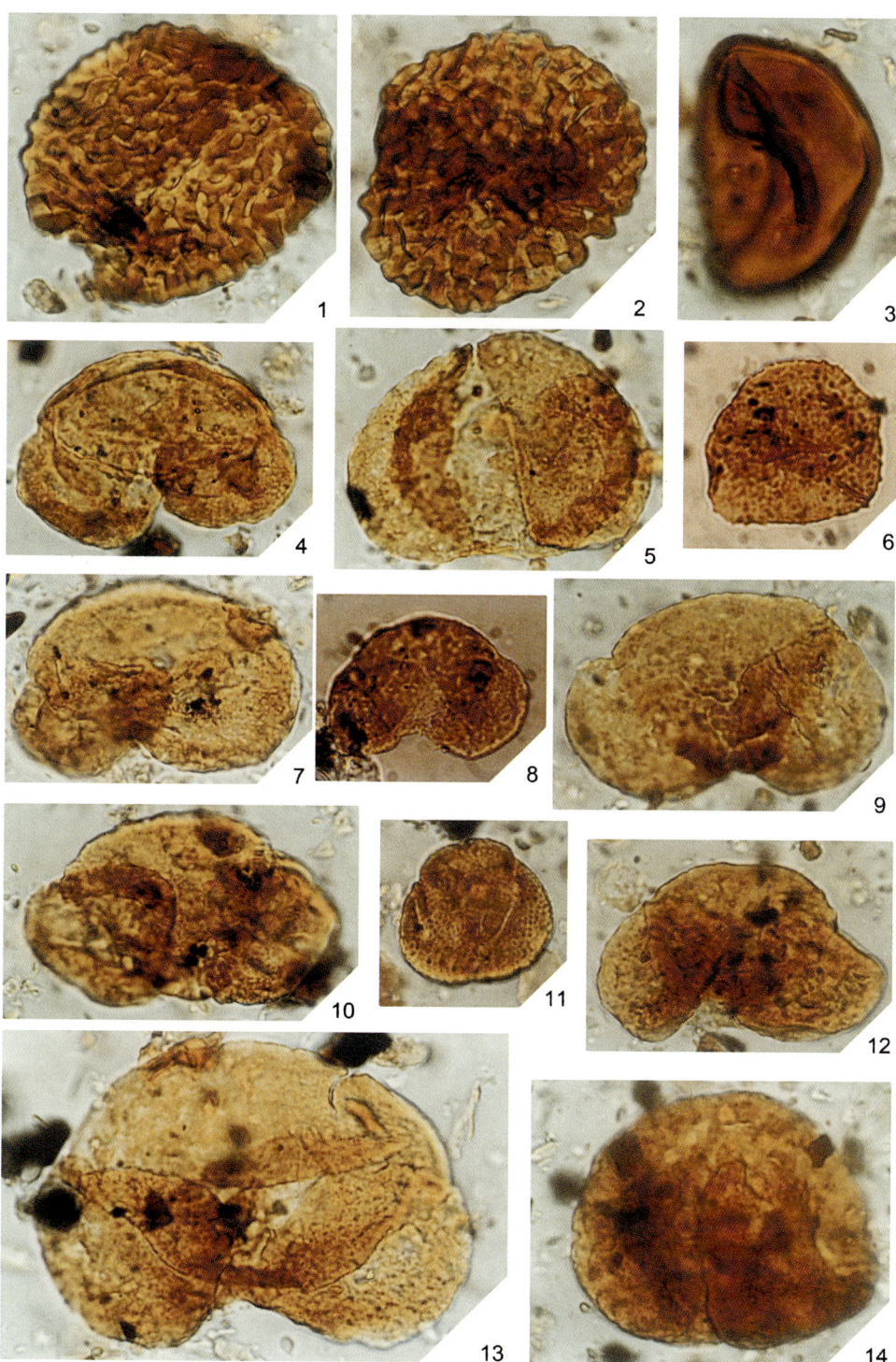

Fossil spores and pollen in crude oils of the Beibu Gulf Basin